PREPARATORY CHEMISTRY

Concepts and Applications
4th Ed.

Jerry Mundell, Ph.D. & Anne O'Connor, Ph.D.

VAN-GRINER

Preparatory Chemistry

Concepts and Applications

4th Edition

Jerry Mundell, Ph.D.

Anne O'Connor, Ph.D.

Printed in the United States of America

10 9 8 7 6 5 4 3 2 1

ISBN: 978-1-61740-197-8

Van-Griner Publishing

Cincinnati, Ohio

www.van-griner.com

Mundell 197-8 Su14

Contents

Introduction to the Book

We have written this *Lecture Exercise Textbook* to help students better understand the concepts and symbolic language of Chemistry, as well as their applications in performing chemical calculations.

For those students who have taken chemistry before, you may well remember that studying chemistry is quite different from other subjects. We always tell our students that chemistry is a **practical** subject that must be **practiced.**

To be able to understand, communicate, and apply such essential concepts as ionization, chemical bonding, electron configuration, and molecular shapes is the proper preparation for more advanced courses in chemistry.

This introductory chemistry book was constructed in the interest of simplicity and a straightforward delivery of the fundamentals of this science. To use this book all that is needed is a pencil, an interested student and a dedicated teacher. For this book to be used successfully requires the full participation of both students and teacher. All assignments must be completed by the individual students alone or working in small Peer-Led groups. It is imperative that the completed assignments be reviewed for errors by the instructor before proceeding to the next section. We have also included additional problems which may be assigned at the instructor's discretion.

Rather than just absorbing facts and directions during the traditional lecture and at a later time attempting to solve assigned problems dealing with the material, the intent of this book is to create hands-on participation of the students during the class meeting.

In this book the facts and application methods of chemistry are presented in a straight-forward style with no digression on historical fact. We believe that the student will be busy enough becoming adept with the fundamentals and applications of chemistry, without the added burden of dealing with historical arguments and anecdotes. Such information, though indisputably important, may be postponed until a later course after the student has established a solid foundation.

We realize that most who purchase this book will not pursue a career in chemistry, but we hope it serves well in making one of the most intriguing subjects more accessible to you.

Anne O'Connor, Ph.D.

And Jerry Mundell Ph.D.

To the Students

For those of you who have never taken a chemistry course before, you may very well be in the dark about the fundamentals of this subject. You may also have some uncertainty about the math involved in this subject, or your own problem-solving abilities.

In the words of Douglas Adams, "**Don't Panic.**"

After having been involved with chemistry as teachers over the years, we have had the realization that learning the fundamentals of chemistry does not have to be thought of as on the same level of experience as getting a root canal.

The text in this book has been set up in a clear, concise, and straight forward fashion. For those who have already had a substantial background in this subject, don't be put off by the simplicity of the presentations. It is all the same traditional textbook information with the new delivery.

Chemistry as any other science has its own vocabulary which enables us to intelligently communicate the questions and concepts relating to this field. For this reason introductory chemistry needs to be taught with the same thoroughness as a language. In order to solve any problem in chemistry the student must not only have a background in the mathematics involved, but also a very clear understanding of all definitions and concepts involved.

This book is a textbook:
This is a straightforward book of information that deals with the introductory concepts, vocabulary, exercises and problems to enable the student to further his or her studies in Chemistry.

This book is a workbook:
The pages of this book are to be read, written on, and sometimes torn out. It's all part of the learning process.

What is Chemistry?

p1

Chapter 1

Chapter Outline

A person walks along the seashore and sees the sand on the beach, the salt water of the waves washing upon the sand. She breathes the fresh air while collecting sea shells along the shore. The chemists on that same beach, as much as they enjoy the same pleasure as you, realize that the sand is silicon dioxide (Figure 1.1), the salt water is a mixture of sodium chloride and dihydrogen monoxide, the air is a mixture of nitrogen and oxygen, and the sea shells are made up of calcium carbonate. These are all forms of matter. **Matter** is anything that has mass and occupies space or volume. In order for chemists to understand the nature of matter, they break it down into two main categories, substances and mixtures of substance. The silicon dioxide, sodium chloride, nitrogen, oxygen, dihydrogen monoxide and calcium carbonate examples of substances. A **substance** is uniform throughout and has a fixed composition. In chapter two, we will learn more about the nature of substances and mixtures.

Figure 1.1

Sand is made up of a pure substance called silicon dioxide.

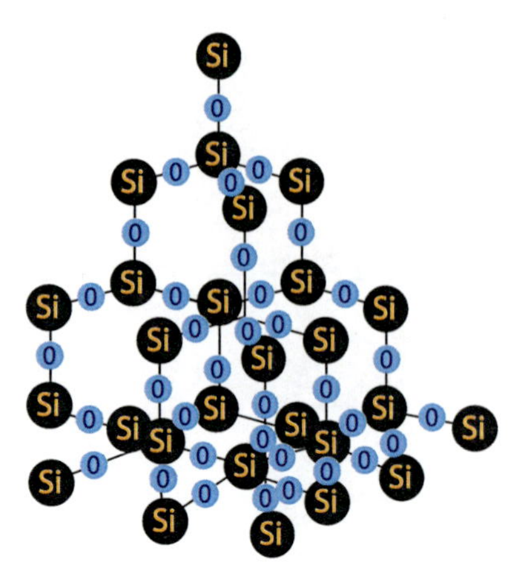

Matter: Anything that has mass and occupies space.

Substance: A form of matter that is uniform throughout and has a fixed composition.

Figure 1.2

Calcium carbonate, the major component in the structure of sea shells.

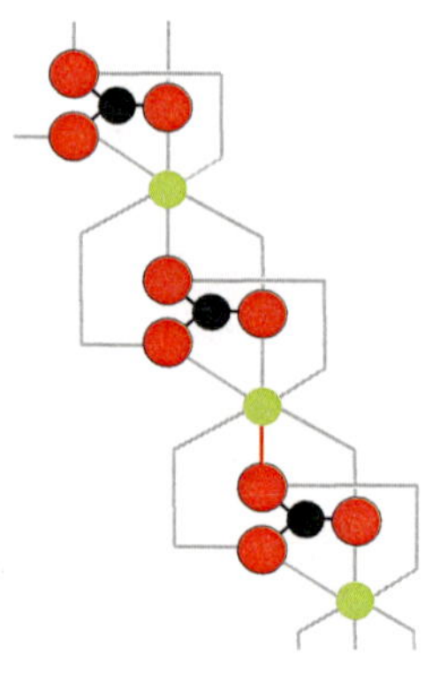

Figure 1.3

Salt water is a mixture consisting, in part, of sodium chloride (NaCl) dissolved in water, dihydrogen monoxide (H_2O).

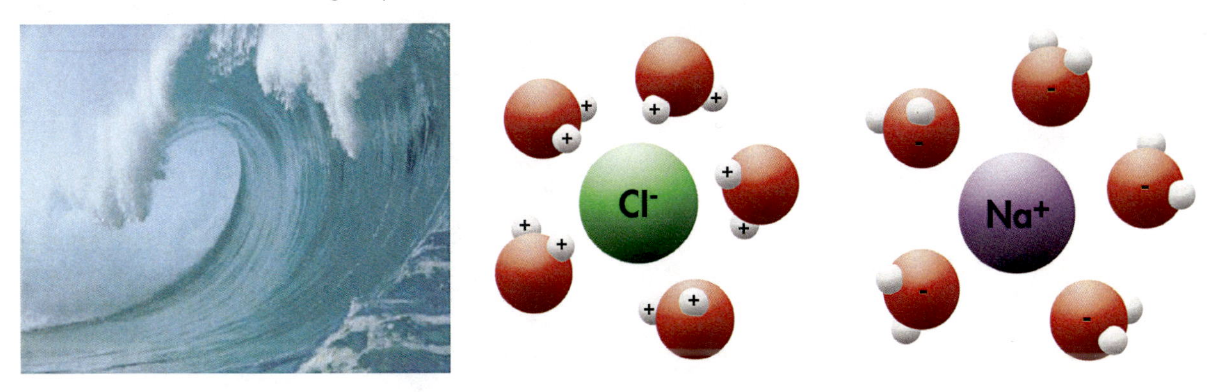

In chemistry the physical and chemical properties of these substances are studied in the hopes of finding how they fit into the world around us. We hear and read much about acid rain and its effect on our environment, Figure 1.4, but the chemist understands the substances sulfur dioxide and nitrogen oxides that are emitted from cars and certain industries and the chemistry that these substances undergo combining with the water in the atmosphere to produce both sulfuric and nitric acid to produce acid rain in the form of diluted sulfuric and nitric acids which effect both aquatic and terrestrial life.

Figure 1.4

The effect of acid rain on the enviroment.

Why Study Chemistry?

Chemistry is often referred to as the central science because it is a foundation to many sciences. It also provides background to students going into health fields. The diagram, Figure 1.5 below shows eight sciences that overlap with chemistry.

Figure 1.5

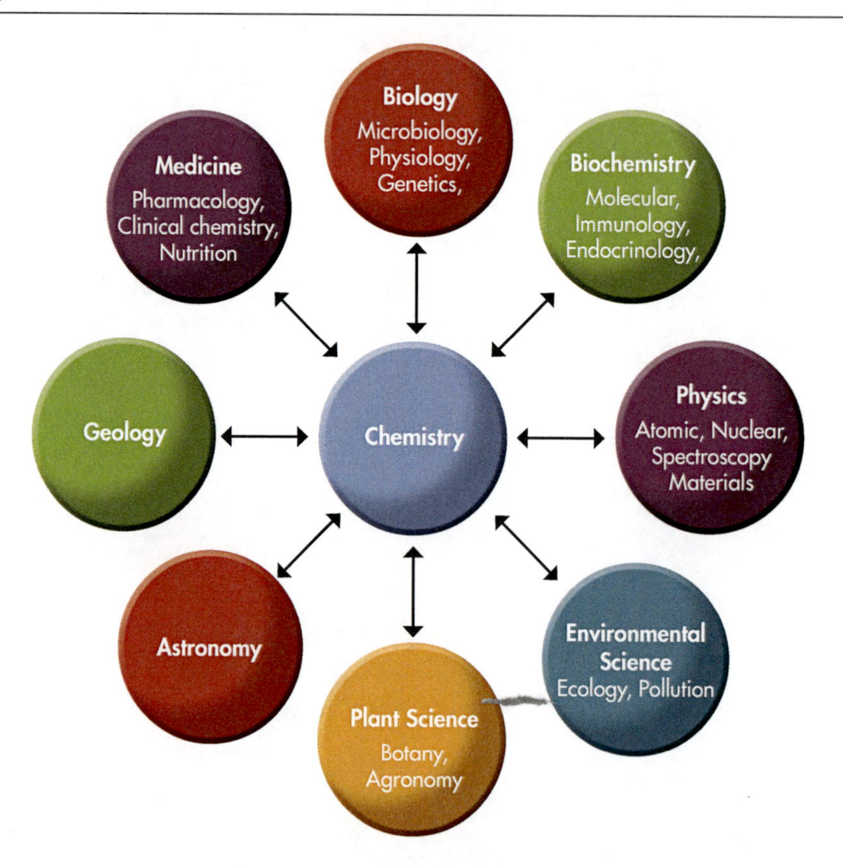

Biology: Understanding life at the molecular level lays the foundation for biologists to study photosynthesis, metabolism, or the replication of DNA. Photosynthesis, (Figure 1.6), both light dependent and light independent involves not only familiar chemical substances, such as oxygen, water, and carbon dioxide, but also more complex biochemical substances such as ATP (adenosine triphosphate), NADP (nicotinamide adenine dinucleotide phosphate), and NADPH.

Figure 1.6

Light dependent and light independent photosynthesis.

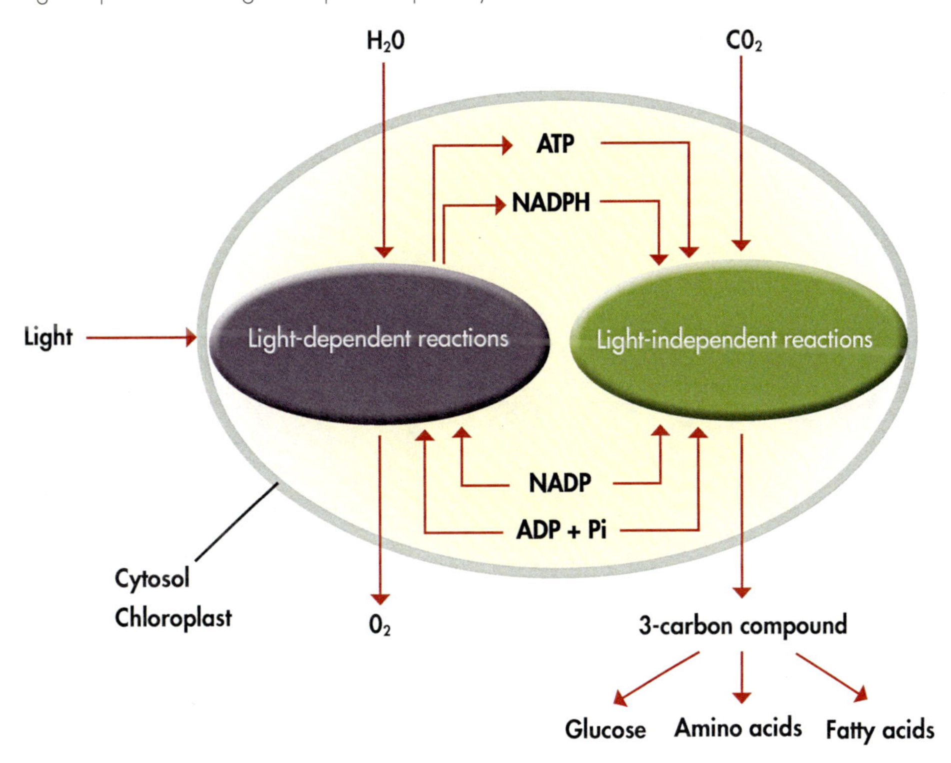

Geology: Through the use of chemical analysis and X-Ray Crystallography geologists know the chemical composition and structure of various gems and minerals Figure 1.7.

Figure 1.7

Rutile is a mineral composed of titanium and oxygen. In its structure, each titanium atom (red spheres) is surrounded by four oxygen atoms (white spheres).

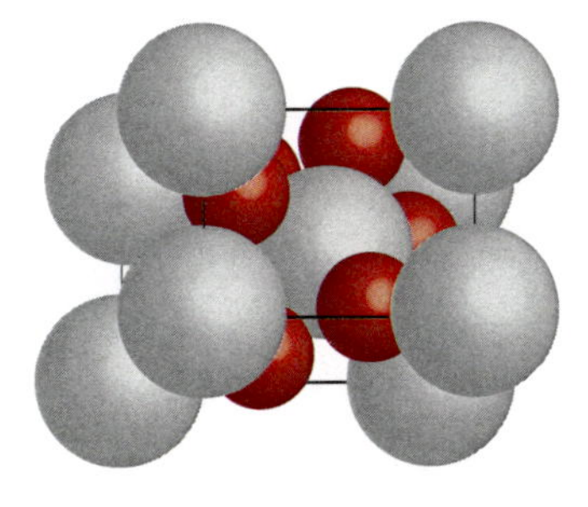

Material Science: The work in the discovery and study of small spherical molecules composed sixty carbon atoms opened the doors to a new and important area of chemistry. The discovery of Buckminsterfullerine (i.e the buckyball Figure 1.8 A) and related compounds composed of larger numbers of carbon atoms has created the nanotechnology (Figure 1.8 B) industry that is currently investigating and developing new materials in areas as diverse as electrodes and building materials stronger than steel at only a fraction of the weight.

Figure 1.8

A) Buckminster fullerene and B) a nanotube.

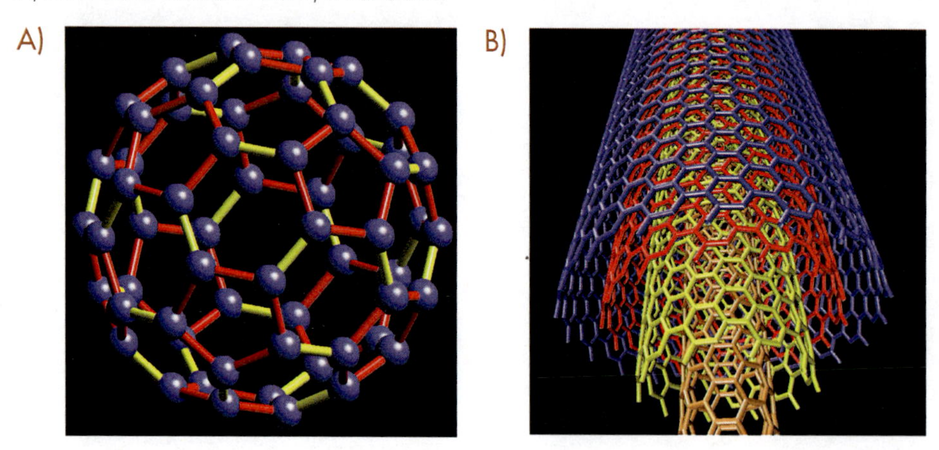

Astronomy: With the knowledge of light spectroscopy, astronomers have the ability to analyze the atmosphere of distant planets and other heavenly bodies. Figure 1.9 is the Butterfly Nebula, located 3,800 light-years away from Earth. The nebula is composed of a central star that is surrounded by hot gases with temperatures that exceed 19,000 °C.

Figure 1.9

The Butterfly Nebula.

For students not intending to pursue careers in the sciences, chemistry enables people to better understand how various chemical substances, both toxic and benign, impact the world around them. Although a steel mill located within a community offers employment to its neighbors, it may not be adhering to proper standards for its emissions. As a result, toxic heavy metals are released into the community resulting in serious health problem to the local population and possibly beyond.

Figure 1.10

Unregulated steel mills can be responsible for releasing oxides of sulfur, resulting in acid rain which impacts environmental balance as well as infrastructure. Emissions can also include toxic heavy metals such as cadmium, lead, and chromium.

1.1 Physical and Chemical Properties

How a substance fits into the world around us can be summed up by the physical and chemical properties of a substance. **Physical properties** are characteristics used to describe a substance and do not involve a chemical change. The physical properties of a substance might include color, hardness, density, and melting point. Chemical properties are characteristics a substance shows when it interacts with another substance to form a new substance. This is called a chemical reaction. A substance's chemical properties might include a vigorous reaction such as when methane gas explodes in the presence of oxygen to produce carbon dioxide and water. Or the surface tarnish formed on silver when it is exposed to air containing hydrogen sulfide. Note that in both of these examples, new substances are formed: the methane gas and oxygen gas react to form carbon dioxide and water, and in the second example the silver metal and hydrogen sulfide react to form tarnish, silver sulfide. Chemical processes will be discussed in much greater detail in later chapters.

Consider the physical and chemical properties of sodium metal (Figure 1.11). Physically, it is so soft it can be cut with a knife. Its other physical properties include a shiny, lustrous appearance, its ability to conduct electricity as well as a melting point of 97.8 °C.

A chemical property of sodium is its reaction with water to produce sodium hydroxide and hydrogen gas.

Figure 1.11

Sodium metal.

Physical Property: Characteristics of matter that describe the matter without changing its composition.

Chemical Property: Characteristics a substance shows when it is converted into or interacts with another substance—a chemical reaction.

Example 1.1 Distinguishing Physical and Chemical Properties

Problem

Indicate if the following are physical or chemical properties.

a) The attraction of iron to a magnet

b) mass of 1 gram of carbon

c) decomposition of vitamin C

d) softness of potassium metal

Solution

a) This is a **physical property.** The iron is attracted to the magnet. Neither the iron or the magnet is changed when brought into contact with one another.

b) Mass is a **physical property.** It describes how much of a substance is present.

c) Decomposition of Vitamin C is a **chemical property.** When a substance decomposes the composition is changed—new substances are formed.

d) The softness of potassium metal is a **physical property.**

Practice Problems

1.1 Identify the following as a physical or chemical property.

a) the melting point of water

b) milk souring

c) solubility of salt in water

d) a bicycle rusts when left outside

1.2 Identify the following as a physical or chemical property

a) electrical conductivity of copper

b) wax melts when heated

c) tarnishing of silver

d) a substance burns in air

Summary of Physical and Chemical Processes Matter is anything that occupies space and has mass.

- A substance is a form of matter that is uniform throughout and has a fixed composition.
- Physical properties are properties that describe matter.
- An example of a chemical property is the reactivity of a substance.

Key Terms

Matter

Substance

Physical Property

Chemical Property

Summary of Physical and Chemical Processes

- Matter is anything that occupies space and has mass.
- A substance is a form of matter that is uniform throughout and has a fixed composition.
- Physical properties are properties that describe matter.
- An example of a chemical property is the reactivity of a substance.

Exercises

1. Name two physical properties of baking soda (sodium bicarbonate).

2. Sucrose, a white crystalline substance, has a density of 1.6 g/cm^3, is very sweet, and decomposes at 186 °C. Using the given information, list the chemical and physical properties of sucrose.

3. Water does not burn. Is this a physical or chemical property of water?

4. Which of the following is a physical property?
   a) flammability
   b) inertness
   c) conductivity

5. Which of the following is not an example of matter (there can be more than one correct answer)?
   a) heat
   b) skin
   c) clothing
   d) sunlight

Worksheet 1.1

1. Name 5 physical properties that a substance can have.

2. Define chemical property and provide an example.

3. Define physical property and provide an example.

4. Indicate if the following are physical or chemical properties.
   a) Coca Cola fizzes when opened

   b) the ripening of a banana

   c) silver tarnishes

   d) the density of water is 1.0 g/mL

   e) sodium chloride is a white crystalline solid

   f) a crystalline substance decomposes at 186 °C

5. Explain how chemical properties differ from physical properties.

6. Which is not an example of matter? Provide a brief explanation.
   a) air
   b) water
   c) electrical current
   d) plants

7. A chemist is asked to analyze an unknown liquid sample. Which of her observations is not a physical property?
   a) The sample is a colorless liquid
   b) The size of the sample is 200 mL
   c) The odor of the sample is very similar to wintergreen
   d) The sample is not flammable

1.2 Physical Process or Chemical Process?

To gain a better understanding of the difference between a chemical and a physical process we first understand that a substance can exist in one of three physical states: solid, liquid, or gas, depending on the conditions such as temperature pressure. Let's consider the melting of sodium, that is when solid sodium metal melts into liquid sodium. Physical processes such as melting (solid to liquid) or boiling (liquid to gas) are called **phase changes**. Whenever a substance melts, whether it be sodium, ice, or iron, nothing new is formed; the substance merely undergoes a phase change. In other words, in the solid state the atoms of sodium are tightly arranged as in Figure 1.12. When the right amount of heat (97.8 °C) is applied, the atoms of sodium gain energy and begin to move; finally breaking away from the rigid structure of the solid to become a liquid. In the liquid state, the sodium atoms, still closely associated, fluidly move about to one another in close proximity. In describing this process of melting, it is apparent that the sodium atoms are unchanged; only the physical state of the sodium has changed.

Figure 1.12

A representation of sodium atoms in the solid state.

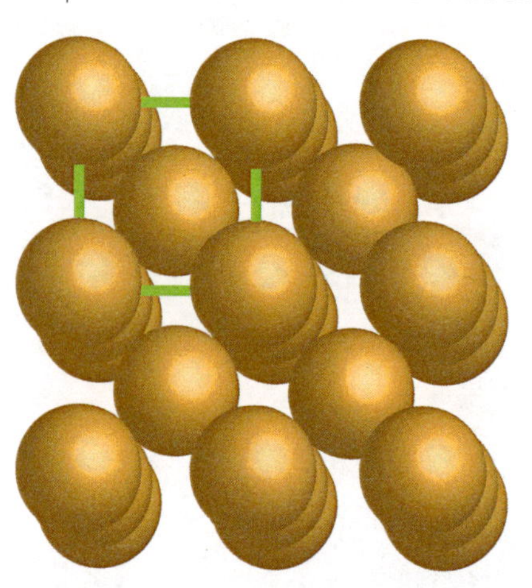

Consider what happens when another substance undergoes the same process. The substance water is a molecule composed of three atoms: one oxygen atom and two hydrogen atoms. Just like sodium, in the solid state, the molecules of water are tightly packed together (Figure 1.13 A). Applying the proper amount of heat, above 0 °C, molecules of water melt and break away from the ice crystalline solid becoming a fluid liquid (Figure 1.13 B). Again we see that the physical process of melting has left the substance unchanged, i.e. each molecule of water is still composed of one oxygen and two hydrogen atoms.

Figure 1.13

A) Water molecules in the solid state, and B) water molecules in the liquid state.

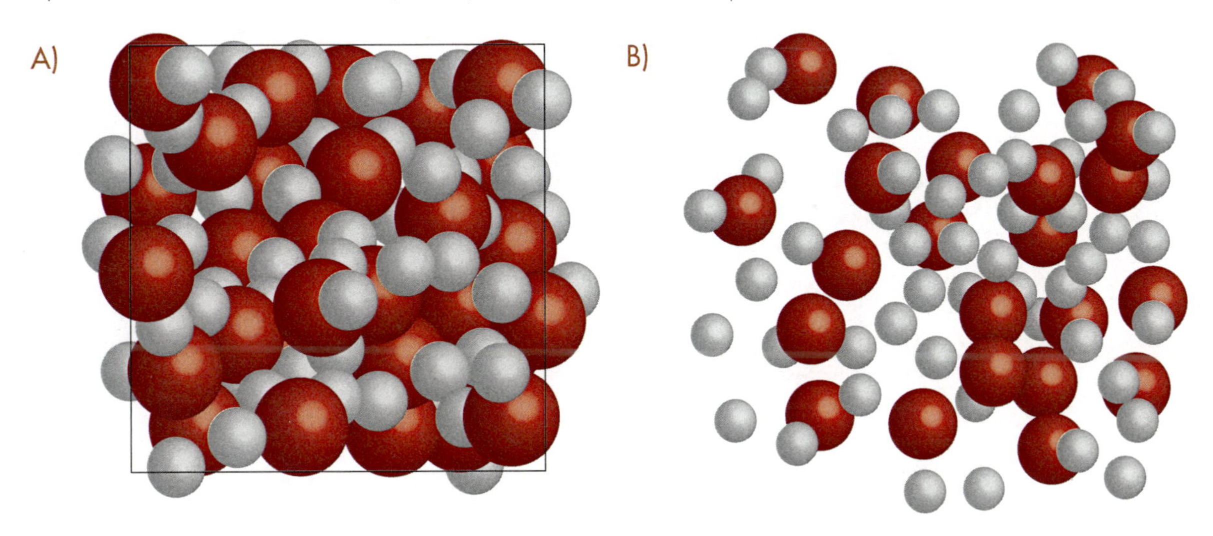

As mentioned above, matter can occur in three physical forms that are called *states*. These are the solid, liquid, and gas states. In the **solid state**, a substance has both fixed shape and volume (Figure 1.4 C). The particles of a solid are held closely together in a regular, three-dimensional arrangement. In the **liquid state** (Figure 1.4 B), the particles are close together but are able to move about one another. Although its volume is fixed, the shape of a liquid will depend on the part of the container in which it is confined. In the **gas state** (Figure 1.4 A), the particles are relatively great distances apart, and not close together as they are in a solid or a liquid. The gas particles are able to randomly move about, taking on both the shape and volume of its container.

Figure 1.14

The three common states of matter.

A) **Solid State**

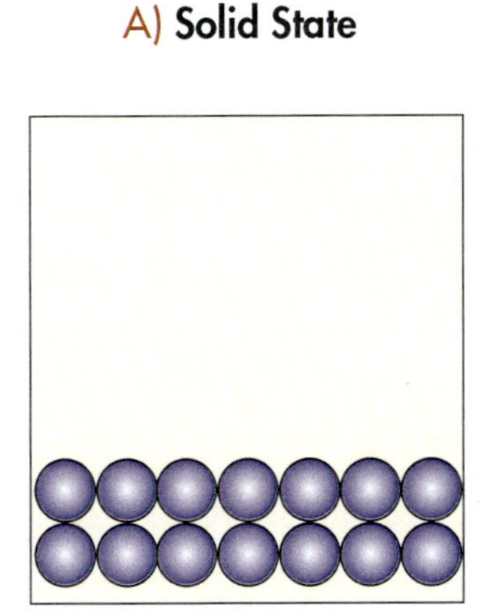

B) **Liquid State**

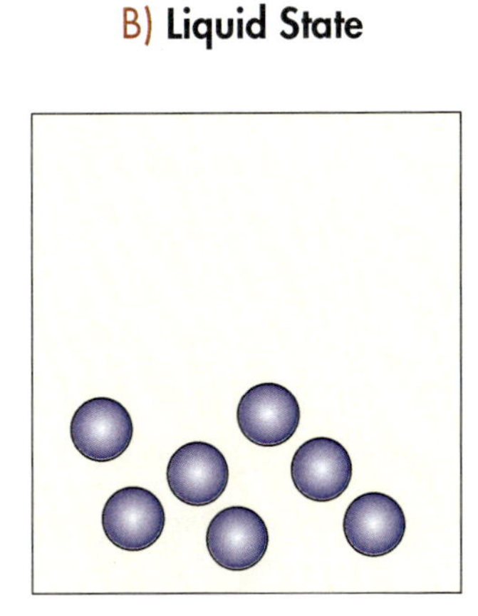

C) **Gas State**

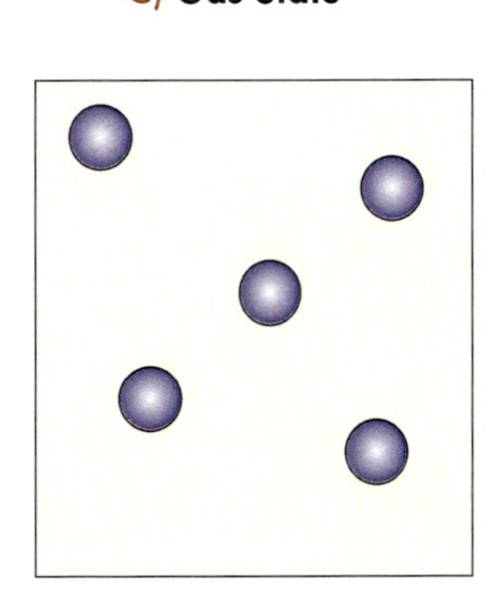

Phase Changes: A physical process in which a substance undergoes a change of physical state.

Solid State: Particles are held closely together in a regular, three-dimensional arrangement. A solid has both fixed shape and volume.

Liquid State: Particles are close together and able to move about in fluid motion. Although its volume is fixed, the shape of a liquid will depend on the part of the container to which it is confined.

Gas State: Particles are relatively great distances apart and moving about in a random motion, taking on both the shape and volume of its container.

Example 1.2 Describing States of Matter

Problem

Describe a cube of ice in terms of:

a) volume

b) shape

c) organization of the water molecules

d) closeness of the molecules

Solution

An ice cube is water in its solid state.

a) the volume is definite

b) the shape is definite and does not conform to the shape of the container

c) the particles are highly organized in a regular three-dimensional shape

d) the molecules are very close together

Practice Problems

1.3 Describe the organization of the particles that comprise a liquid. Comment on the shape (how the particles fill the container).

1.4 A substance has a definite volume. The particles that comprise the substance are very close together but not in an organized three-dimensional shape. Is this a liquid, solid, or gas?

1.5 An unknown substance has a variable volume and it takes on the shape of its container. Is this a solid, liquid, or a gas?

Summary of Physical Processes or Chemical Processes

In a physical process, matter changes in form, but still retains its chemical identity.

- The common three states of matter are solid, liquid, and gas.
- Solids have both invariable shape and volume, liquids have variable shape and invariable volume, and gases have both variable shape and volume.

Key Terms

Phase Change

Solid State

Liquid State

Gas State

Exercises

1. Describe the organization of gas particles and comment on the shape.

2. A substance has a variable volume and fixed shape. Is this a liquid, solid, or gas?

3. An unknown substance has an invariable volume and variable shape. Is this a solid, liquid, or a gas?

4. Give two examples of phase changes.

5. Is melting a physical or chemical change?

6. Explain why phase changes are physical processes.

Worksheet 1.2 Physical and Chemical Processes

1. Which of the following is a chemical process?
   a) ethanol melts at –114 °C
   b) ethanol mixes with water
   c) ethanol burns to form water and carbon dioxide
   d) ethanol vaporizes at 78.4 °C

2. Fill in the blanks.
   A solid has ____________ shape and ____________ volume.
   A liquid has ____________ shape and ____________ volume.
   A gas has ____________ shape and ____________ volume.

3. The conversion of a solid to a liquid is called ____________.

4. The conversion of a liquid to a gas is called ____________.

5. Draw diagrams that represent a substance in the solid, liquid, and gas states.

1.3 Changes of State

Any *change of state* is defined as a physical process. If we allow a cube of ice to melt in a sauce pan and bring the pan of water to a boil on the stove the fluid water molecules will quickly gain enough energy to overcome the liquid state and escape into the gas state as steam, maintaining the one oxygen atom, two hydrogen atom structure. Figure 1.15 illustrates the six changes of state substances undergo. Notice that for each change of state, such as freezing there is an opposite process, in this case it would be melting. Freezing, melting, evaporation, and condensation are physical processes with which we are all familiar. Most substances will pass from a solid state to a liquid state and finally into a gas state.

Figure 1.15

The three states of matter and the various phase changes.

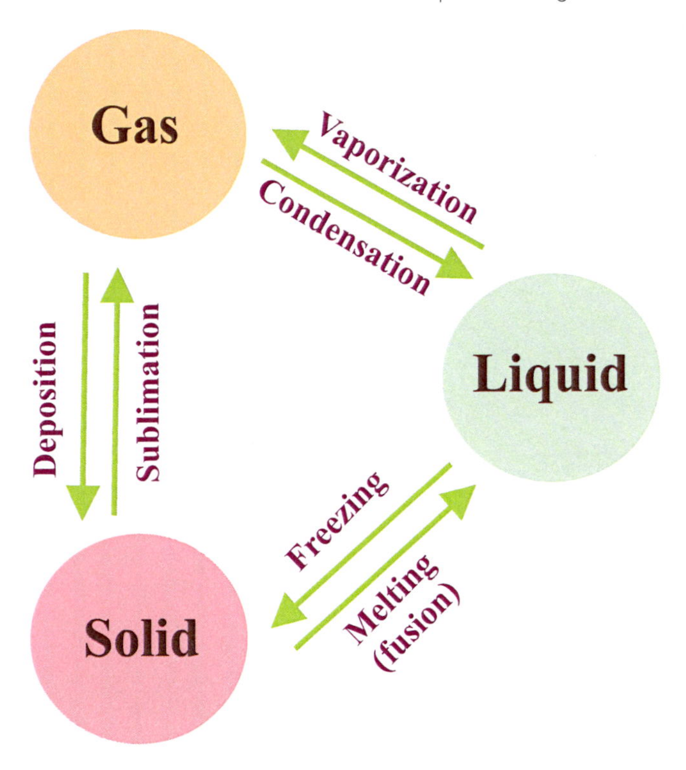

However another process exists in which a substance passes directly from the solid state directly into the gas state. This process is called **sublimation** (the reverse process is resublimation or **deposition**). Dry ice, Figure 1.16, is a substance that under normal atmospheric conditions passes directly from a solid to a gas. Dry ice is actually carbon dioxide in the solid state. The carbon dioxide gas that we exhale will undergo deposition at –78 °C. The liquid state of carbon dioxide can be formed only under high pressure. In Figure 1.10, the fog is not carbon dioxide, but condensed water vapor.

Figure 1.16

Dry ice passes direct from the solid state into the gas state.

Sublimation: The process in which a substance passes from the solid state directly to the gas state.

Deposition: The process in which a substance passes from the gas state directly to the solid state.

Example 1.3 Identify Phase Changes Problem

What is the name of the transformation when a substance is converted from a gas to a liquid?

Solution

This is called condensation. It is the reverse process of vaporization.

Practice Problems

1.6 Which of the following is not a phase change?

a) melting b) fusion c) mixing d) deposition

1.7 Which transformation is sublimation?

a) liquid ⟶ solid
b) liquid ⟶ gas
c) solid ⟶ liquid
d) solid ⟶ gas

What Causes Changes of State?

Molecules and individual atoms have varying amounts of attractive forces for each other. These attractive forces are called **intermolecular forces**. Molecules of water under normal temperature and pressure are strongly attracted to one other by a special type of intermolecular force called hydrogen bonding (Figure 1.17). Intermolecular **hydrogen bonding** occurs when a molecule has a hydrogen bonded to an oxygen, nitrogen, or a fluorine atom. Molecules of methane under the same conditions have very weak attraction to each other—methane has hydrogen bonded to carbon as a result intermolecular hydrogen bonding is not possible for methane molecules.

Figure 1.17

Hydrogen bonding between molecules of water.

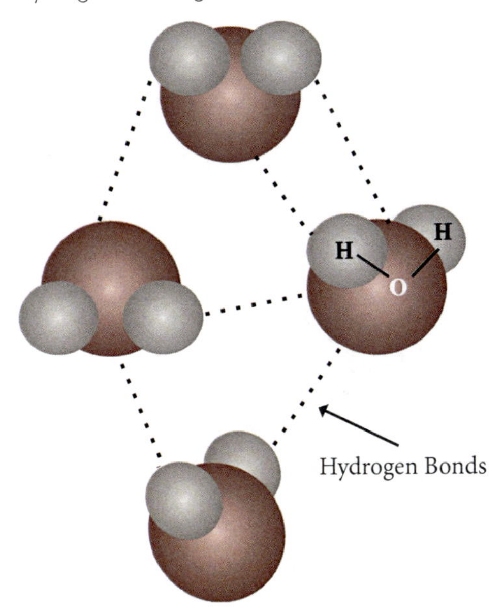

Substances such as water that have strong intermolecular forces tend to stay closely associated. Because of this, these substances melt and also boil at relatively high temperatures. For example, ice will melt at 0 °C and with additional heating, the water will boil at 100 °C. Methane, a gas at room temperature, can be cooled down into a liquid and with continued cooling it will freeze into a solid. Just as ice will melt at 0 °C, solid methane will melt at −182.5 °C, and with continued heating the liquid methane will boil into a gas at −164 °C. Just as the strong intermolecular forces result in substances with relatively high melting and boiling points, weak intermolecular forces result in substances existing as gases at room temperature which will condense into liquids and freeze into solids at very low temperatures.

Intermolecular Forces: The attractive forces between molecules.

How Does Temperature Affect Changes in State?

Temperature can be thought of as an indication of heat transfer. On a hot summer day, as heat is transferred into a house, the temperature within the house increases. On a cold winter night, as heat is transferred out of the house, the temperature within the house decreases. **Heat** is the flow of kinetic energy from a hotter body to a colder one. This form of kinetic energy is called **thermal energy**. Consider what happens when an ice cube is taken out of the freezer and placed in a pan on a counter. Since the ice cube is colder (approximately 0 °F or –17 °C) than its new surroundings (72 °F or 22 °C), the thermal energy immediately surrounding the ice cube will flow into the ice. Thermal energy is partly **kinetic energy**—energy of motion—and as the thermal energy flows, kinetic energy will start to build within the ice, and the molecules of water will begin to break away from their tightly held structure. And as such, the cube of ice commences to melt.

Figure 1.13

Under normal conditions, most substances will undergo a solid to liquid phase change.

Heat: The flow of kinetic energy from a warmer body to a cooler one.

Thermal Energy: Heat energy.

Kinetic Energy: The energy of motion.

If the pan is placed on a lit burner, heating will rapidly increase, transferring more kinetic energy to the H_2O molecules, and quickly melt the remainder of ice into liquid water. In the liquid state the kinetic energy of the water molecules is much greater than the kinetic energy of those same water molecules locked into solid ice. In order to melt, the water molecules in the rigid ice structure must take on enough kinetic energy to overcome the strong attractive forces between them and begin moving about.

In the liquid state, even though kinetic energy is much greater than before, the attractive forces still keep the molecules in close proximity, tumbling around one another. As more heat is added, the kinetic energy of the molecules begins to overcome the intermolecular forces of the liquid. With enough energy, the molecules of water begin to break away into the gas state as in Figure 1.19.

Figure 1.19

As water is heated, water molecules break away into the gas state.

So far, we have discussed physical processes (melting and evaporation) in which energy is absorbed. These processes are called **endothermic** because in order to travel from a lower energy state to a higher energy state energy must be absorbed. The processes opposite to melting and evaporation are freezing and condensation. These processes are **exothermic** which are appropriately named because in order to travel from a high energy state, such as the gas state, to a lower energy state, the liquid state, energy must be released. Figure 1.20 indicates the various physical states and the respective heat flow (i.e. endothermic or exothermic) for the respective change of state.

Figure 1.20

Endothermic and exothermic phase changes.

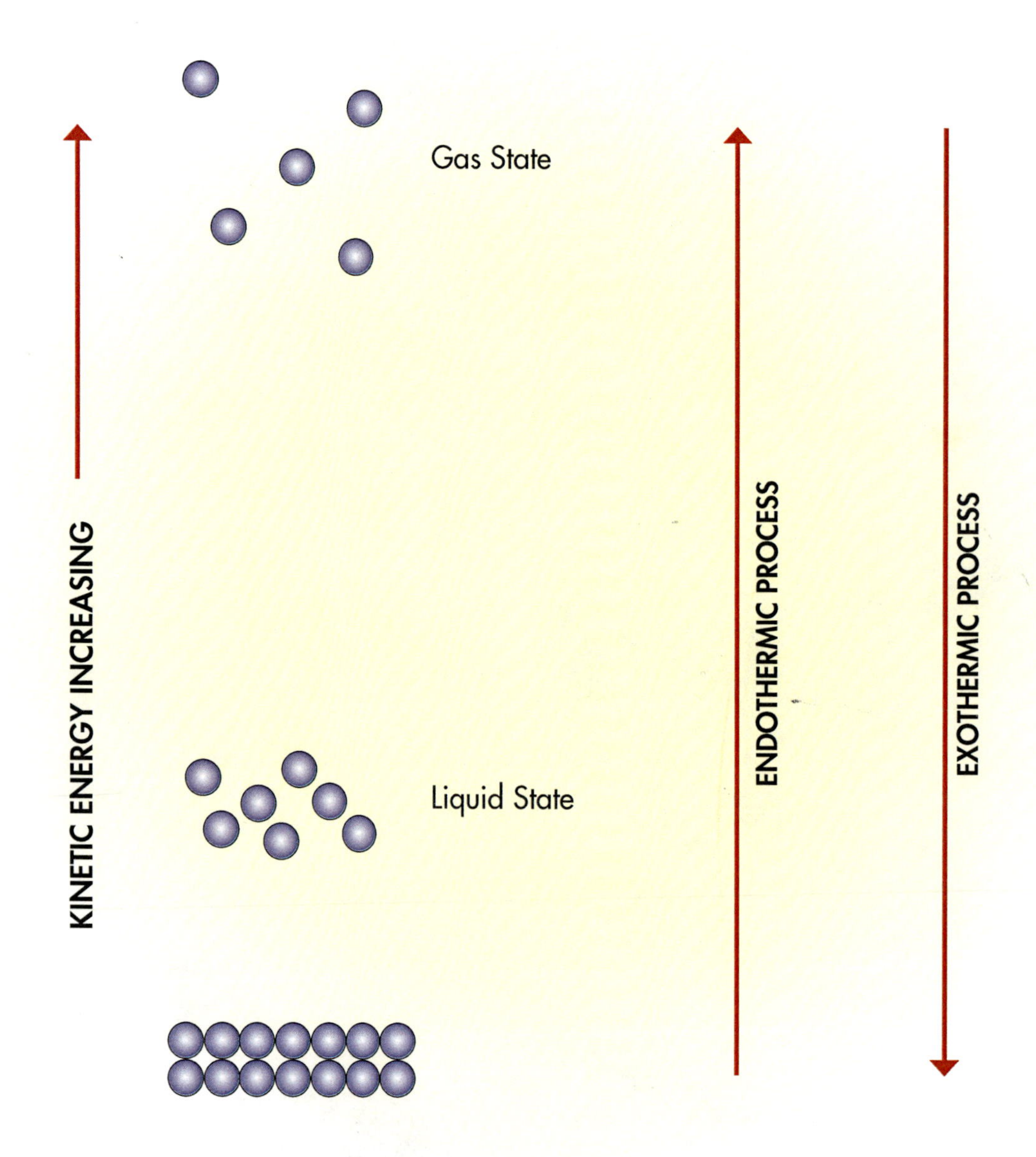

Endothermic Process: A process that requires the absorption of heat energy from the surroundings.

Exothermic Process: A process that involves the release of heat into the surroundings.

Summary of States of Matter

The description of the physical states of matter, i.e., solids, liquids, and gas, on the atomic or molecular level is known as the *kinetic-molecular theory*. According to this theory, the physical state of a substance is dependent upon the kinetic energy (the amount of heat) and intermolecular forces (IMF) within the system. In the solid state, the particles have minimal kinetic energy, have very strong intermolecular forces, and are highly ordered. In the liquid state, the particles possess high kinetic energy, strong intermolecular forces, but the particles are less ordered and can move randomly about one another. In the gas state, the kinetic energy of the particles is very strong, the intermolecular forces or attractions are very weak, and the particles are far apart and move randomly about. Figure 1.21 summarizes the relationship between the state of a substance and kinetic energy and intermolecular forces.

Figure 1.21

States of matter, intermolecular forces, and kinetic energy.

A) Solid State

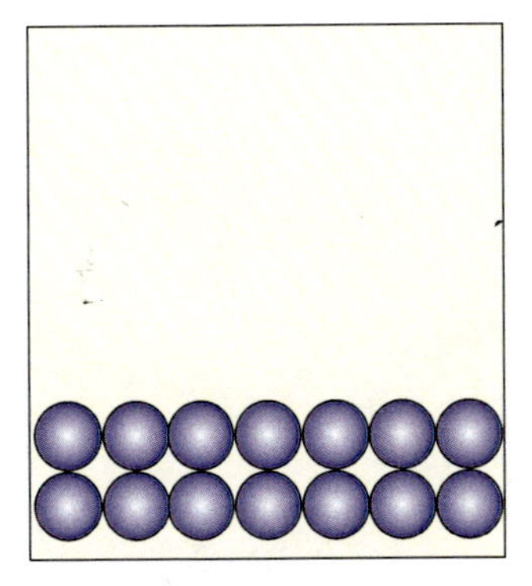

Minimal Kinetic Energy
Very Strong IMF
Highly Ordered

B) Liquid State

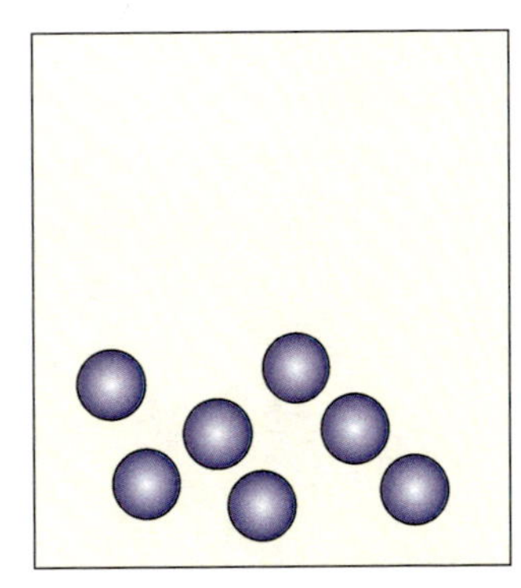

Strong Kinetic Energy
Strong IMF
Less Ordered, More Random

C) Gas State

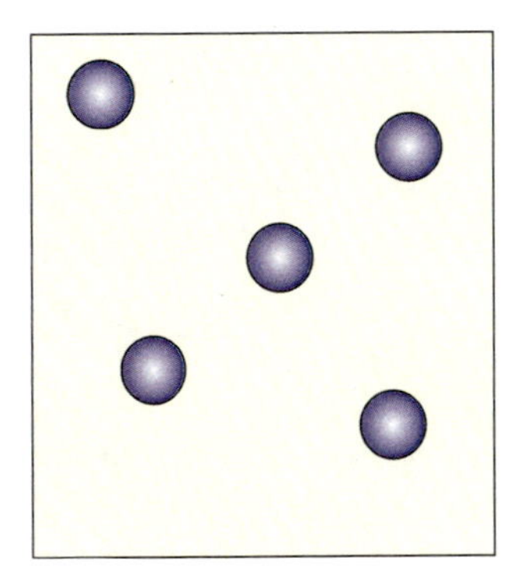

Very Strong Kinetic Energy
Very Weak IMF
No Order, Chaotic

Example 1.4 Determine the State of a Substance When Given Melting Point and Boiling Point

Problem

Calcium carbonate is an odorless white crystalline solid that melts at 825 °C and decomposes at temperatures higher than 898 °C. Calcium carbonate would be in what state at 560 °C?

Solution

Calcium carbonate melts at 825 °C which means it is in the solid state at temperatures below 825 °C. At 560 °C calcium carbonate is a solid.

Practice Problems

1.8 Which process is exothermic?

a) solid ⟶ liquid b) liquid ⟶ solid c) solid ⟶ gas d) liquid ⟶ gas

1.9 The state in which a substance exists at normal temperatures is determined by what two factors?

1.10 Is the process of vaporization endothermic or exothermic?

Summary of Changes of State A change of state is a physical process.

- Each change of state has a reverse process.
- The physical state of a substance is dependent on the strength of intermolecular forces and the temperature.
- Changes of state are either endothermic or exothermic.

Key Terms

Sublimation

Deposition

Intermolecular Forces

Hydrogen Bonding

Heat

Kinetic Energy

Thermal Energy

Endothermic Process

Exothermic Process

Exercises

1. Methanethiol, a compound added to natural gas giving it that distinct odor, freezes at –123 °C and boils at 5.95 °C. What phase is it in at room temperature, 25 °C?

2. Name two factors that determine the physical state of a substance.

3. Acetone, when left in an open container, evaporates very quickly. Is this an endothermic or exothermic process?

4. A piece of solid iron is melted. Does this process require energy or is energy released? Is this endothermic or exothermic?

5. Some air fresheners contain a scented gel. The package is opened and placed in a room. After a month or so, the gel disappears. What is the phase change? Is the process exothermic or endothermic?

Worksheet 1.3 Changes of State

1. If formaldehyde melts at –92 °C and boils at –19 °C, what state is it in at –50 °C?

2. If methane boils at –162 °C and melts at –182 °C, what state will it be in at –200 °C?

3. If the bromine melts at –7.2 °C and boils at 58.8 °C, what is its state at room temperature?

4. A substance sublimes at 25 °C. Is this an exothermic or endothermic process?

5. Identify the following processes as exothermic or endothermic.
   a) Condensation ____________
   b) Melting ________________
   c) Boiling ________________
   d) Deposition ____________
   e) Freezing _______________

6. What are the two phase changes that can occur between a liquid state and the solid state of a substance?

7. A cube of ice is removed from a freezer and placed on the kitchen counter. Describe the phase change it undergoes and why does it occur?

8. Mercury is a liquid at 25 °C. Butane is a gas at the same temperature. Explain why these two substances exist in these states at the same temperature.

9. Water is found to condense on the outside of a cold bottle of soda. Is this an endothermic or exothermic process?

End of Chapter Problems

Section1.1 Physical and Chemical Properties

1. Define chemistry.

2. Define the following:
   a) matter

   b) substance

   c) physical property

   d) chemical property

3. Which of the following is not an example of matter.
   a) rain
   b) leaves
   c) salad dressing
   d) heat

4. Name 4 physical properties that a substance can have.

5. Classify each of the following as a physical or chemical property.
   a) water is a liquid at room temperature
   b) aluminum is shiny
   c) gasoline is flammable
   d) chlorine is a reactive gas

6. Classify each of the following properties of hydrogen as physical or chemical.
   a) is the lightest element on the periodic table
   b) has a melting point of –259.16 °C
   c) will burn in air
   d) has a density of 0.0899 g/L at 0 °C

7. Potassium metal is soft and can be cut easily with a knife. It is shiny and lustrous, conducts electricity, and has a melting point of 63.5 °C. It's boiling point is 659 °C and it reacts violently with oxygen and water in air to produce peroxides and hydroxides. It does not react with most hydrocarbons and readily dissolves in liquid ammonia. From the information given, list the physical and chemical properties of potassium metal.

8. Aluminum metal is a silvery white, soft, metal that is ductile and malleable. It is one of the most abundant metals found in the earth's crust. It has a density of 2.7 g/ cm^3 and it forms aluminum oxide when exposed to air. From the given information, list the physical and chemical properties of aluminum.

9. Name three intensive properties of water.

Section 1.2 Physical or Chemical Processes

10. Name 3 physical changes.

11. Define phase change.

12. Describe the solid, liquid, and gas states in terms of volume and shape.

13. Describe liquid water in terms of
    a) volume
    b) shape
    c) organization of water molecules
    d) closeness of molecules

14. An unknown substance takes the shape of its container and has a fixed volume. Is this a solid, liquid, or a gas?

15. Answer True or False
    a) a solid has variable shape and invariable volume
    b) in a physical process, matter changes in form and in chemical identity
    c) in a chemical process matter can change in form and chemical identity
    d) a gas has both variable volume and shape
    e) phase changes are chemical processes

Section 1.3 Changes of State

16. Butter melts on a very hot day. Is this a chemical or physical change?

17. Fluorine has a freezing point of –219.7 °C and a boiling point of –188.1 °C. Is the element fluorine a liquid, solid, or a gas at 25 °C?

18. Is a change of state a physical or chemical process. Provide an example.

19. What is the reverse process for each of the following?
    a) melting
    b) freezing
    c) condensation
    d) sublimation

20. Which transformation is condensation?
    a) liquid $\longrightarrow$ solid
    b) solid $\longrightarrow$ gas
    c) gas $\longrightarrow$ liquid
    d) liquid $\longrightarrow$ gas

21. Which of the following is not a phase change?
    a) condensation of water
    b) a popsicle melts
    c) salt dissolves in water
    d) nail polish remover evaporates

22. Define each of the following:
    a) heat
    b) thermal energy
    c) kinetic energy
    d) endothermic process
    e) exothermic process

23. Name three phase changes that are exothermic.

24. Name three phase changes that are endothermic.

25. Name two factors that determine the state of a substance.

26. Describe the solid state in terms of kinetic energy, intermolecular forces, and arrangement of particles that make up the solid.

27. The particles of a substance in the gaseous state experience very weak intermolecular forces. Would the kinetic energy of the particles be described as minimal or very strong.

28. Hydrogen peroxide has a melting point of –0.43 °C and a boiling point of 150 °C. Is hydrogen peroxide a solid, liquid, or a gas at 0 °C?

29. Carbon monoxide is an odorless gas that is toxic to humans. It freezes at –205 °C and boils at –191.5 °C. Indicate if carbon monoxide exists as a gas, liquid, or solid at each of the following temperatures.

    a) 25 °C

    b) –210 °C

    c) 100 °C

    d) 300 °C

30. Water freezes at 0 °C. What is the melting point of ice?

Summation Questions

31. Which characteristic describes a physical change?

    a) change in composition

    b) no change in composition

    c) no change in form

32. Name three intensive properties of water.

34. Nickel is a silvery-white lustrous metal that is hard and ductile. It has a melting point of 1,455 °C, a boiling point of 2,730 °C, and a density of 8.91 g/cm^3. Nickel will react with acids to produce hydrogen gas. Nickel does not react with sodium hydroxide.

35. Which of the following is a chemical change? Explain your answer

    a) water evaporating in a dish

    b) sulfur burning to form sulfur dioxide

    c) alcohol freezing at −114 °C

    d) dry ice sublimating at room temperature

36. Figure 1.5 shows the physical sciences that are related to chemistry. Choose one of these sciences and explain in your own words how it uses chemistry.

37. The element iodine melts at 115 °C, but at 100 °C solid iodine can easily sublimate. Describe the change of state that iodine goes through when it sublimates in regards its shape, volume, and the kinetic energy of its particles. Is sublimation an endothermic or exothermic process?

38. The element chlorine is a gas at room temperature; at the same temperature, the elements bromine and iodine are liquid and gas, respectively. Based on this observation, comment on how the intermolecular forces differ between these three substances.

39. A cold bowl of soup is warmed in a hot oven, and later taken from the oven and placed on a table to cool for a few minutes. Describe the heat flow when the soup is first put into the oven, and later after it was placed on the table.

40. A sample of ethane is at –195.0 °C. The temperature of the ethane is changed to –86.3 °C. What are the initial and final physical states of the ethane? Ethane has a melting point of –182.8 °C and a boiling point of –88.5 °C. Is the phase change an endothermic or exothermic process?

How are Substances Classified?

p1

Chapter 2

Chapter Outline

In the previous chapter we referred to methane gas, water, and sodium as substances. We can now characterize these substances into elements and compounds. Chemistry is the study of all of the "stuff" in the universe. In chemistry we refer to this "stuff" as matter. Recall, that matter is anything that has mass and occupies space. Flowers, trees, sand, water, methane gas, sodium, gold, people, DNA, are all examples of matter.

Matter can be classified into two separate groups. The first group are **pure substances** which consists of elements and compounds. An **element** is a pure substance that cannot be chemically broken down into simpler substances. Elements are composed of atoms. An **atom** is the smallest particle of an element that still retains all of the element's characteristics. A **compound** is a pure substance that can be broken down chemically into simpler substances. The second group consists of *mixtures of pure substances* and may be considered an extension of the first group, Figure 2.1. For now, we will consider elements and compounds.

Pure substances Elements and compounds.

Element: A pure substance that cannot be chemically broken down into simpler substances.

Atom: The smallest particle of an element that still retains all of the element's characteristics.

Compound: A pure substance that can be chemically broken down into simpler substances by chemical means.

Figure 2.1

A) Structure of DNA—a pure substance B) calcium carbonate, the major component in the structure of seashells, and C) gold jewelry which is a mixture of gold alloyed with another metal(s), usually silver and copper.

A)

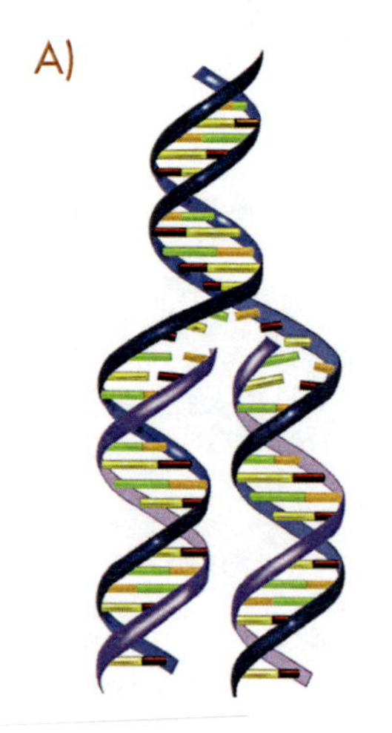

B)

C)

2.1 Elements

There are over 114 known elements in the universe. Most of us are already familiar with some of their names. We are well acquainted with the names carbon, gold, copper, iron, oxygen, and hydrogen to name a few. These are all **elements**, each of which is made up of its own unique type of atom. The Periodic Table, Figure 2.2, lists the names and symbols of all the known elements.

Figure 2.2

The periodic table of the elements. Each element is represented by a chemical symbol.

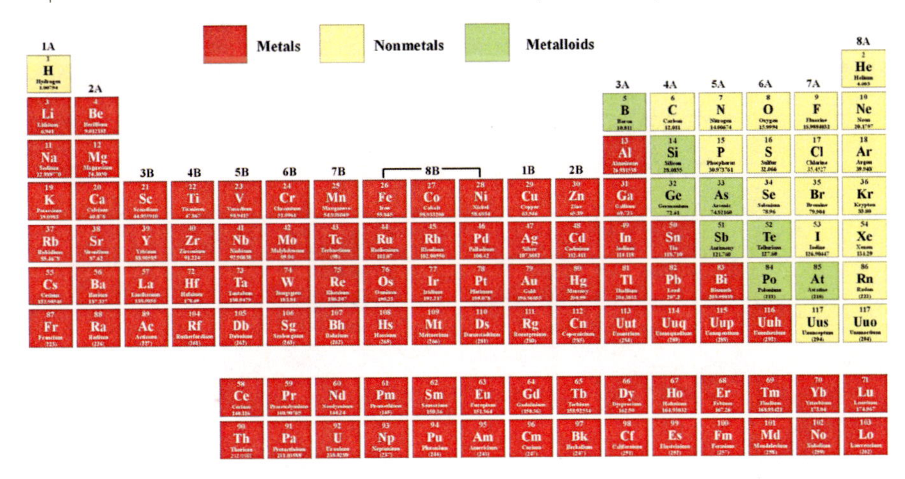

Each element has its own position on the periodic table. The element hydrogen is located in the upper left-hand corner of the periodic table. The elements are referred to by one and two letter symbols. One letter symbols are always capitalized while only the first letter of a two letter symbol is capitalized and the second letter is lowercase. The element symbol for hydrogen is the capital letter H and the element symbol for sodium is Na. There is a number directly above each element symbol. This number is called the *atomic number* and is represented by an uppercase **Z**. The atomic number for hydrogen is 1. The element symbol for gold is Au, and the atomic number is 79. The elements are arranged on the periodic table by atomic number.

Exercise 2.1 Using your periodic table as a reference, fill in the missing information in the following table.

Element	Symbol	Atomic Number
Magnesium	Mg	12
Sodium	Na	11
Molybdenum	Mo	42
carbon	C	6
Nitrogen	Ni	7
Chromium	Cr	24
Bromine	Br	35
Antimony		
mercury	Hg	80
Neon	Ne	10

Exercise 2.2 Write the symbol and name for the element that has atomic number 13.

Al aluminum

Exercise 2.3 What is the name and atomic number for Sn?

Tin 50

Atom vs. Elements

At the beginning of their chemistry studies, many chemistry students confuse the terms *atoms, elements, and elemental.* Are they the same in essence? Are the terms interchangeable? You can think of an element as the name of a certain type of atom. For example the element carbon is made up of carbon atoms, and the element nitrogen is made up of nitrogen atoms.

The term **elemental** is used to describe the most stable physical state that an element exists in under normal conditions. For example, the most stable form for chlorine atoms is as discrete diatomic structures, i.e. two chlorine atoms bonded together comprise one unit or molecule of elemental chlorine.

Figure 2.3

Chlorine in its elemental form.

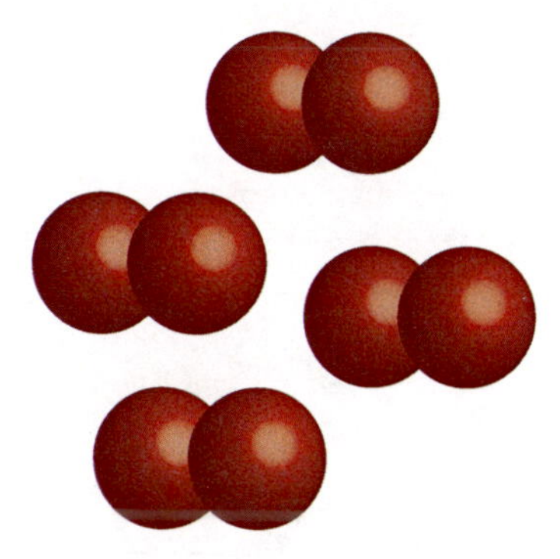

Elemental: A term used to describe the most stable physical state that an element exists under normal pressures and temperatures.

On the periodic table, the symbol for chlorine is Cl (the *monoatomic* species). When referring to the elemental form of chlorine, Figure 2.3, the chemical symbol is Cl_2, the *diatomic* species. Elemental chlorine occurs as a gas in nature.

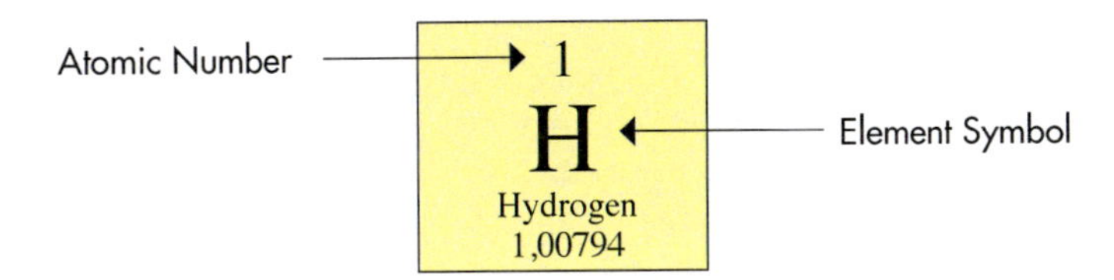

In contrast to *elemental chlorine*, carbon atoms in the most stable state occur in the form of solid graphite. Pencil lead is an example. This form of carbon occurs as 6-membered rings that are fused together producing extensive sheets which form layers, Figure 2.4. Even though this structure of elemental carbon seems complex, its chemical symbol is a single C.

Figure 2.4

Elemental carbon.

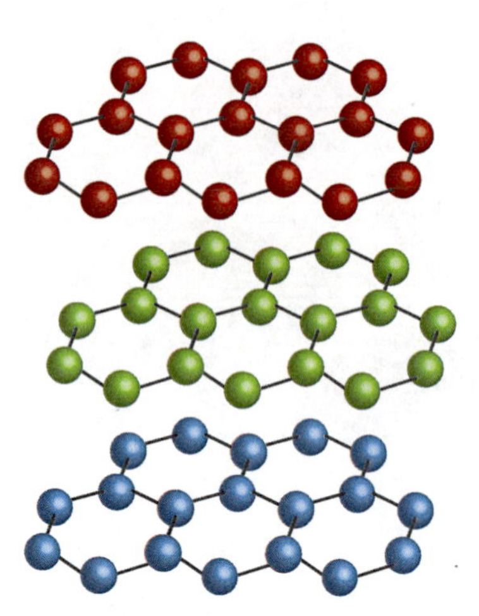

Regardless of how atoms bond together or whether they exist as a gas, liquid, or a solid in an elemental state, the primary difference between different elements is the type of atoms involved. For example, the element carbon is made up of carbon atoms, whereas the element gold is made of gold atoms. We will address this difference in detail later, but for now understand that any atom can be broken down into a collection of its subatomic building blocks—protons, electrons, and neutrons which will be discussed in more detail later—by doing so, the atom would lose its elemental identity, and that both the carbon and gold atoms would become just piles of subatomic particles. An **atom** is the smallest particle of an element that still retains all of the characteristics of that element.

Atom: The smallest particle of an element that still retains all of the element's characteristics.

More About Elements: Elements vs. Atoms

Remember, when talking about elements, we are referring to those atoms in their most stable form. That is, elemental sodium, Figure 2.5, occurs as a collection of individual sodium atoms. Elemental sodium is monotomic whereas elemental oxygen, Figure 2.6 occurs as a collection of two oxygen atoms bonded together. Oxygen is a diatomic species. Phosporous, in its most stable elemental form, will occur in groupings of four phosphorus atoms, Figure 2.7. The symbol for elemental phosphorus is P_4. This is called a tetratomic species.

Figure 2.5

Representation of elemental sodium, Na.

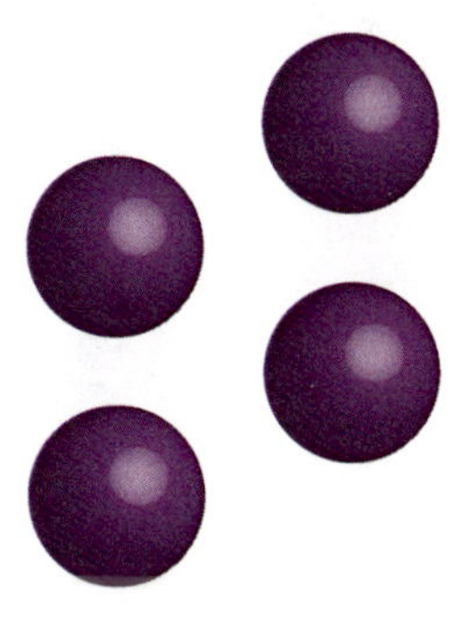

Figure 2.6

Representation of elemental oxygen, O_2, in its diatomic state.

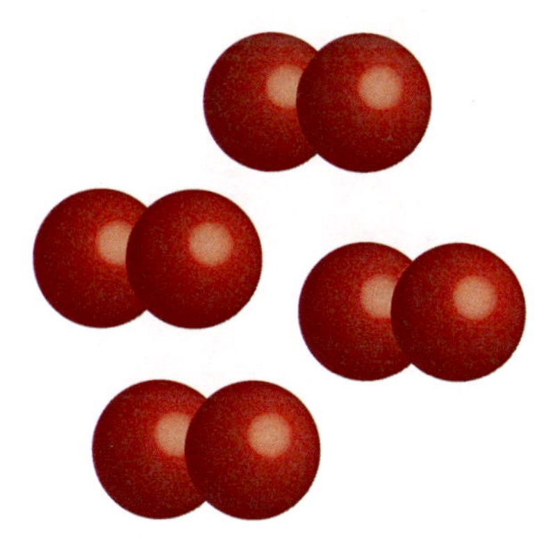

Figure 2.7

Representation of the most stable form of phosphorus, P_4.

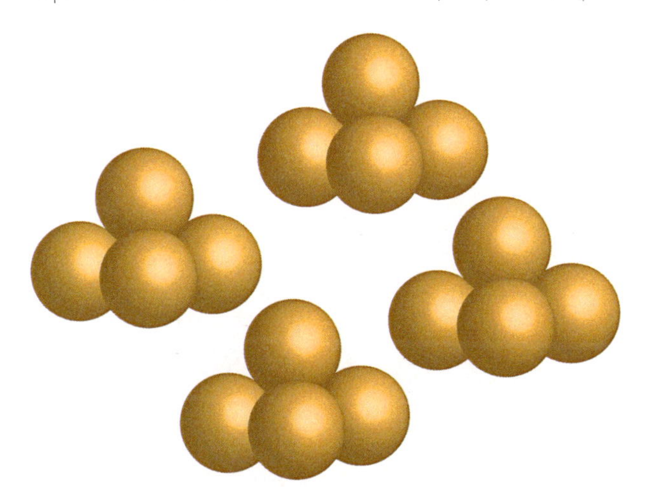

Keep in mind that when referring to elements, the atoms occur in their most stable forms; that is for example, as monatomic, diatomic, or tetratomic species. For now, just to keep things simple, we will refer to the elements as being only monatomic or diatomic species. This simplifies matters greatly since only seven elements occur as diatomic species: hydrogen, nitrogen, oxygen, fluorine, chlorine, bromine, and iodine, Figure 2.8.

Figure 2.8

Periodic table of the elements. Elements highlighted in yellow exist as diatomic species. Notice that, if you leave out hydrogen, the remaining six elements form an upside down L.

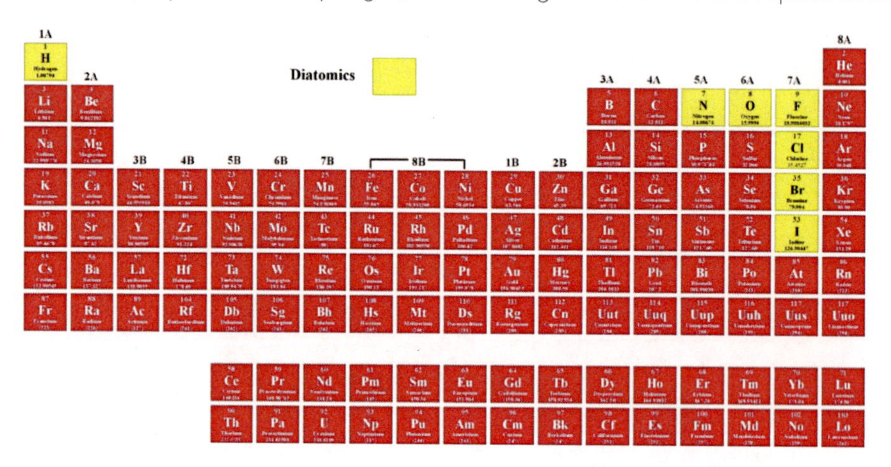

2.2 Design of the Periodic Table

As you already know, the periodic table is a list of all known elements; some of which have only been synthesized in laboratories and have fleeting lifetimes. At first glance, the periodic table seems to be merely the listing of these elements, but it is within the arrangement of these elements in the table that a wealth of information is stored. The key to learning chemistry can be found in the structure of the periodic table. We can predict physical and chemical properties of an element by its position on the table.

The most obvious structure in the periodic table is the arrangement of atomic numbers. We will learn in the next section that these numbers indicate the number of protons within an atom. For example, the carbon atom has an atomic number of six and it contains six protons, whereas the oxygen atom has an atomic number of eight and it contains eight protons. The atomic numbers of the elements increase going from left to right and from top to bottom on the periodic table.

Periods and Groups

The horizontal rows of the periodic table are called **periods**. The period number is indicated outside of the periodic table on the left. In Figure 2.9, the 3rd period is highlighted in yellow. The uppermost row contains the two elements hydrogen and helium. This row is referred to as the first period. *The number of the period increases from top to bottom*. For example, the element sodium, Na, is in the third period as is the element aluminum, Al.

Period: The horizontal rows of the periodic table.

Figure 2.9

The elements highlighted in yellow are in the third period.

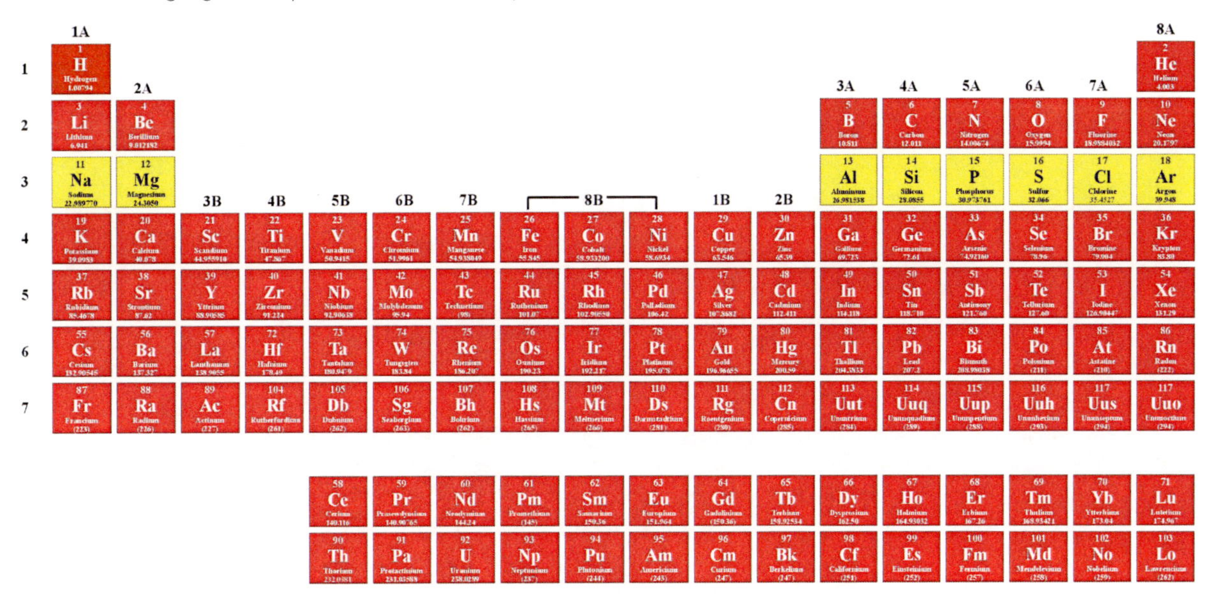

The periodic table is also divided up into vertical columns called groups. There are 2 numbering systems used: 1) from left to right the groups are numbered 1 to 18 and 2) from left to right the groups are listed numerically, that is 1A, 2A, 3B, etc. Elements within the same group display similar chemical activity. In Figure 2.10, group 2 A is highlighted in yellow. The A groups are known as the main group elements (Groups 1A to 8A) shown in Figure 2.12. **The main group elements** are also known as the *representative elements.*

Figure 2.10

Elements highlighted in yellow are in Group A.

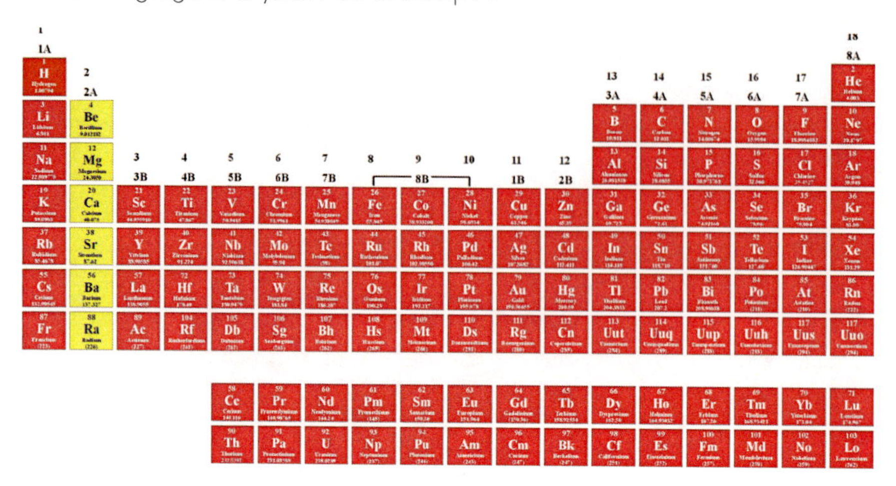

Main Group Elements: The elements in Group 1A through Group 8A (Groups 1,2,13,14,15,16,17, and 18).

Figure 2.11

Marine group elements (representative elements) are highlighted in yellow.

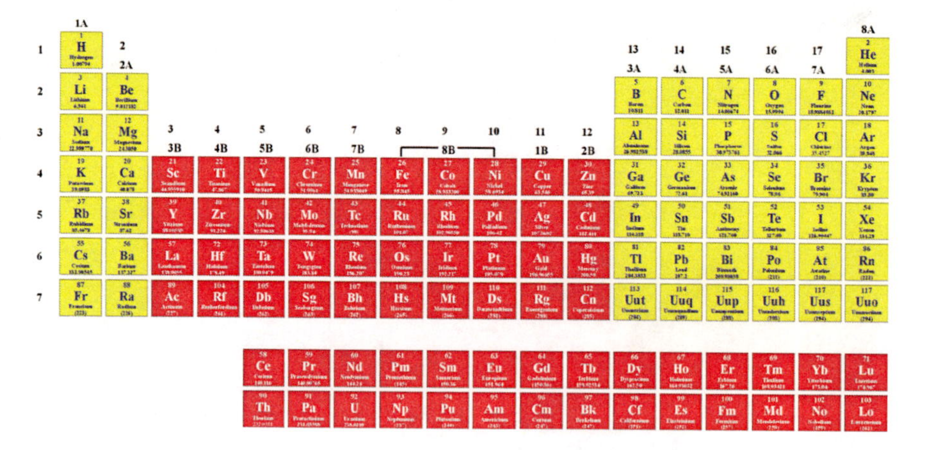

Exercise 2.4 Periods and Groups

Fill in the following table. Refer to your periodic table.

Element	Symbol	Period	Group
Sodium	Na	3	1A
magnesium	Mg	3	2A
silicon	Si	3	4A
Chromium	Cr	4	6B
sulfur	S	3	6A
Chlorine	Cl	3	7A
Oxygen	O	2	6A
Selenium	Se	4	6A
Cesium	Cs	6	1A
palladium	Pd	5	1B

Main Group Elements

Main Group Elements contain metals, nonmetals, and metalloids (discussed later in this chapter). Most exist as solids, several as gases, and several as liquids at normal temperatures and pressures. (0°C, 1 atm). When investigating the various trends of the elements, our concentration will be on Main Group Elements. Of the eight groups making up the main group elements, four of the groups have special names.

Group 1 A:	Alkali Metals
Group 2 A:	Alkaline Earth Metals
Group 7 A:	Halogens
Group 8 A:	Noble Gases

Elements in groups 1A and 2A, including helium, are in the region known as the s-block. Elements in groups 3A through 8A are in the region known as the p-block.

Transition Elements

The transition metals are located in the central block (d-block) of the Periodic Table, within periods 4 through 7 highlighted in Figure 2.12. All transition metals are solids with the exception of mercury, Hg, which is a liquid at normal pressures and temperatures. Transition metals tend to form brightly colored chemical compounds due to the nature of their outermost electrons. Both industrial catalysts and biological enzymes contain transition metals.

Figure 2.12

The transition elements highlighted in yellow.

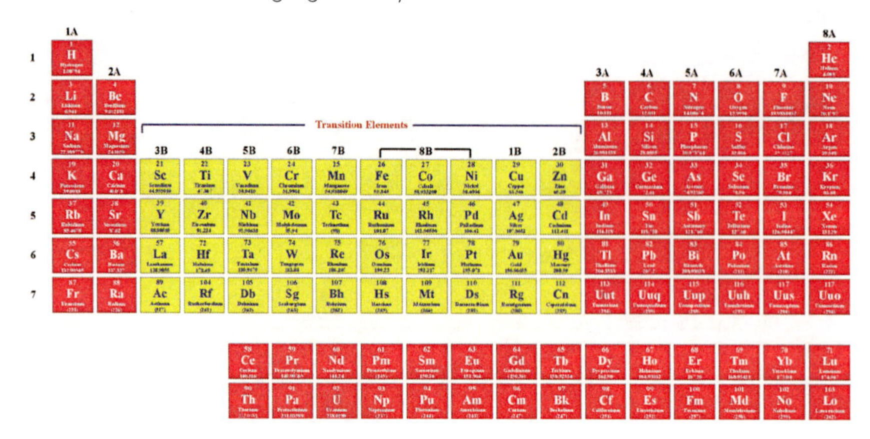

Inner Transition Elements

These other transition metals are located separate from and at the bottom of the main Periodic Table; this region is known as the f-block (Figure 2.13). These metals are composed of the **Lanthanide Series** which immediately follows the element lanthanum and includes the elements from cerium to lutetium, and the **Actinide Series** which immediately follows the element actinium and includes the elements from thorium through lawrencium. Of the Actinide Series, only the first three elements (thorium, protactinium, and uranium) naturally occur. All the others are synthesized in a laboratory.

Figure 2.13

Inner transition elements highlighted in yellow. The first row is called the lanthanide series. The second row is reffered to the actinide series.

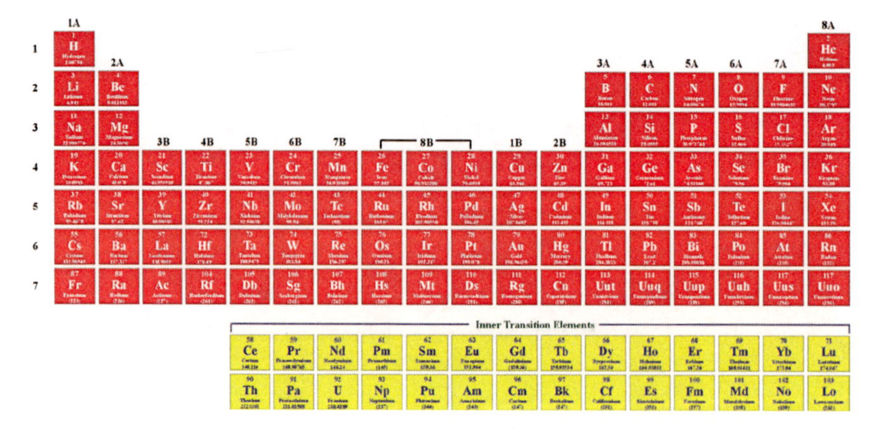

Metals, Nonmetals and Metalloids

All the elements in the periodic table can be divided into three regions according to their properties: metals, nonmetals, and metalloids (also known as semi-metals).

Figure 2.14

Metals, metalloids, and nonmetals.

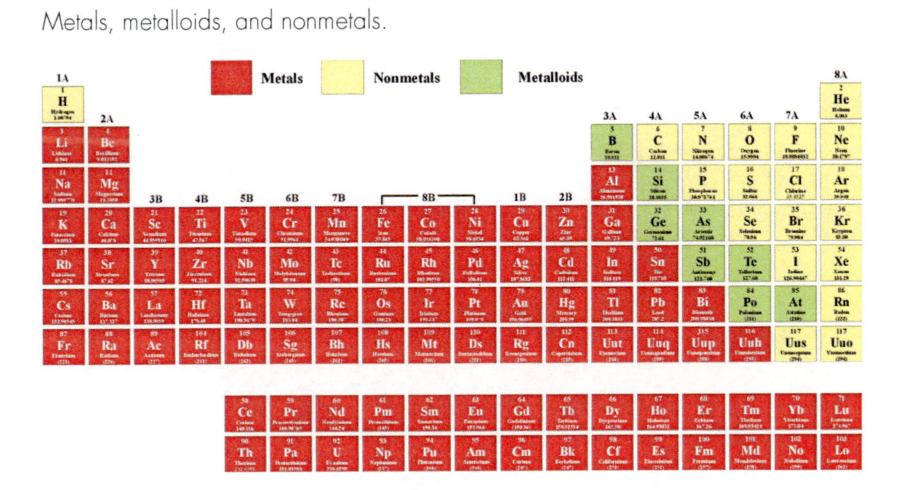

From the shaded area of the above periodic table, it is obvious that most of the elements fit into the category of metal. Their properties include an illustrious surface, the ability to conduct heat and electricity, and they are both malleable and ductile. With the exception of mercury (Hg) which is a liquid, all of the metal elements are solids.

Unlike metals, which exist as solids (with the exception of liquid mercury), nonmetals exist as solids, liquid, and gases at room temperature. For example, in the halogen group the element chlorine exists as a gas, the element bromine exists as a liquid, and iodine exists as a solid. The properties of nonmetals are also quite different from those of metals. Nonmetals tend to have dull surfaces. They are not conductors of electricity or heat, and they are brittle.

The remaining six elements are called metalloids because their characteristics and properties overlap in both the metals and the nonmetals, that is, physical properties of metals and chemical properties of nonmetals. The metalloids are also referred to as semi- metals. An important property of metalloids is their variable conductivity. Because of this property, they also go by the well-known name of semiconductors, which has played a big part of putting Silicon Valley on the map.

Exercise 2.5 Answer the following questions.

1. In the following group, neon, argon, selenium, and helium, which element does not belong?

   selenium

2. Of the following elements, fluorine, nitrogen, phosphorus, and bromine, which are in the same period? fluorine + Bromine

   nitrogen + phosphorus

3. What group are the elements copper, silver, and gold in?

   1B

4. Identify the element found in period 5 and group 5A.

   Sb Antimony

5. Identify the element found in period 3 and group 4A.

   Si Silicone

6. Which of the following elements is not a metal?
   chromium, calcium, phosphorus, and palladium

7. Identify the metals, nonmetals, and metalloids that exist in group 5A.

8. Which of the following elements is not considered Main group?
   cesium, osmium, xenon, and nitrogen

9. Which of the following elements is not diatomic?
   hydrogen, chlorine, nitrogen, and boron

   boron

10. Which of the following elements have similar chemical properties?
    chlorine, phosphorus, sulfur, and bromine

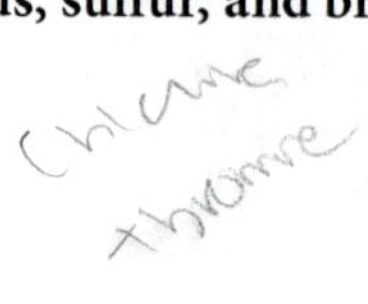

2.3 Compounds

Now that we have defined an element as a collection of just one type of atom, we shall now consider the second type of pure substance, the compound. Unlike an element, a compound is a collection of more than one type of atom. Recall that both elemental nitrogen (N_2) and elemental oxygen (O_2) occur as diatomic species. However, elemental nitrogen and elemental oxygen can be mixed together under certain conditions to form a new species in which one oxygen atom is bonded to one nitrogen atom (Figure 2.15). This new species is called a compound. By definition, **compounds** *are pure substances consisting of more than one type of atom.* As can be seen in the above reaction, a new chemical substance is formed, nitric oxide, NO, a compound composed of both one nitrogen atom and one oxygen atom.

Figure 2.15

Elemental oxygen and nitrogen react to form nitric oxide, a compound.

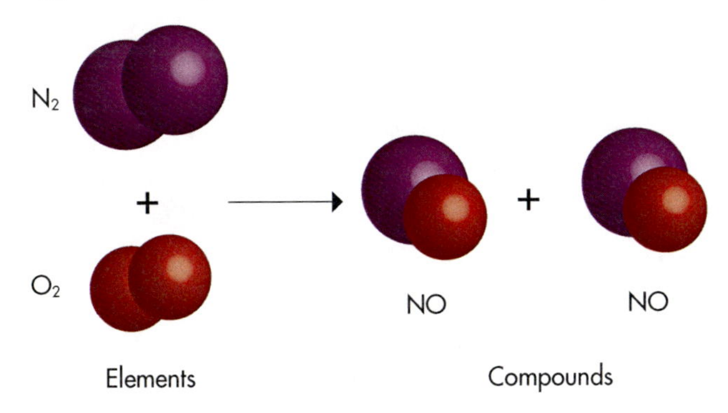

Some common examples of compounds are carbon monoxide (CO), sodium bicarbonate ($NaHCO_3$), dihydrogen monoxide (H_2O), sodium chloride (NaCl), and calcium carbonate ($CaCO_3$), Figure 2.16. The symbolic notations, NO, CO, and $NaHCO_3$, are chemical formulas. Chemical formulas indicate the type and number of each atom in the compound.

Figure 2.16

Common compounds. A) carbon monoxide, CO, B) sodium bicarbonate, $NaHCO_3$, also known as baking soda, C) dihydrogen monoxide, H_2O which is water, D) sodium chloride, NaCl, which is table salt, and calcium carbonate, E) $CaCO_3$ also called chalk.

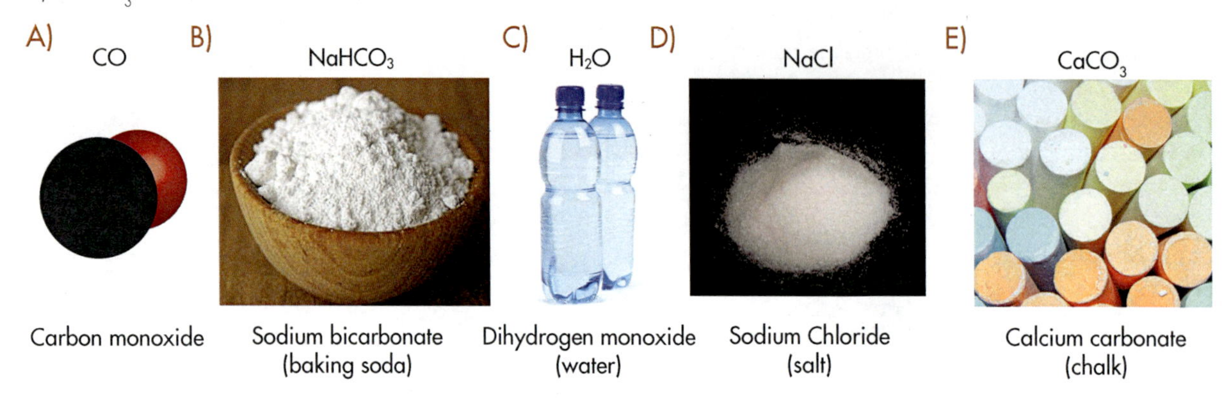

The compound nitric oxide has the chemical formula NO. This compound contains 1 nitrogen atom and 1 oxygen atom. Water has the chemical formula H_2O and contains 2 hydrogens and 1 oxygen. Elemental oxygen, O_2 contains two oxygen atoms. Baking soda (sodium bicarbonate) has the formula $NaHCO_3$ and contains 1 atom of sodium, 1 hydrogen atom, 1 carbon atom, and 3 oxygen atoms.

Compounds are also defined as having **constant composition**. That is, the compound carbon monoxide, CO, must always consist of one carbon atom and one oxygen atom. To add an additional oxygen atom to this compound would yield a new substance, carbon dioxide (CO_2). By the same token, to add a second oxygen to water, H_2O, would yield hydrogen peroxide, H_2O_2.

Compounds can be chemically broken down into their elemental components. For example, hydrogen peroxide, H_2O_2, can be broken down into elemental hydrogen and elemental oxygen. Hydrogen and oxygen gas can be produced from water, H_2O, when an electrical current is passed through it. Recall that unlike compounds, elements cannot be chemically broken down into simpler substances.

Exercise 2.6 Write the chemical formula for caffeine if it contains eight carbon atoms, ten hydrogen atoms, four nitrogen atoms, and two oxygen atoms.

C_H_N_O_

Exercise 2.7 The chemical formula for ibuprofen is $C_{13}H_{18}O_2$. Indicate how many of each type of atom are in this compound.

Carbon ____ Hydrogen ____ Oxygen ____

Assignment 2.8 Indicate if the pure substance is an element or a compound.

Pure Substance	Element or Compound
$NaCl$	compound
CH_4	compound
Pb	element
N_2	element
$C_6H_{12}O_6$	compound
P_4	elemen
$Fe(OH)_2$	compound
H_2	element
Na_2SO_4	compound
H_2O	compound
CO_2	element

2.4 Mixtures of Pure Substances

Having discussed the various aspects of both elements and compounds let's now look at the properties of these same substances when mixed together. A **mixture** is a combination of two or more substances. Substances in a mixture do not chemically react and each substance retains its own chemical identity. Suppose we mixed together the two pure substances, water (H_2O) and salt ($NaCl$). The result, to no surprise, is a mixture we commonly refer to as saltwater. If you are told to make the mixture of salt and water, how much salt would you add to the water? Given these directions, you could make a saltwater mixture containing very little salt or a lot of salt. We need to be told how concentrated to make the mixture. *In mixtures, the amount of any component may vary.* That is, mixtures have variable composition.

Describing Mixture Composition

At 20°C, a 100 g saltwater mixture at room temperature may contain up to about 36 grams of sodium chloride. If 100 g of a salt water mixture contains 10 g of sodium chloride it is described as a 10% mixture of sodium chloride. If the mixture contains 20 grams of sodium chloride in 100 g of salt water mixture, it is described as a 20% solution. This represents just one of several ways to describe mixture or solution composition. More about this later.

Increasing the concentration of sodium chloride in the water will not only increase the electrical conductivity of the water, the boiling point of the water will also increase accordingly. A 100 gram sample of pure water boils at 100°C under normal atmospheric pressure. If 11.32 grams of sodium chloride are dissolved into the water, its boiling point will increase to 101°C. If an additional 11.32 grams of sodium chloride are added to the mixture, the boiling point will increase another 1°C. It is also interesting to note that while the boiling point of saltwater mixture increases, the freezing point of the same mixture decreases to −0.054°C and −0.108°C, respectively. This is why we use salt on icy roads. The dependence of the physical properties of mixtures on the concentration of their individual components are called colligative properties. **Colligative properties** occur due to the physical interactions of the components in the mixture and not because of any chemical reactions between the components.

Most people know what happens when a saucer of saltwater is left outside in the sun for an extended period of time. The water will evaporate away from the mixture, leaving behind a saucer full of salt, Figure 2.17. In other words, the salt that was added to water can be recovered. In fact, using the proper apparatus (a distillation apparatus) both the salt and the water can be recovered. *Each component can be recovered from a mixture.* This is a very important property of mixtures. The fact that the various components of a mixture can be recovered intact, tell us that while in the mixture, these same components maintain their chemical identities. *In mixtures, the components do not chemically react with each other.*

Figure 2.17

Water has evaporated away from the salt.

Many people are not aware that the pure substance water does not conduct electricity. It's all the extra stuff, like metal ions, that give tap water its electrical conductivity. Salt is one such substance that produces electrical current in water. By increasing the concentration of salt in a given volume of water the electrical conductivity of the mixture is also increased.

Consider another physical property that is affected by changing the amount of salt in the water. When adding salt to a pot of boiling water on the stove the bubbling momentarily stops. This is because the addition of the salt raises the boiling point of the water. *Properties of mixtures (conductivity, boiling point...) vary with the amount of substances.*

Types of Mixtures

With final regards to the salt water mixture, let's discuss the ways in mixtures are described. Into 100 ml of water, mix 1 g of salt. The salt will completely dissolve, resulting in a clear, colorless mixture. In this mixture, the salt and the water are uniformly mixed. This is called a **homogeneous mixture**, which is also known as a **solution**. To say homogeneous solution is redundant and unnecessary. When considering solutions, most of us think in terms of liquids. In reality, homogeneous mixtures or solutions may consist of solids, liquids, or gases. Some examples are given in Table 2.1. A solution can be composed of one or more solutes dissolved in a solvent. The **solute** is present in the lesser amount while the **solvent** is present in the larger amount. For example, when salt is added to water it dissolves in the water resulting in a solution. The salt is the solute and the water is the solvent. When water is a solvent, we refer to the mixture as an **aqueous** solution.

Table 2.1

Example of Homogeneous Mixtures.

Medium	Name of Solution	Composition
Gas in gas	Air	21% O_2 78% N_2
Liquid in liquid	Rubbing alcohol	70% isopropyl alcohol 30% water
Solid in solid	Brass	70% copper 30% zinc
Liquid in liquid	Gasoline	Mixture of hydrocarbons
Solid in liquid	Seawater	NaCl and water
Gas in liquid	Soda water	Co_2 gas in water

Salt dissolves easily in water, that is, it's very soluble in water. But in reality, there is a limit of how much salt will dissolve in the water. **Solubility** is the amount of solute that dissolves in a given amount of solvent usually given in grams of solute per 100 mL of solution (g/100 mL). The solubility of salt is 35.7 g/100 mL at 20°C. In 100 ml of water, at 20°C, about 35 g of salt will easily dissolve resulting in a homogeneous mixture. This mixture is said to be **unsaturated**. When additional salt is added to the solution, a precipitation of salt will occur. That is, no additional salt can be dissolved in the water. The solution is said to be saturated. A **saturated** solution contains the maximum amount of solute that can dissolve in a given amount of solute.

A **heterogeneous mixture** is not uniform throughout. You can see the components of the mixture, whereas a homogeneous mixture is one that is uniform throughout—you are not able to see the components with the naked eye. Examples of heterogeneous mixtures include Italian salad dressing, paint, granite, sand and water, seawater, chocolate chip cookies, orange juice with pulp, etc. Examples of homogeneous mixtures include sugar dissolved in water, gold alloys, shaving cream, milk, gasoline, etc.

2.5 Nomenclature of Compounds I

The IUPAC (International Union of Pure and Applied Chemistry) convention is used to assign names to chemical compounds. Scientists from around the world formed IUPAC because they recognized the need for standardization in chemistry. IUPAC sets the guidelines for how compounds are named.

There are two main divisions of compounds. Covalent compounds which are composed of only nonmetal species. Some examples are CO, NO_2, SCl_2, CCl_4. The second division are ionic compounds which are composed of both metal and nonmetal species such as NaCl, MgO, K_2SO_4, and NaOH. The differences in the bonding characteristics of the two groups of compounds will be discussed in detail in later sections of this book. For now you need to focus on differentiating between covalent and ionic compounds when presented with a chemical formula and learn the basic rules for naming these compounds.

In naming **covalent** compounds, the prefixes in Table 2.2 are used to indicate the number of each atom in the compound.

Table 2.2

Prefixes used to indicate the number of atoms in a compound.

Number of Atoms	Prefix	Number of Atoms	Prefix
1	mono	6	hexa
2	di	7	hepta
3	tri	8	octa
4	tetra	9	nona
5	penta	10	deca

There are three rules in naming binary covalent compounds:

1. Never use the prefix mono in naming the first atom of a covalent compound

2. If an atom in the compound starts with a vowel, drop the vowel at the end of the prefix.

3. Modify the suffix of the second atom to have an –ide ending.

In naming the compound CO, all three rules are demonstrated.

CO is called carbon monoxide. The prefix mono is used for the second atom but not for the first atom. The second atom, oxygen, becomes oxide. Because oxygen begins with a vowel we drop the o at the end of mono. We have mono + oxide = monoxide.

NO_2 is named nitrogen dioxide, SCl_2 is named sulfur dichloride, and CCl_4 is named carbon tetrachloride.

Exercise 2.9 Name the following covalent compounds.

N_2O_5, SF_6, and ClF_2

2.6 Nomenclature of Compounds II: Metal–Nonmetal Compounds: Ionic Compounds

Whenever a chemical formula contains a metal and a nonmetal, the result is an ionic compound, and just as the name indicates, this chemical species is composed of ions. These compounds are structurally very different from covalent compounds. But before continuing, a brief introduction to ions and their nomenclature is in order. In an atom, the number of protons, which have positive charges, equals the number of electrons which have negative charges. All positive and negative charges cancel out. **Ions** are atoms that have either lost one or more electrons. The sodium atom, Na, loses an electron to form the sodium ion (Na^+) and the atom calcium, Ca, loses 2 electrons to form the calcium ion (Ca^{2+}). Notice the charge is positive and the number superscript corresponds to the number of electrons lost. Some atoms gain one or more electrons to become negatively charged. For example, the fluorine atom, F, gains one electron to form the fluoride ion (F^-) and the oxygen atom gains two electrons to form the oxide ion (O^{2-}). When atoms gain electrons the charge is negative and the superscript number corresponds to the number of electrons gained. If only one electron is lost or gained, the number "1" is assumed.

Atoms such as sodium and fluorine will form just one ion, Na^+ and F^-, while other atoms can form more than one ion. For example, iron can lose 2 electrons to form Fe^{2+} or 3 electrons to form Fe^{3+} ions. These simple ions, which are derived from single atoms, are called monatomic ions. The rules for naming **monatomic** ions are straightforward.

1. Ions with negative charges are named by:

   The element name (drop ending) + add –ide + ion

   Some examples are: Cl^- is the chloride ion, O^{2-} is the oxide ion and S^{2-} is the sulfide ion

2. Metals that form positively charged ions of only one charge are named as:

   The element name + ion

   Some examples are: Na^+ is the sodium ion, Ca^{2+} is the calcium ion, and K^+ is the potassium ion.

3. Metals that form cations of more than one possible charge are named as:

   The element name + roman numeral which indicates the charge + ion

Fe^{2+} is the iron(II) ion, and Fe^{3+} is the iron(III) ion. These are pronounced as the iron two ion and iron three ion. Ions that can form two or more different cations are also referred to by their common names. For example the common name for the iron(II) ion is the ferrous ion and for iron(III) is the ferric ion. The suffixes –ic and –ous are added to the latin names of the elements. The suffix –ic is used for the cation with the smaller charge while –ous is used for the cation with the higher charge. Table 2.3 lists both the systematic and common names for some common metals that can form two different cations. In this text we will use the systematic method for naming these ions.

Table 2.3

Common and systematic names for metals that can form more than one cation.

Metal	Ion Symbol	Systematic Name	Common Name
Iron	Fe^{2+} Fe^{3+}	Iron(II) ion Iron(III) ion	Ferrous ion Ferric ion
Tin	Sn^{2+} Sn^{4+}	Tin(II) ion Tin(IV) ion	Stannous ion Stannic ion
Copper	Cu^{+} Cu^{2+}	Copper(I) ion Copper(II) ion	Cuprous ion Cupric ion
Chromium	Cr^{2+} Cr^{3+}	Chromium(II) ion Chromium(III) ion	Chromous ion Chromic ion

Since the sodium atom only forms one ion, Na^{+}, it is not necessary to indicate the sodium ion as sodium(I) ion. The same holds for the calcium and potassium ions. The method in determining the charge on these simple monatomic ions employs use of the Periodic Table. Group 1A atoms form ions with a +1 charge, Group 2A atoms form ions with a +2 charge, Aluminum in Group 3A forms an ion with a +3 charge, Group 6A atoms form ions with a –2 charge and Group 7A atoms form ions with a –1 charge. The periodic table in Figure 2.19 shows the common monotomic ions.

Figure 2.19

Common monatomic ions.

1A	2A	3A	4A	5A	6A	7A	8A
Li^+				N^{3-}	O^{2-}	F^-	
Na^+	Mg^{2+}	Al^{3+}		P^{3-}	S^{2-}	Cl^-	
K^+	Ca^{2+}				Se^{2-}	Br^-	
Rb^+	Sr^{2+}				Te^{2-}	I^-	
Cs^+	Ba^{2+}						

Ionic Compounds

Unlike the covalent compounds, prefixes are not used in naming ionic compounds. For example $BaCl_2$ is named barium chloride (not barium dichloride). Here is why: With covalent compounds there can be more than one combination of two atoms. A few examples of nitrogen and oxygen compounds:

NO	NO_2	N_2O_5
nitrogen monoxide	nitrogen dioxide	dinitrogen pentoxide

With ionic compounds, there is only one combination of any two species.

As the name **ionic compound** indicates, these compounds are made up of ions, which is the result of combining metals and nonmetals. Later we will see that some ions do not contain metal atoms. Recall that ions are charged atoms. That is, atoms that have either picked up additional electrons, negatively charged ions called **anions** or lost electrons, positively charged ions called **cations**—pronounced cat-ion. As demonstrated above, there are many combinations of nonmetals, however in dealing with charged species, such as ions, only one combination is possible. This will make more sense in light of one more rule: *All compounds, both covalent and ionic, have a zero net charge.* Since covalent compounds are composed of neutral nonmetallic atoms, this rule isn't apparent, but since ions are charged species, all charges must cancel in ionic compounds. For example, there is only way to combine ionic calcium, Ca^{2+} and ionic fluorine, F^- one calcium ion containing two positive charges and two fluorine ions each containing one negative charge, CaF_2. Two positive charges plus two negative charges sum to zero. Since there is only one combination of these two ions, calcium fluoride is the proper name (not calcium difluoride). Yes, there are multiples of this combination, which would also yield a neutral species, Ca_2F_4, Ca_3F_6, and $Ca_{10}F_{20}$ to name a few. In determining the proper combination of the ions in the formula of an ionic compound, *the simplest ratio of ions is the rule.*

Other examples of ionic compounds are sodium chloride, NaCl w... Na^+ ion and one Cl^- ion. Magnesium oxide, MgO, is composed of one M... and one O^{2-} ion. Two Na^+ ions and one S^{2-} ion combine to form sodium sulfi... Na_2S. Again, ionic compounds are neutral and the charges must add up to zero. In writing formulas and names of ionic compounds, the positive ion always comes before the negative ion.

When the Fe^{3+} ion is combined with Cl^- ions, the formula is $FeCl_3$. The name of this compound is iron(III) chloride. Recall that iron is a metal that can form more than one ion. Combining Fe^{2+} ions and Cl^- ions gives iron(II) chloride, $FeCl_2$. The combination of Cu^+ and F^- ions will give CuF called copper(I) fluoride. Cu^{2+} ions and Cl^- ions form $CuCl_2$ which is called copper(II) chloride.

Exercise 2.10 Given the following ions, determine the formulas and names of the ionic compounds. In writing the formulas and names of compounds, the positive ion always comes before the negative ion.

Ions	Formula	Name
Al^{3+} Br^-	$AlBr_3$	Aluminium Bromide
Ca^{2+} P^{3-}	Ca_3P_2	Calcium Phosphide
Li^+ O^{2-}	Li_2O_2	Lithium Oxide
Al^{3+} Cl	$AlCl_3$	Aluminum Chloride
Al^{3+} O^{2-}	Al_2O_3	Aluminum Oxide

Compounds vs Polyatomic Ions

The nitrate ion, NO^{3-}, is an example of what is called a polyatomic ion. A **polyatomic ion** is an ion that contains more than one atom. Polyatomic ions can be negatively or positively charged. By inspection, its formula resembles a normal molecule. In the section, Chemical Bonding and Structure, a closer look will be taken at such ionic structures.

Since a polyatomic ion is a charged species, it is not regarded as a molecule. Molecules do not possess a net charge. The polyatomic ions are to be treated in the same respect as the simpler monatomic ions. Just as one sodium ion balances one chloride ion,

...hloride

...one nitrate ion.

NO^{3-} → $NaNO_3$

sodium nitrate

...ate ions, SO_4^{2-}, form copper(II) sulfate, $CuSO_4$, and Cu^+ and ...ne to form copper(I) sulfate, Cu_2SO_4. Listed in Table 2.4 are some ...tomic ions. Memorize not only the names, but also the formulas ... Note that all but the last two are anions.

...2.4

...mon polyatomic ions. Other polyatomic ions can be derived from those listed in the ...e.

Name	Formula
Acetate ion	$C_2H_3O_2^-$
Carbonate ion	CO_3^{2-}
Nitrate ion	NO_3^-
Oxalate ion	$C_2O_4^{2-}$
Permanganate ion	MnO_4^-
Phosphate ion	PO_4^{3-}
Sulfate ion	SO_4^{2-}
Cyanide ion	CN^-
Hydroxide ion	OH^-
Chlorate ion	ClO_3^-
Bromate ion	BrO_3^-
Iodate ion	IO_3^-
Ammonium ion	NH_4^+
Hydronium ion	H_3O^+

There are additional polyatomic ions that are derived from the ones in the table. Rather than memorizing one or two more tables of formulas and names, four simple rules that follow will help to extend the list of polyatomic ions. These rules pertain to the **oxo anions**, which are negatively charged polyatomic ions that contain a metal or nonmetal atom combined with one or more oxygen atom.

Rule 1

With the addition of a hydrogen ion, H $^+$, the word hydrogen is inserted into the name of the polyatomic ion.

SO_4^{2-}	+	H^+	$\longrightarrow$	HSO_4^-
sulfate ion		hydrogen ion		hydrogen sulfate ion

The addition of the hydrogen ion to the sulfate ion reduces the negative charge on the hydrogen sulfate ion from negative two to a negative one. Since the hydrogen sulfate ion contains a net charge of negative one, it can still accept one more hydrogen ion:

HSO_4^-	+	H^+	$\longrightarrow$	H_2SO_4
hydrogen sulfate ion		hydrogen ion		sulfuric acid

By accepting the second hydrogen ion, the net charge has become zero, and this compound is no longer defined as an ion; it has become the parent acid, sulfuric acid.

Rule 2

As the acid, the –ate ending is dropped, and an –ic ending is used along with the word acid.

Another example is nitrate ion:

NO_3^-	+	H^+	$\longrightarrow$	HNO_3
nitrate ion		hydrogen ion		nitric acid

By taking on the single hydrogen ion, the net charge is reduced to zero and the acid, HNO_3, is formed. For this reason, the term hydrogen nitrate is not used. If the polyatomic ion retains a net charge after the second hydrogen ion is added, dihydrogen is used.

AsO_4^{3-}	+	H^+	$\longrightarrow$	$HAsO_4^{2-}$
arsenate ion		hydrogen ion		hydrogen arsenate ion
$HAsO_4^{2-}$	+	H^+	$\longrightarrow$	$H_2AsO_4^-$
hydrogen arsenate ion		hydrogen ion		dihydrogen arsenate ion
$H_2AsO_4^-$	+	H^+	$\longrightarrow$	H_3AsO_4
dihydrogen arsenate ion		hydrogen ion		arsenic acid

Rule 3

When an oxo anion contains one less oxygen, but retains the charge of its parent polyatomic ion, the –ate ending is changed to –ite.

For example, sulfate ion has the formula SO_4^{2-}. The sulfite ion, SO_3^{2-} has one less oxygen and retains the charge of the sulfate. Another example is the nitrite ion, NO^{2-}, which has the same charge and one less oxygen than the nitrate ion, NO^{3-}. The rules for the addition of hydrogen ion(s) also apply to the polyatomic ions with the *–ite* ending.

SO_3^{2-}	+	H^+	$\longrightarrow$	HSO_3^-
sulfite ion		hydrogen ion		hydrogen sulfite ion

Rule 4

As the acid, the –ite ending is dropped, and an –ous ending is used along with the word acid.

HSO_3^-	+	H^+	$\longrightarrow$	H_2SO_3
hydrogen sulfite ion		hydrogen ion s		sulfurous acid
NO_2^-	+	H^+		HNO_2
nitrate ion		hydrogen ion		nitric acid

Now that you know how to name the polyatomic ions, let's do a few more examples of naming and writing formulas for ionic compounds that contain polyatomic ions. When Na^+ ions combine with HCO_3^- ions we get sodium hydrogen carbonate, $NaHCO_3$, also called sodium bicarbonate which is baking soda. Notice the charges balance. Combining Al^{3+} and SO_4^{2-} ions we have aluminum sulfate, $Al_2(SO_4)_3$. Parenthesis are used around the polyatomic ion when the charges of both ions are different. It would be incorrect to write $Al_2S_3O_{12}$. The combination of Mg^{2+} and OH^- ions give magnesium hydroxide, $Mg(OH)_2$.

Exercise 2.11 Given the following polyatomic ions, write both the formulas and names of the pecies that result from the addition of hydrogen ions.

	Formula	Name
SO_4^{2-}	HSO_4^-	Hydrogen Phosphate ion
	H_2SO_4	Phosforic acid
CO_3^{2-} Carbonate ion	HCO^-	Hydrogen carbonate ion
	H_2CO	Carbonic acid
PO_4^{3-}	HPO_4^{-2}	Hydrogen Phosphate ion
	$H_2PO_4^-$	Di Hydrogen Phosphate ion
	H_2PO_4	Phosphoric acid
BrO_3^-	$HBrO_3$	Bromic acid
$C_2H_3O_2^-$	$HC_2H_3O_2$	Acedic Acid

Additional Forms of Oxo Anions

The halogen-containing oxo anions have two additional forms. For example, as with the other oxo anions, not only can they exist as chlorate ion, ClO_3^-, and chlorite ion, ClO_2^-, but a perchlorate or a hypochlorite ion can form. The perchlorate ion, ClO_4^-, has one more oxygen that the chlorate ion. The hypochlorite ion, ClO^- has one less oxygen than the chlorite ion. The formula for hypoiodite ion is written as IO^- and periodate ion has the formula IO_4^-. These four oxo anionic states will apply to chlorine, bromine, and iodine. Fluorine cannot form an oxo anion.

The combination of Na^+ ions and ClO_4^- ions gives $NaClO_4$, sodium hypochlorite. When BrO_3^- ions are combined with sodium we have sodium bromate, $NaBrO_3$. Potassium iodite is formed from the combination of potassium ions, K^+, and iodite ions, IO_2^-, and its formula is written as KIO_2.

Exercise 2.12 Fill in the following table.

Name	Formula
Carbonic Acid	
	H_2PO_4
	NO_2^-
Iodite Ion	
	BrO^-
Hydrogen Sulfite Ion	
	$H_2C_2O_4$
Hypobromite Ion	
Phosphorous Acid	
Hydrogen Carbonate Ion	
	H_2SO_4

Exercise 2.13 Fill in the table.

Formula	Name	Ion, Ionic Compound, or Covalent Compound
SF_4		
$BeCl_2$		
SO_3^{2-}		
$CuCl_2$		
NO_2		
$AlCl_3$		
N_2O_4		
XeF_4		
NO_3		
SnO_2		
SO_3		
CS_2		
ClF_3		
$AgCl$		

Exercise 2.14 Fill in the table.

Ions	Formula of Ionic Compound	Name of Compound
Sr^{2+} OH^{-}	Sr OH^{+}	
Ca^{2+} HPO_4^{2-}		
K^{+} SO_3^{2-}	K_2 SO_3	
Ba^{2+} NO_2^{-}	Ba NO_2(2)	
Fe^{3+} ClO_4^{-}	Fe ClO_4(3)	
Na^{+} ClO^{-}	Na ClO	
Cu^{+} CO_3^{2-}	Cu_2 CO_3	
Al^{3+} PO_3^{3-}	Al_3 PO_3(3)	
Sr^{2+} N^{3-}	Sr_3 N_2	
Sn^{4+} O^{2-}	Sn_2 O_4	

Exercise 2.15 Identify the following acids and ions.

Acid Formula	Acid Name	Ion Formula	Ion Name
HI	Hydrogen Iodate	SO_3^{2-}	Sulfite ion
HClO	Hypochlorous acid	P^3	Phosphous ion
HNO_2	Nitritous acid	S^{2-}	Sulfur ion
H_2SO_3	Sulfurous acid	NO_2^-	Nitrate ion
HIO_2	Iodous acid	IO_4^-	Periodate ion
$HClO_4$	Perchloric acid	HPO_3^{2-}	Hydrogen phosphite ion
H_3PO_4	Phosphoric acid	Fe^{3+}	Iron (III) ion
HBr	Hydrogen Bromate	I^-	Iodide ion
H_2CO_3	Chloric acid	HSO_4^-	Hydrogen Sulfate
HNO_3	Nitric acid	NH_4^+	Ammonium ion

p1

Structure of Atoms

Chapter 3

The atom was earlier defined as the smallest particle of an element that retains all the characteristics of the element. An atom is made up of subatomic particles called protons, neutrons, and electrons. Each proton possesses a single positive charge and each electron has a single negative charge. In a neutral atom there are an equal number of protons and electrons. Whereas protons and electrons are charged particles, neutrons have no charge.

Atoms are composed of two regions: the nuclear region includes the **nucleus,** where the protons and neutrons are combined; the **extranuclear region** is the area surrounding the nucleus, where the electrons are found (**Figure 3.1**). Although the nuclear region contains most of the atom's mass, it is only approximately 1/10,000 the size of the total volume of the atom. You can think of the nucleus as a small bead in the center of a football stadium and the extranuclear region as the rest of the stadium.

proton:
A subatomic particle that has a positive charge.

neutron:
A subatomic particle that has no charge.

electron:
A subatomic particle that has a negative charge.

Figure 3.1

The nuclear and extranuclear regions of an atom (not drawn to proportion).

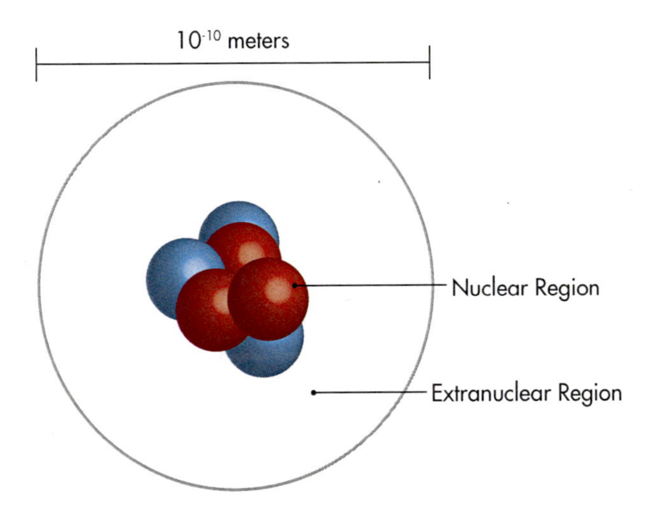

nucleus:
The dense region at the center of the atom that contains the protons and neutrons.

extranuclear region:
The region surrounding the nucleus that contains the electrons.

Figure 3.2

Copper metal, shown in the photo on the left, is composed of copper atoms, represented in the diagram on the right.

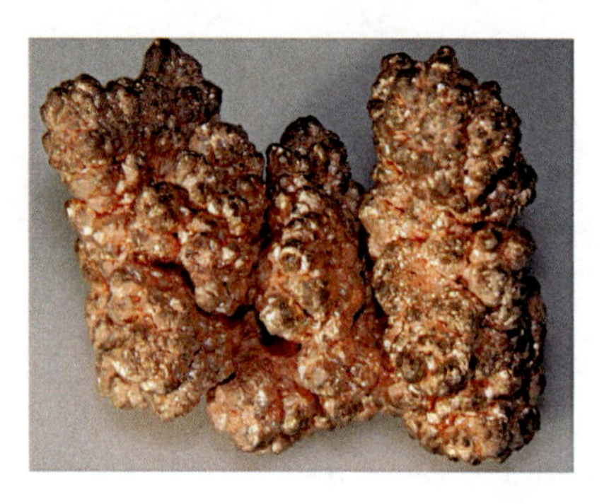

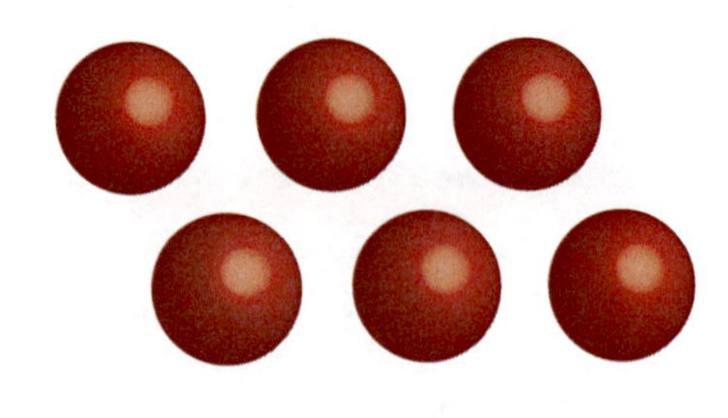

An atom's characteristics are based on the arrangement and number of subatomic particles. For example, copper is an element that is made up of copper atoms (**Figure 3.2**). The smallest particle of copper that still retains the identity of copper is the copper atom. However, if we broke the copper atom down into its subatomic particles, we would no longer be able to identify it as a copper atom—it would just be a pile of protons, neutrons, and electrons.

We begin this chapter by considering the nuclear region. Then we will turn to the extranuclear region, where we will learn how the electrons that are found in the extranuclear region play a most important role in chemical structure and reactivity. To better understand how the electrons behave chemically we need to know the arrangement of the electrons about the nucleus of an atom. We will startthis discussion with a brief introduction to the first substantial model of the hydrogen atom and move into the modern theories that support how the electrons are configured about the nucleus.

The Nuclear Region

As we have already stated, the nuclear region contains the nucleus, which accounts for over 99% of the atom's mass, but less than 1% of the total volume of the atom. The nucleus consists of protons and neutrons (**Figure 3.3**). Because the protons carry a positive charge and neutrons are neutral, the nucleus of an atom always has a positive charge. Both the protons and the neutrons in the nucleus are referred to as *nucleons.*

Figure 3.3

The nucleus of an atom contains protons and neutrons.

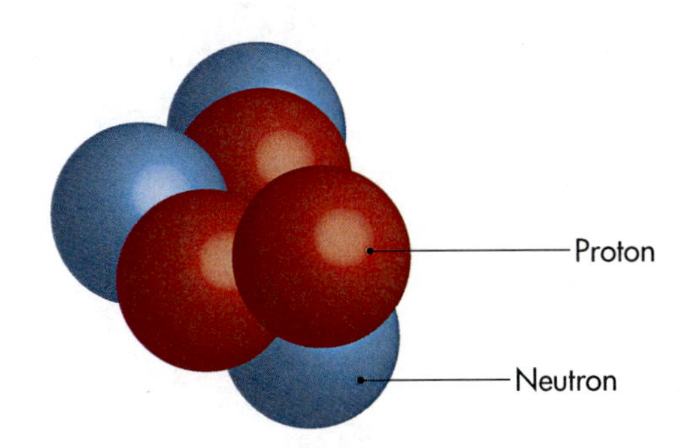

Atomic Mass Units: Dealing with Small Particles

Atoms, protons, neutrons, and electrons have infinitesimally small masses. For example, a carbon atom containing six protons and six neutrons has a mass of 1.9926×10^{-23} grams, and a hydrogen atom containing a single proton in its nucleus has a mass of 1.6605×10^{-24} grams. Dealing with such small numbers can be quite difficult both to remember and compare. We need a more convenient way of expressing these masses.

A much easier way to express the masses of subatomic particles and atoms is with the **atomic mass unit (amu)**, which is defined as one-twelfth the mass of a carbon atom that has six protons and six neutrons. Using atomic mass units instead of grams, the masses of the atoms described above are 1 amu for hydrogen and 12 amu for carbon. Atomic mass unit measurements are determined in an instrument called a *mass spectrometer,* which can give us the mass of an individual atom or molecule. Atomic mass units are called *relative masses* because all measurements of mass are recorded relative to the mass of the carbon atom. The mass spectrometer is calibrated using a carbon atom that contains six protons and six neutrons, setting the mass of this atom at 12.0000 amu. All other masses that the spectrometer records are relative to this carbon atom. As it turns out, the single proton hydrogen atom is one-twelfth the mass of the carbon atom containing six protons and six neutrons. The fact that the relative mass of carbon is twelve times the relative mass of hydrogen can be verified by comparing the actual masses of these atoms:

$$\frac{\text{Mass of hydrogen atom}}{\text{Mass of carbon atom}} = \frac{1.6737 \times 10^{-24}\text{ grams}}{1.9945 \times 10^{-23}\text{ grams}} = 0.0839 \approx \frac{1}{12}$$

atomic mass unit, amu:
A unit of mass that is equal to one-twelfth the mass of a carbon atom that has 6 protons and 6 neutrons.

Example 3.1 Calculate Relative Atomic Mass

Problem

If a cobalt atom is 4.9111 times the mass of a carbon atom that contains 6 protons and 6 neutrons in its nucleus, what is the relative mass of a cobalt atom in atomic mass units?

Solution

The mass of one atom of carbon that has 6 protons and 6 neutrons is 12.000 amu. We are asked to calculate the mass of a cobalt atom that is 4.9111 times heavier than the carbon atom.

Mass of cobalt atom = 12.000 amu × 4.9111 = 58.933 amu

The cobalt atom has a relative mass of 58.933 amu

Practice Problems

3.1 A copper atom is 5.2955 times the mass of a carbon atom that has 6 protons and 6 neutrons. What is the relative mass of a copper atom? pencast

3.2 A helium atom is only 0.33355 times the mass of a carbon atom that has 6 protons and 7 neutrons. What is the relative mass of a helium atom? pencast

The use of atomic mass units will not only be used in describing the relative masses of atoms and molecules, but is also used in comparing masses of subatomic particles. In **Table 3.1,** we see that a proton possesses a charge of +1 and is about the same size as a neutron, which does not carry a charge. The electron is about 1/10,000 the mass of either a proton or a neutron and carries a –1 charge.

Table 3.1

Charges and Masses of the Subatomic Particles

Subatomic Particle	Charge	Mass (g)	Mass (amu)
proton	+1	1.673×10^{-24}	1.0073
neutron	0	1.675×10^{-24}	1.0087
electron	-1	9.1094×10^{-28}	5.486×10^{-4}

Before leaving this section, it should be mentioned that to interconvert between grams and atomic mass units, the following relationship is used:

$$1 \text{ amu} = 1.6605 \times 10^{-24} \text{ grams}$$

This relationship will be used in later calculations. Note that both a proton and a neutron have a mass of approximately 1 amu.

Example 3.2 Calculate the Mass of a Nucleus

Problem

The nucleus of a carbon atom contains 6 protons and 7 neutrons. Calculate the mass of this nucleus in both grams and amu.

Solution

We are given the number of protons and neutrons in the carbon nucleus. Each proton has a mass of 1.673×10^{-24} g and each neutron has a mass of 1.675×10^{-24} g (see again Table 3.1).

Mass (g) = $6 \times (1.673 \times 10^{-24})$ g + $7 \times (1.675 \times 10^{-24})$ g = 2.176×10^{-23} g

In Table 3.1, we see that each proton has a mass of 1.0073 amu and each neutron a mass of 1.0087 amu.

Mass (amu) = 6×1.0073 amu + 7×1.0087 amu = 13.1047 amu

Practice Problems video

3.3 Calculate the mass, in amu, of an oxygen nucleus that contains 8 protons and 8 neutrons.

3.4 A chlorine nucleus has a mass of 35.2807 amu. The mass of the 17 protons is 17.1241 amu. What is the total mass of the neutrons?

Atomic Number and Mass Number

An atom's chemical identity is determined by the number of protons in its nucleus, called the **atomic number, Z.** For example, if an atom's nucleus contains 11 protons, it is a sodium atom. All carbon atoms contain six protons. All phosphorus atoms contain 15 protons. The atomic number is located on the periodic table just above the element's symbol (**Figure 3.4A**). There are over 115 different types of atoms listed in the periodic table in order of increasing atomic number (**Figure 3.4B**).

> **atomic number, *Z*:**
> The number of protons contained in the nucleus of an atom.

Figure 3.4

A) The periodic table lists the atomic number for each element, along with its symbol and name. **B)** Elements are ordered on the periodic table by increasing atomic number, Z.

A)

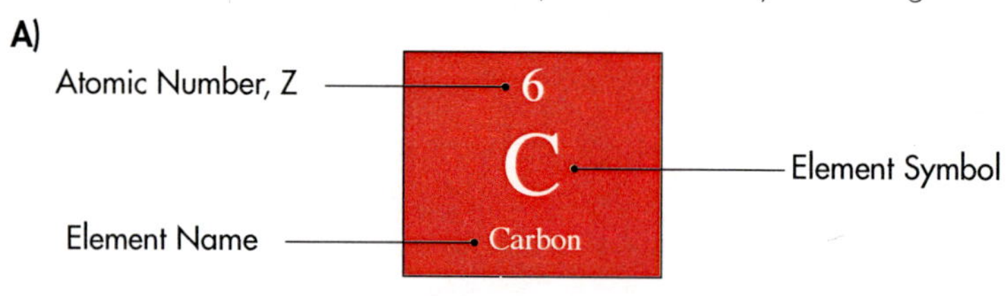

B)

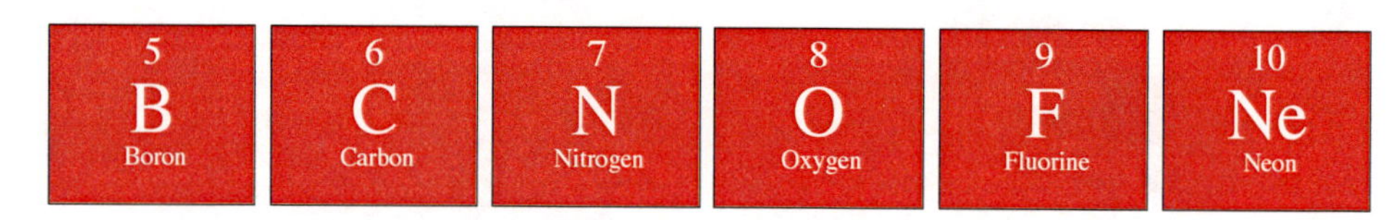

The total number of protons and neutrons in a nucleus is called the **mass number, *A*.** For example, a lithium atom contains 3 protons and 4 neutrons. The sum of protons and neutrons in this atom equals four. We say the lithium atom has mass number, *A*, equal to 4. The mass number is always a whole number because it is the sum of protons and neutrons—it is a particle count and we can't have half a proton or neutron. The mass number is not the same as the atomic mass, which is the average mass of the atoms in an element. We will go into more detail about how the atomic mass is determined later in this chapter. For now we are concerned with counting the number of protons and neutrons that reside in the nuclear region.

> **mass number, *A*:**
> The total number of protons and neutrons in the nucleus of an atom.

Knowing the atomic number of an atom also tells us about the number of electrons in the atom. For an atom to be neutral, the positive charge from the protons must be balanced by an equal, negative charge from the electrons, which means that a neutral atom has equal numbers of protons and electrons. For example a neutral oxygen atom has 8 protons and 8 electrons. This means that the number of electrons in a neutral atom is also equal to *Z*.

So, how do we determine the number of neutrons in the nucleus of an atom? If you know both Z and the mass number, A, you can determine the number of neutrons in the nucleus. Remember, A is the sum of neutrons and protons in the nucleus. For example, the name of the element with $Z = 14$ is silicon (look up $Z = 14$ on the periodic table). A silicon atom that has 14 protons and 29 neutrons has a mass number of 43 ($A = Z + \#$ neutrons $= 14 + 29 = 43$).

Example 3.4 Composition of the Atomic Nucleus

Problem

a) What is the name of the element with $Z = 28$?

b) If $A = 64$, how many neutrons are in the nucleus of the atom?

Solution

a) We are given Z, which corresponds to the number of protons in the atom's nucleus.

In this case, $Z = 28$. Look up atomic number 28 on the periodic table to determine the identity of the element.

The name of the element with $Z = 28$ is nickel.

b) We are asked how many neutrons are in the nucleus of a nickel atom that has $A = 64$.

We know the mass number is equal to the sum of the neutrons and protons. We know the number of protons is equal to 28. We can determine the number of neutrons from the following:

$A = Z + \#$ of neutrons

If the mass number is equal to the sum of protons and neutrons, then the number of neutrons will be equal to the mass number $-Z$.

$A - Z = \#$ of neutrons

$64 - 28 = 36$ neutrons

A nickel atom with $Z = 28$ and $A = 64$ has 36 neutrons in its nucleus.

Practice Problems video

3.5 An atom has 11 protons and 12 neutrons in its nucleus. **(a)** What is the mass number, A? **(b)** What is the identity of the element?

3.6 An atom contains 7 neutrons and has a mass number of 13. What is the identity of the element?

Summary of the Nuclear Region

- The nucleus of an atom contains protons and neutrons, which account for most of an atom's mass.
- The atomic number, *Z*, is the number of protons, and it determines the identity of the atom.
- The mass number, *A*, is the sum of protons and neutrons contained in the nucleus of an atom. It is a particle count; therefore *A* is always a whole number.
- The number of neutrons in the nucleus is determined from both *A* and *Z*. $A - Z$ = # neutrons.
- The masses of atoms and subatomic particles are given in atomic mass units, amu.
- Both protons and neutrons have a relative mass of approximately 1 amu.
- A neutral atom contains equal numbers of protons and neutrons.

Worksheet 3.1 Atomic Number, Z (Section 3.1)

Use the periodic table to fill in the table below.

Z	Name of Atom
6	carbon
3	
	fluorine
22	
	argon
62	
7	
73	
	hydrogen

3.2 Isotopes: Same Atom, Different Mass

Although each atom of an element has the same atomic number (*Z*), different atoms of the same element may have different mass numbers (*A*), meaning they have different numbers of neutrons. For example, the two carbon nuclei shown in **Figure 3.5** have the same number of protons but a different number of neutrons in the nucleus. There are 6 protons and 6 neutrons, $A = 12$, in one nucleus and 6 protons and 7 neutrons in the other, $A = 13$. Such atoms are called **isotopes.**

Figure 3.5

All carbon atoms have six protons (shown in red), but carbon-12 has six neutrons (shown in blue),whereas carbon-13 has seven.

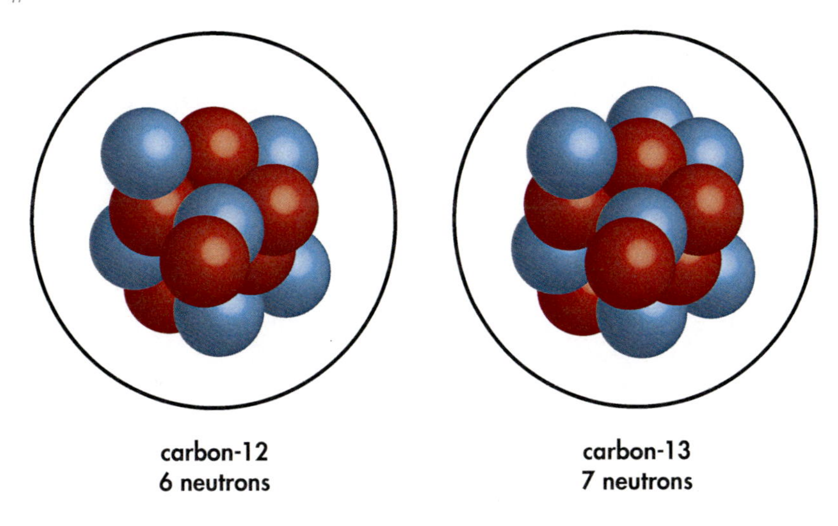

carbon-12
6 neutrons

carbon-13
7 neutrons

> **isotopes:**
> Atoms that have the same number of protons, *Z*, but a different mass number, *A*.

A chlorine atom (Cl) has 17 protons within the nucleus ($Z = 17$), but one chlorine isotope has a mass number of $A = 35$ and another chlorine isotope has a mass number of $A = 37$. We call the chlorine isotope with mass number $A = 35$ chlorine-35, or Cl-35. Chlorine-35 contains 17 protons and 18 neutrons. Chlorine-37 contains 17 protons and 20 neutrons.

When writing an isotope symbolically, we use the element symbol along with its atomic and mass numbers. The symbol for chlorine-35 is written as:

Mass Number ⟶ 35

Atomic Number ⟶ 17

$$^{35}_{17}\mathrm{Cl}$$

At this point we have defined the nuclear region of the atom as the central part containing protons and neutrons. Each type of atom contains a certain number of protons (atomic number, Z), but each type of atom may exist as more than one isotope.

Example 3.5 Isotopes

Problem

a) Nitrogen has two isotopes: nitrogen-14 and nitrogen-16. How many neutrons are contained in the nucleus of each isotope?

b) Write the symbols for the two isotopes of nitrogen.

Solution

a) Nitrogen-14 has mass number, $A = 14$. The atomic number, Z, of nitrogen is 7. The number of neutrons in the nucleus is $A - Z$.

$14 - 7 = 7$ neutrons in the nucleus of nitrogen-14

Nitrogen-16 has mass number, $A = 16$. The number of neutrons is $A - Z$

$16 - 7 = 9$ neutrons in the nucleus of nitrogen-16

b) Isotope symbols are written with the mass number in the upper left corner of the elemental symbol and and the atomic number is located in the lower left corner.

$^{14}_{7}N \quad ^{16}_{7}N$

Practice Problems

3.7 Consider the isotope symbol for the unknown element X, $^{26}_{12}X$. **a)** How many protons and neutrons are contained in the nucleus? **b)** What is the identity of X? pencast

3.8 Write the isotope symbol for a nucleus that contains 38 protons and 48 neutrons. video

3.9 How many neutrons are in the nucleus of the oxygen-18 isotope? pencast

We now know that most elements are composed of various isotopes, that is, atoms containing the same atomic number but different mass numbers. Another important fact is that these isotopes occur in specific amounts called *percent abundances.* For example the element magnesium has three isotopes: magnesium-24, magnesium-25, and magnesium-26. For any number of magnesium atoms, 78.70% are magnesium-24, 10.13% are magnesium-25, and 11.17% are magnesium-26. These percentages are fixed for all the isotopes of magnesium, and this is true for the isotopes of other elements. **Table 3.2** shows more examples of percent abundances of isotopes for other elements. Notice for each element, the percent abundances add to 100.

Table 3.2

Examples of Isotope Mass and Percent Abundance

	Isotope	Mass, amu	% Abundance
Chlorine Isotopes	^{35}Cl	34.9688	75.77
	^{37}Cl	36.9650	24.23
Neon Isotopes	^{20}Ne	19.9924	90.48
	^{21}Ne	20.9938	0.27
	^{22}Ne	21.9914	9.25
Oxygen Isotopes	^{16}O	15.9949	99.75
	^{17}O	16.9991	0.04
	^{18}O	17.9992	0.21

Worksheet 3.2 Isotopes, Mass Number, and Atomic Number (Section 3.2)

Fill in the missing information in the following table.

	Name	Mass Number, *A*	Atomic Number, *Z*	Number of Neutrons
	Lithium-7	7	3	4
${}^{16}_{8}O$				
	Uranium-235			
${}^{4}_{2}He$				
			13	6
	Neon-22	22		
		56	26	
Pt			114	
	Magnesium-26			

Isotopes: Atomic Mass Versus Mass Number

At this point you know that most elements are composed of isotopes, which occur in certain amounts called percent abundances. When discussing the mass of an element, we do not refer directly to the individual masses of the isotopes, but instead to the average of the isotope masses, which is called the **atomic mass.** The atomic mass is located directly below the name of the element on the periodic table, and its units are expressed in atomic mass units, amu. For example, the element carbon has an atomic mass of 12.011 amu (**Figure 3.6**).

Figure 3.6

The element carbon as it appears on the periodic table.

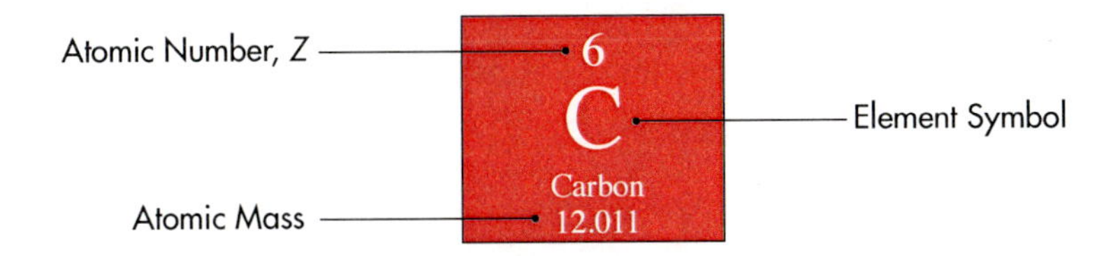

As we discussed earlier, an element may have two or more isotopes, each with different mass numbers. Consider the three isotopes of magnesium:

magnesium-24 isotope (relative mass = 23.985 amu)

magnesium-25 isotope (relative mass = 24.986 amu)

magnesium-26 isotope (relative mass = 25.983 amu)

A simple average of these masses would be 24.98 amu. However, we cannot simply average together the masses, because the three magnesium isotopes occur in different abundances in nature, each having its own contribution to the average mass. Instead we need to calculate the *weighted average atomic mass*, which takes into account the relative abundance of each isotope.

atomic mass:
The average mass, in amu, of the naturally occurring isotopes of an element.

To properly calculate the atomic mass of magnesium, we must use the percent abundance of each isotope along with its mass. Along with the masses of the magnesium isotopes we will incorporate the following percent abundances: 78.70% of all magnesium atoms occur as magnesium-24, 10.13% occur as magnesium-25, and 11.17% occur as magnesium-26. We will use this information to calculate the weighted average atomic mass of magnesium in the following example. In these calculations we will be using fractional abundance, which is obtained by dividing the percent abundance by 100 and dropping the percent symbol. For example, a percent abundance of 65.35% can be expressed as a fractional abundance of 0.6535.

Example 3.6 Calculating the Weighted Average Atomic Mass

Problem

Using a mass spectrometer it is determined that magnesium has three isotopes, with the following masses and abundances. Calculate the weighted average atomic mass of magnesium.

Isotope	Mass (amu)	% Abundance	Fractional Abundance
Mg-24	23.985	78.70	0.7870
Mg-25	24.986	10.13	0.1013
Mg-26	25.983	11.17	0.1117

Solution

Multiply each isotopic mass by its fractional abundance, and then add all three products together. The fractional abundance is calculated by dividing the percent abundance by 100.

fractional abundance (Mg-24) = 78.70/100 = 0.7870

$$\left.\begin{array}{l} 23.985 \text{ amu} \times 0.7870 = 18.88 \text{ amu} \\ 24.986 \text{ amu} \times 0.1013 = 2.531 \text{ amu} \\ 25.983 \text{ amu} \times 0.1117 = 2.902 \text{ amu} \end{array}\right\} \text{Add these}$$

18.88 amu + 2.530 amu + 2.902 amu = 24.31 amu

Compare the result with the atomic mass on the periodic table.

Practice Problems

3.9 Gallium has two isotopes, ^{69}Ga and ^{71}Ga. Ga-69 has an atomic mass of 68.527 amu with a percent abundance of 60.27% and Ga-71 has an atomic mass of 70.9249 amu with a percent abundance of 39.73%. Calculate the weighted average atomic mass of gallium.

3.10 The imaginary element X has three isotopes. Calculate the weighted average atomic mass of element X using the following information: pencast

Isotope	Mass (amu)	% Abundance
X-20	19.992	90.5
X-21	20.994	0.27
X-22	21.991	9.2

Averages and Weighted Averages

If you Google the word *average*, you will read a definition that states the average is the result obtained by adding several quantities together and then dividing this total by the number of quantities. In the above discussion about calculating atomic masses, we determined something called a *weighted average*. What do we mean by a weighted average, and how is it different from what we consider the normal way of finding an average?

Rather than stating definitions, let's look at an example in which we will calculate the "normal" average and also the weighted average.

Example 3.7 Calculate an Average and a Weighted Average

Problem

In an exercise class, there are three weight groups of people. In the first weight group, the people all weigh 125 pounds, in the second weight group, people weigh 165 pounds, and in the third weight group, people weigh 210 pounds. Everyone has to sign in for the class and write down their weight group. At the end of the class, the sign in log has this information:

Bill (210 lbs)	Matt (165 lbs)	Tom (165 lbs)	LaToya (125 lbs)
Amber (125 lbs)	Frank (125 lbs)	Mary Beth (125 lbs)	
Pete (210 lbs)	Lynda (125 lbs)	Toni (165 lbs)	

a) Find the average mass of all of the people in the exercise class.

b) Find the weighted average mass of all of the people in the exercise class.

Solution

a) To find the average weight of all the people in the exercise class, we add the weights together, 210 + 165 + 165 + 125 + 125 + 125 + 125 + 210 + 125 + 165 = 1,540 lbs and divide by the total number of people in the class, 10.

1,540 lbs/10= 154.0 lbs

b) Now, let's find the weighted average, and we will see that it gives the same result!

To use this method, we will need to use percentages, which are discussed in more detail in the Application part of the book. Find the percentage of people in each weight group Out of a total of 10 people, 5 were in the 125 lbs group:

$5/10 \times 100 = 50\%$ (125 lb group)

Out of a total of 10 people, 3 were in the 165 lbs group:

$3/10 \times 100 = 30\%$ (165 lb group)

Out of a total of 10 people, 2 were in the 210 lbs group:

$2/10 \times 100 = 20\%$ (201 lb group)

In finding a weighted average, instead of adding up all the weights and dividing by the total number of people, we just multiply the weight group number by its percentage/100.

$$\left.\begin{array}{l} 125 \text{ lbs} \times 0.50 = 62.5 \text{ lbs} \\ 165 \text{ lbs} \times 0.30 = 49.5 \text{ lbs} \\ 210 \text{ lbs} \times 0.20 = 42.0 \text{ lbs} \end{array}\right\} \text{add these}$$

62.5 lbs + 49.5 lbs + 42.0 lbs = 154 lbs

Note the two results are identical.

Practice Problems

3.11 A basket contains 12 oranges. There are 4 small oranges, 6 medium-sized oranges, and 2 large oranges contained in the basket. What is the weighted average mass of the basket of oranges?

3.12 Ten students were asked to record the weight of their book bags. Sixty-five percent of the book bags weighed 10 lbs, 20% weighed 8 lbs, and 15% weighed 5 lbs. What is the weighted average mass of the book bags?

If the same result is obtained by either method, which method is better to use? For this particular problem (Example 3.6), because the amount of data was so small, it doesn't make much difference which method we used. But for very large amounts of data, such as when the numbers of isotopes number in the thousands of trillions, a weighted average is the way to go. But how do we find the percentages of all these isotopes? The mass spectrometer provides the percentages of the isotopes as well as the isotopic masses.

Summary of Isotopes

- Isotopes have the same atomic number, *Z*, but different mass number, *A*. They differ in the number of neutrons contained in the nucleus.
- Isotope symbols are written with the element symbol, the atomic number, *Z*, and the mass number, *A*.
- The isotopes of an element occur in amounts called percent abundance.
- The weighted average atomic mass is calculated by multiplying the atomic mass of each isotope by its fractional abundance and then summing the products.

Worksheet 3.3 Calculate Weighted Average Atomic Mass (Section 3.2)

Given the following information about the element X, answer the following questions:

Isotope	Atomic Mass (amu)	Percent Abundance
X-16	15.9949	99.757
X-17	16.9991	0.038
X-18	17.9992	?

a) What is the mass number of X-16? ________________

b) What is the mass number of X-17? ________________

c) What is the percent abundance of X-18 (to three digits)? ________________

(Hint: the percent abundances add up to 100%)

d) What are the fractional abundances of

i) X-16

ii) X-17

iii) X-18

e) Calculate the average weighted atomic mass of the element X.

The average weighted atomic mass of element X = _________ amu.

3.3 Charge Balance: When Atoms Become Ions

One of the most fundamental chemical phenomena is the formation of ions from atoms. As mentioned earlier, the overall charge of an atom is determined by the positive charge of the nucleus (each proton contributes a +1 charge) and the negative charge of the extranuclear region (each electron contributes a −1 charge). We refer to atoms containing an equal number of protons and electrons as being *neutral*, meaning they have no net charge. For example, the neutral lithium atom has an equal number of protons and electrons, so the positive and negative charges of the subatomic particles cancel each other out (**Figure 3.7**).

Figure 3.7

In neutral lithium, the +3 charge in the nucleus is balanced by the −3 charge of the extranuclear region.

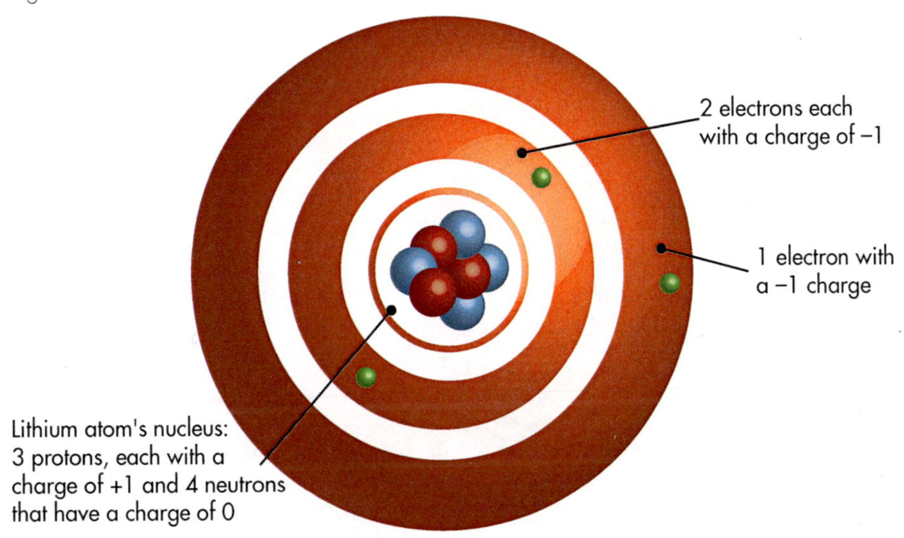

Recall that metal atoms tend to lose electrons, whereas nonmetal atoms tend to gain extra electrons. When electrons are removed from a neutral species or extra electrons are added, the atom is regarded as an **ion** with either a positive or negative net charge. A positively charged ion is called a **cation**, and a negatively charged ion is called an **anion**. If a single electron were removed from the extranuclear region of a neutral lithium atom, this would result in a cation containing three protons and only two electrons, resulting in a net charge of +1 (**Figure 3.8**). The symbol for the lithium ion is written as

$$Li^{+}$$

If the charge is either +1 or −1, the 1 is not used and only the positive or negative sign appears.

Figure 3.8

A +3 charge in the nuclear region and a −2 charge in the extranuclear region results in a net charge of +1 for the lithium ion.

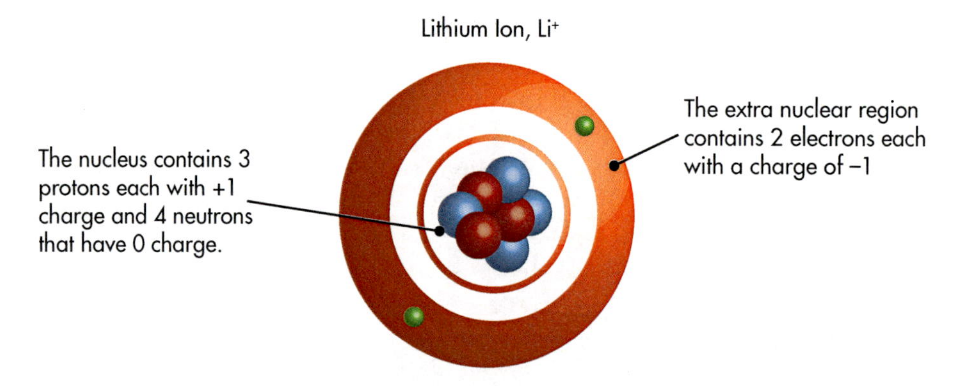

ion:
A charged atom with an unequal number of protons and electrons.

cation:
A positively charged ion.

anion:
A negatively charged ion.

To see how an ion gets a +2 charge, consider the neutral calcium atom ($Z = 20$), which contains 20 protons and 20 electrons. If this atom lost two electrons, it would contain 20 positive charges and only 18 negative charges. This would result in a net charge of +2 for the cation. The calcium cation is symbolically written as

$$Ca^{2+}$$

Now let's see how an atom gets a −2 charge when an oxygen atom gains 2 electrons to become the oxide anion. The neutral oxygen atom contains 8 protons and 8 electrons. When oxygen gains 2 electrons, it contains 8 protons and 10 electrons. Therefore it has a +8 charge in the nuclear region and a −10 charge in the extranuclear region, which results in a net charge of −2. The oxide anion is written as

$$O^{2-}$$

Example 3.7 Writing Ion Symbols

Problem

A potassium atom loses one electron. **a)** Is this a cation or an anion? **b)** Write the symbol of the ion.

Solution

a) A neutral potassium atom contains 19 protons and 19 electrons. If potassium loses one electron, there will be 19 protons and 18 electrons resulting in a net charge of +1. Cations are positively charged. A neutral sodium atom that loses one electron is a cation.

b) When a potassium atom loses one electron, the net charge is +1. The symbol is written as K^+.

Recall, the one is not used and only the positive sign appears.

Practice Problems

3.13 Write the symbol for a sulfur atom that has gained 2 electrons. Is this a cation or an anion? video

3.14 Write the symbol for an aluminum atom that loses three electrons. Is this a cation or an anion?

Example 3.8 Composition of Ions

Problem

Indicate the number of protons, neutrons, and electrons in the following.

a) $^{84}_{38}Sr^{2+}$ **b)** $^{37}_{17}Cl^-$

Solution

a) $Z = 38$ and $A = 84$. The ion has a +2 charge, which means it contains 2 more protons than electrons (electrons have been lost).

The Sr^{2+} ion contains 36 electrons, 38 protons, and 46 neutrons.

b) $Z = 17$ and $A = 37$. The chloride ion has a net charge of –1, which means it has one more electron than the neutral atom (an electron has been gained).

The Cl^- ion contains 17 protons, 20 neutrons, and 18 electrons.

Practice Problems

3.15 What are the total number of electrons in a Fe^{2+} ion ($Z = 26$)? pencast

3.16 What is the charge on the nucleus of a neutral copper atom ($Z = 29$)? pencast

Summary of Charge Balance: When Atoms Become Ions

- Ions have unequal numbers of electrons and protons, which results in a net charge. A cation is a positively charged ion and an anion is a negatively charged ion.
- The overall charge on an atom is determined by the positive charge of the nucleus and the negative charge of the extranuclear region.
- If you know how many electrons an atom has gained or lost, you can write the symbol of the ion.

Worksheet 3.4 The Nuclear Region (Sections 3.1, 3.2, and 3.3)

1. Fill in the missing information in the following table.

	Z	A	# neutrons	# electrons
Ne	10	24	14	10
S	15	32	17	15
Fe	26	55	29	30
$^{1}_{33}As$				
	3	1		
Br^-		79		
	13	27		10

2. What is the mass, in amu, of the gallium-69 nucleus?

protons ____________

neutrons ____________

Mass of proton, amu ____________

Mass of neutron, amu ____________

Mass of nucleus ____________ amu

3. It was found that an unknown element, X, exists as three isotopes. Using the information below, calculate the average weighted atomic mass of unknown element X. Identify the unknown element X. (Hint: compare atomic masses.)

Isotope	Mass, amu	Percent Abundance
^{28}X	27.976927	92.2297
^{29}X	28.976495	4.6832
^{30}X	29.973770	3.0872

The weighted average atomic mass of X is ____________________ amu

X is most likely to be the element ____________________.

4. Fill in the following information for Pb-204.

mass number = ______________ # protons ______________

Z = ______________ # neutrons ______________

atomic mass ____________ # electrons ______________

5. Consider $^{40}Ca^{2+}$ to answer the following:

protons __________ # electrons __________ # neutrons ________

6. A lead atom (Pb) loses two electrons. How many electrons remain? ____________

3.4 The Extranuclear Region

Besides their role in determining the overall charge of an atom, electrons are also responsible for all chemical bonding. To understand how this works, we need to learn about how the electrons are arranged about their nucleus. This arrangement is called the **electron configuration**, and the rest of this chapter will focus on understanding electron configuration and how it applies to the formation of chemical bonds. But before we move into this, we need to know something about the energies that electrons experience when they are within an atom, and how this differs from the normal energies that we experience.

> **electron configuration:**
> The arrangement of electrons about a nucleus.

Electron Energetics

In our everyday world, we can think of energy as continuous, much as the volume levels on a radio can be continuously increased or decreased. As an example of continuous energy, consider the potential energy (PE) that results from placing a weight on an inclined ramp (Figure 3.10). We can see in the figure that as the weight is moved up the ramp, its distance from the ground increases along with its potential energy. With this example in mind, imagine placing the weight at the bottom of the ramp and slowly moving it towards to top. As the weight is continuously moved up the ramp, its potential energy continuously increases. In this way, we can think of the energy we encounter as continuous, much the same as the weight being moved up, or down, the ramp.

Figure 3.10

Potential energy (*PE*) increases continuously as a weight moves up an inclined ramp.

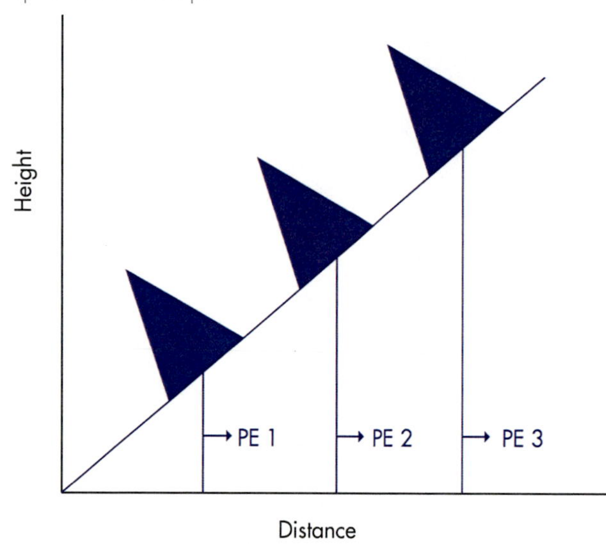

But the energy that an electron experiences within an atom is quite different. For electrons, energy is not continuous, but instead occurs in energy levels that we can think of as steps. For example, instead of using the analogy of a weight on a ramp, consider standing before a staircase, and gently tossing a ball so that it lands and remains on one of the steps. By adjusting your toss that is, increasing or decreasing the energy of the toss—you can make the ball land on a different step, with each step representing a different potential energy level. Now imagine an electron within an atom experiencing similar step-like energy levels. That is, this electron can only do work in whole steps of energy. In other words, just like the ball, the electron must have the right amount of energy to land on the first, second, or third step; it cannot land between steps (**Figure 3.11**). This type of energy is called *quantized energy*, and the steps are called *quantized energy levels.*

Figure 3.11

Each step corresponds to the available energy levels that an electron may occupy. Electrons can be located in energy levels 1 and 2 but not in energy level 1.5.

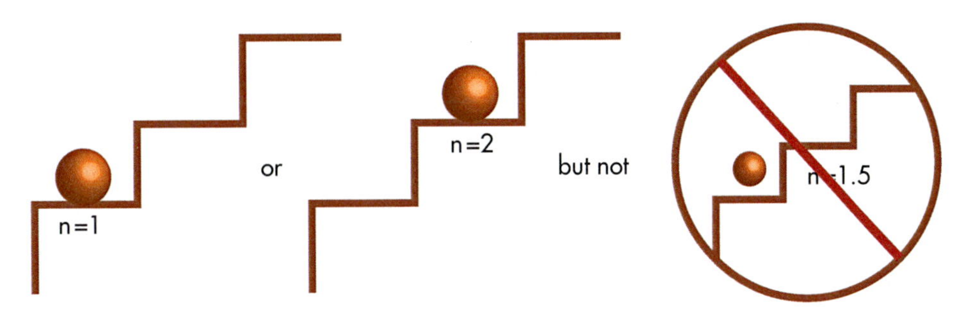

In the following discussion, we will see an early representation of the atomic structure of the hydrogen atom, which shows how these quantized energy levels work.

The Bohr Model of the Hydrogen Atom

In 1915, Neils Bohr, a Danish physicist, applied the idea of quantized energy levels to help explain the information that had been discovered up to that time regarding subatomic particles. In his model of the hydrogen atom, the electron circles the nucleus in an orbit, much the same way that a planet circles its sun (**Figure 3.12**). For this reason his model is also called the *planetary model.*

Figure 3.12

The Bohr orbit model for a hydrogen atom. In the diagram on the left, hydrogen's one electron occupies a single orbit around the nucleus. Empty orbits are shown on the right.

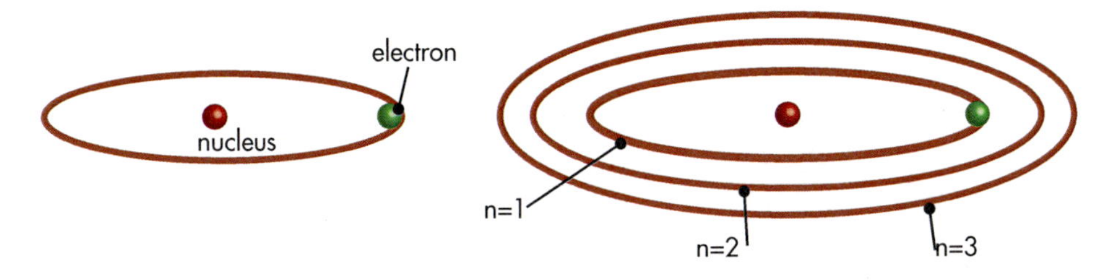

In his model, Bohr also postulated that other empty orbits exist, and as the orbits increase in size, so do their quantized energies. The orbits were labeled $n = 1$ for the first orbit, $n = 2$ for the second orbit, $n = 3$ for the third orbit, and so on. Bohr also stated that the electron could either pick up additional energy and move from the first orbit, $n = 1$, into a higher energy orbit, or release energy and move back into the lower energy orbit. For example, if an electron in a hydrogen atom moves from $n = 1$ to $n = 2$ (this process is written as $n_1 \rightarrow n_2$), energy is absorbed by the electron because it is moving from a lower energy orbit to a higher energy orbit.

Example 3.10 Electron Transitions in a Hydrogen Atom

Problem

An electron in a hydrogen atom moves from $n = 1$ to $n = 3$. Is energy absorbed or released by the electron in this process?

Solution

From Bohr's description, we know that the energy levels of the orbits increase as the value of n increases, therefore $n = 1$ is lower energy than $n = 3$. For the electron to move to $n = 3$, it would have to absorb extra energy.

In the process $n_1 \rightarrow n_3$ energy is absorbed by the electron.

Practice Problem

Determine in these transitions, if energy is absorbed or released.

a) $n_2 \rightarrow n_5$ ________________________

b) $n_4 \rightarrow n_3$ ________________________

c) $\mathrm{n}_1 \rightarrow n_2$ ________________________

d) $n_4 \rightarrow n_1$ ________________________

Bohr was awarded the Noble Prize in 1922 for the calculations he developed in his model of the hydrogen atom. The Bohr model was a good first approximation of electron energetics of the simple hydrogen atom, but it couldn't answer the more complicated questions of electron behavior. By 1926, research produced a more sophisticated model, based on complex mathematical equations, to represent the arrangement and behavior of electrons in atoms called the *quantum mechanical model.* In the following pages, we will use this model to describe the arrangement of electrons in all the elements of the periodic table, and go on to discuss how these arrangements account for chemical bonding.

In the following discussions and exercises, we will learn how to determine the arrangement of electrons within an atom, the electron configuration, using modern theories. These theories go far beyond what Neils Bohr first proposed in 1913. We will no longer think of electrons circling the nucleus in orbits, but rather of electrons surrounding the nucleus within specific regions of space, or energy levels, about the nucleus. All of this will be presented in an easy-to-follow method, one step at a time.

Summary for The Extranuclear Region

- The arrangement of electrons about the nucleus of an atom is called the electron configuration.
- The energy of an electron is quantized.

3.5 Electron Shells

Although electrons do not move about the nucleus in precise orbits, as Bohr imagined, neither do they move randomly about the nucleus. Instead they are confined to regions in space called **principle energy levels,** or **shells**. Each electron of an atom resides in a specific shell which is dependent on 1) the energy of the electron and 2) the number of electrons in the atom. The shells are numbered 1, 2, 3, 4, and so on with the first shell being closest to the nucleus. The electrons in the first shell are lower in energy because they are closer to the positively charged nucleus and are therefore held more strongly to the nucleus. The electrons in the second shell are further away from the nucleus and are not held as strongly. They are higher in energy than those electrons in the first shell.

shell (principle energy level):
A region in space about the nucleus that contains the electrons.

In describing the arrangement of electrons surrounding a nucleus we use quantum numbers. Instead of saying shells number one, two, and three, we say $n = 1$, $n = 2$, $n = 3$. The lower case letter n indicates a shell and is called the **principle quantum number, *n***. Notice that this is the same letter that Bohr used to indicate his orbits. Other quantum numbers will be introduced later in the chapter.

principle quantum number, *n:*
The number that corresponds to the principle shell number.

Because the shells surround the nucleus, it isn't hard to imagine that in regard to size and volume,

$$n = 4 > n = 3 > n = 2 > n = 1$$

and in accordance with this increase in volume as the number of the shell increases, so does the number electrons that the shell may contain. The maximum number of electrons within a shell can be determined by the equation

$$\text{maximum number of electrons in a shell} = 2n^2 \text{ with } n = \text{shell number}$$

Example 3.11 Electrons and Principle Energy Levels

Problem

What is the maximum number of electrons that can occupy the shell with principle quantum number 2?

Solution

We can use the following equation

maximum number of electrons in a shell = $2n^2$ where n = the shell number

The principle quantum number corresponds to the second shell, $n = 2$

$2 \times 2^2 = 8$

A maximum of 8 electrons can occupy the shell with principle quantum number 2.

Practice Problems

3.17 What is the maximum number of electrons that can occupy $n = 3$?

3.18 A shell can hold a maximum of 32 electrons. What is the principle quantum number (n)?

Thus far, we have addressed the fact that the electrons of an atom surround the nucleus in a series of shells of increasing volume and energy. We have focused on the first 4 shells ($n = 1$ to $n = 4$), but in reality the value of n can go from 1 to infinity. In discussing the normal energies (ground states) of atoms, we will employ values of $n = 1$ to $n = 7$. In regard to chemical activity, bond formation, and bond breaking, of all the atom's electrons, the ones in the outermost shell are the most important. The outermost shell is called the **valence shell** and the electrons within this shell are called the **valence electrons**. The first step in determining the arrangement of the electrons in an atom, the electron configuration, is to understand how the shells are filled. First, determine the total number of electrons in the neutral atom. This can be obtained from the atomic number (Z). Second, knowing the maximum number of electrons each shell can hold, start filling the shells with the first shell ($n = 1$) and proceed to $n = 2$, $n = 3$, etc.

valence shell:
The outermost shell of an atom.

valence electron:
An electron that occupies the valence shell.

An atom of carbon, $Z = 6$, contains 6 electrons. Two of the electrons occupy $n = 1$ and the remaining 4 electrons occupy $n = 2$. The valence shell is $n = 2$, and the number of valence electrons is equal to 4.

Example 3.12 Electrons, Shells, and Valence Electrons

Problem

Consider the fluorine atom. **a)** Determine the number of filled shells in a fluorine atom. **b)** Determine the number of valence electrons in the fluorine atom.

Solution

a) The fluorine atom, $Z = 9$, contains 9 electrons. We know the first shell, $n = 1$, can hold a maximum of 2 electrons. That leaves 7 electrons that can go into the second shell, $n = 2$. The second shell can hold a maximum of 8 electrons.

$n = 1$ 2 electrons

$n = 2$ 7 electrons

9 electrons total

There are 2 electrons in $n = 1$ and 7 electrons in $n = 2$.

b) From the definition of the valence shell, we can see that the outermost shell is $n = 2$, which contains 7 electrons.

There are 7 valence shell electrons.

Practice Problems

3.19 A potassium (K) atom contains how many valence electrons?

3.20 What is the valence shell number for an atom of oxygen?

3.21 An atom has 6 valence electrons in $n = 2$. What is the identity of the element?

You might have noticed the equivalences of 1) the valence shell number and the period number and 2) the number of valence electrons and the group number. The valence shell corresponds to the period number on the periodic table. For main group elements, Groups 1A to 8A, the number of valence electrons corresponds to the group number. To determine the number of valence electrons for a *main group element*, all you have to do is look up the group number. For example, nitrogen is located in Group 5A and has 5 valence electrons. Nitrogen is located in period 2 of the periodic table and its valence shell is $n = 2$. Sodium is located in Group 1A, period 3. It has one valence electron which is located in $n = 3$.

Subshells

Determining the arrangement of the electrons within an atom, the electron configuration, would be very simple if it ended with the listing of only shells. It gets a bit more complicated because within each shell there exist further energy divisions called **subshells.** The number of subshells available varies from shell to shell. There is a very simple rule to determine the number of subshells within a given shell. The number of subshells within a shell is equal to the number of the shell, *n*, as shown in **Table 3.4.** For example, the second shell is $n = 2$ and contains 2 subshells.

subshell:
Energy sublevels that are contained in a principle shell.

Table 3.3

Number of Subshells Contained in a Principle Shell

n	Number of Subshells
1	1
2	2
3	3
4	4

Summary of Electron Shells

- Electrons are arranged according to their energy in regions around the nucleus called principle energy levels or shells. Shells are designated by the principle quantum number, *n*.
- The maximum number of electrons that can be contained in a shell is $2n^2$ where n is equal to the shell number.
- For main group elements, valence electrons are in the outermost shell. For transition elements we include the outermost s subshell electrons and the outermost d subshell electrons.
- Each shell is contains sublevels called subshells.
- The number of subshells is equal to *n*.

Worksheet 3.5 Electron Arrangement in the Extranuclear Region (Section 3.5)

Fill in the missing information in the following table. (Ne is completed as an example.)

Atom	Number of Electrons	*n* 1	*n* 2	*n* 3	Valence Shell Number	Number of Valence Electrons
Ne	10	2	8	0	n=2	8
Li	3	2	1	6	n=1	1
Be	4	2	2	0	n=2	2
B	5	2	3	0	n=3	3
C	6	2	4	0	n=4	4
N	7	2	5	0	n=5	5
O	8	2	6	0	n=6	6
F	9	2	7	0	n=7	7
Mg	12	2	8	2	n=2	2
					n=3	2

3.6 Electron Configuration

Each subshell has 1) a quantum number designation and 2) a spectroscopic notation designated by the letter *s*, *p*, *d*, or *f* (**Table 3.4**).

Table 3.4

Quantum Numbers and Letter Designations

Quantum Number	Spectroscopic Notation
$l = 0$	*s*
$l = 1$	*p*
$l = 2$	*d*
$l = 3$	*f*

For subshell designation, the lowercase letter l is used. It is called the *azimuthal quantum number*, l. The quantum number, l, can have values of $l = 0, 1, 2$, or 3, with 0 being the lowest energy level and 3 being the highest energy level The letters used in spectroscopic notation have their origin in *spectroscopy*, which is defined as the study of the interactions between matter and light. The first shell contains one subshell, $l = 0$ that is designated by the letter *s*. The second shell contains both an *s* subshell and a *p* subshell, $l = 1$. The third shell contains an *s*, *p*, and a *d*, subshell, $l = 2$ (**Table 3.5**). The fourth shell contains an *s*, *p*, *d*, and an *f* subshell, $l = 3$.

Table 3.5

Distribution of Electrons Within Subshells

Azimuthal Number	Subshell	Maximum Number of Electrons a Subshell Can Hold
$l = 0$	*s*	2
$l = 1$	*p*	6
$l = 2$	*d*	10
$l = 3$	*f*	14

Just as the number of electrons within a shell increases with the value of n, so does the number of electrons increase with the value of l. An *s* subshell can hold a maximum of 2 electrons, a *p* subshell 6 electrons, a *d* subshell 10 electrons, and an *f* subshell 14 electrons (**Table 3.6**). So when describing the energy of an electron, we also need to designate its subshell ($l = 0, 1, 2, 3, 4\ldots$). For example the electron in a hydrogen atom can be described by the quantum numbers $n = 1$ and $l = 0$, which corresponds to 1*s* in spectroscopic notation, where the number 1 is the principle shell number, n.

Table 3.6

Distribution of Electrons in Principle Shells and Subshells

n	Subshells Contained in Shell	Maximum Number of Electrons Held in Shell
1	1*s*	2
2	2*s*, 2*p*	$2 + 6 = 8$
3	3*s*, 3*p*, 3*d*	$2 + 6 + 10 = 18$
4	4*s*, 4*p*, 4*d*, 4*f*	$2 + 6 + 10 + 14 = 32$

Example 3.13 Quantum Numbers and Spectroscopic Notation

Problem

An electron has the following quantum numbers, $n = 5$ and $l = 0$. Translate the information into spectroscopic notation.

Solution

The principle shell has $n = 5$, and $l = 0$ corresponds to an *s* subshell. In spectroscopic notation we have

5*s*

Practice Problem

3.22 Fill in the following table.

Quantum Numbers	Spectroscopic Notation
$n = 2$ $l = 0$	2*s*
$n = 4$ $l = 2$	4d
$n = 3$ $l = 1$	3p
$n = 4$ $l = 1$	4*p*

S 0
P 1
D 2
F

Electron configurations are written with the shell number, the letter designation of the subshell, and a superscript that indicates the number of electrons.

principle shell (n=1) → $1s^2$ → number of electrons in subshell; → subshell

Electrons occupy the lowest energy principle shells and subshells that are available starting with the 1*s* subshell. The energies of the subshells within each principle shell increase in order of *s*, *p*, *d*, and *f*. Above the 3*p* level are energy crossovers that occur between the subshells. For now we will concentrate on the electron configurations for the 1st and 2nd period elements. For example, a neutral fluorine atom contains 9 electrons. Two electrons are assigned to the first shell and occupy a 1*s* subshell. The remaining 7 electrons are assigned to the second principle shell with 2 electrons in the 2*s* subshell and 5 electrons in a 2*p* subshell. The electron configuration would be written as

$$1s^2 2s^2 2p^5$$

The following example shows how to write an electron configuration for the electrons of a carbon atom.

Example 3.14 Electron Configurations

Problem

Consider the neutral carbon atom with $Z = 6$. Use the table to determine the electron configuration of the carbon electrons.

		Number of Electrons
$n = 3$	$l = 2$ $l = 1$ $l = 0$	
$n = 2$	$l = 1$ $l = 0$	
$n = 1$	$l = 0$	

Solution

The neutral carbon atom has 6 electrons. We need to determine the number of electrons in each shell and subshell. Two of the electrons occupy $n = 1$ in an *s* subshell, $l = 0$. The remaining 4 electrons occupy $n = 2$. Because it is lower in energy, the 2*s* subshell will fill before the 2*p* subshell. Two of the four electrons are assigned to the *s* subshell, $l = 0$, and the remaining two electrons to the *p* subshell, $l = 1$. Let's fill in the table using this information.

		Number of Electrons
$n = 3$	$l = 2$ $l = 1$ $l = 0$	
$n = 2$	$l = 1$ $l = 0$	2 2
$n = 1$	$l = 0$	2

From the table we can write the electron configuration.

For n = 1 and l =0 $1s^2$

For n = 2 and l =0 $2s^2$

For n = 2 and l =1 $2p^2$

The electron configuration is $1s^2 2s^2 2p^2$

Practice Problems

3.23 Fill in the following tables to determine the electron configuration of **a)** neon and **b)** silicon. Write the electron configurations using spectroscopic notation.

a)

Neon Z= 10 Total number of electrons 10		
		Number of Electrons
	= 2 d	______
n=3	= 1 p	______
	= 0 s	______
n=2	= 1 p	6
	= 0 s	2
n=1	l=0 s	2
Electron Configuration	$1s^2 2s^2 2p^6$	

$2n^2$ 10 8

b)

Silicon Z = 11 Total number of electrons 11		
		Number of Electrons
n=3	= 2	___
	= 1	___
	= 0	1
n=2	= 1	6
	= 0	2
n=1	l=0	2
Electron Configuration	$1s^2 2s^2 2p^6 3s^1$	

Up until now we have been using the following filling order

1s→2s→2p→3s→3p→3d→4s…and so on.

This order can only be used in the case of single electron systems such as the hydrogen atom. Most of the time, we will be dealing with atoms that contain more than one electron. For elements in the first and second period, we do not go above the 3p level. Due to the effects of repulsive electron–electron interactions, some of the shell/subshell positions shift as indicated in **Figure 3.13**. Notice that above the 3p level, the 4s subshell is lower in energy than the 3d subshell. Because of this, the 4s subshell will fill before the 3d subshell. The 4d subshell fills after the 5s subshell is filled. This crossover of energies continues as n increases. The order that shells and subshells fill is

1s→2s→2p→3s→3p→4s→ 3d→4p→5s→4d→5p→6s→4f→5d→6p …and so on.

Energy of subshells ⟶

Figure 3.13

Electrons occupy the lowest energy shells and subshells. Note the crossover in subshell energies above the 3p subshell.

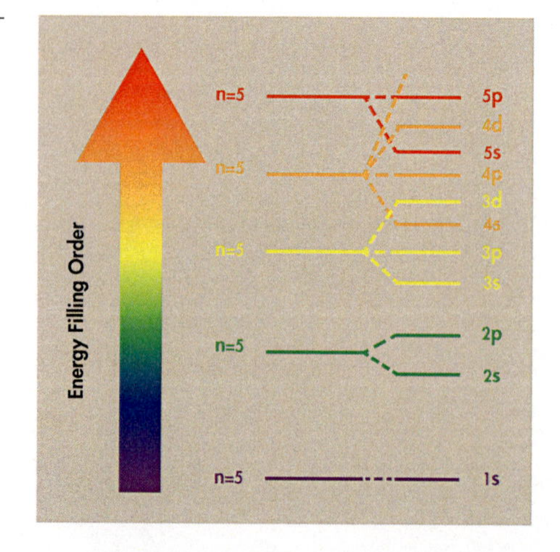

Example 3.15 Electron Configurations Beyond the Second Period

Problem

Write the electron configuration for an atom of calcium.

Solution

First we find Z for calcium. $Z = 20$. We have 20 electrons to distribute. Recall that the lowest energy shells and subshells fill first.

For $n = 1, l = 0$,
 2 electrons will occupy the 1s subshell.

For $n = 2, l = 0$ and $l = 1$,
 2 electrons will occupy the 2s subshell and 6 electrons will occupy the 2p subshell.

For $n = 3, l = 0, l = 1$ and $l = 2$,
 we have 10 electrons left; 2 electrons will occupy the 3s subshell, and 6 electrons will occupy the 3p subshell.

For $n = 4, l = 0, l = 1, l = 2$, and $l = 3$,
 we have two electrons left; they will occupy the 4s subshell.

The electron configuration is written as
 $1s^2\ 2s^2\ 2p^6\ 3s^2\ 3p^6\ 4s^2$
 You can check your answer by summing the superscripts in the electron configuration. The sum is equal to 20, the number of electrons in the neutral atom.

Practice Problem

3.24 Use the table to write the electron configuration for an atom of iron, Fe.

Iron Z = ______ Total number of electrons ______		
		Number of Electrons
n=3	=	______
	=	______
	=	______
n=2	=	______
	=	______
n=1	l=0	______
Electron Configuration		

Recall that for main group elements, the number of valence electrons is equal to the group number found on the periodic table. The situation is different for transition elements. The valence electrons for a transition element include both the electrons in the outermost *s* subshell and the outermost *d* subshell. For example, vanadium has the electron configuration 1*s* 2*s* 2*p* 3*s* 3*p* 4*s* 3*d* and has 5 valence electrons. We include the 2 electrons in the 4*s* subshell and the 3 electrons in the 3*d* subshell.

Orbitals

We have established that the *s* subshell ($l = 0$) can accommodate up to 2 electrons; the *p* subshell ($l = 1$) holds up to 6 electrons; the *d* subshell ($l = 2$), 10 electrons; and the *f* subshell ($l = 3$) has a capacity of 14 electrons. In addition to the arrangement of electrons into shells and subshells, we will now consider the next ordering of electrons: **orbitals**. An orbital is a region in space in an atom where there is a high probability of finding a specific electron. Within each subshell, electrons are grouped into orbitals.

orbital:
A region of space in an atom where a specific electron is most likely to be found.

Each orbital can hold a maximum of 2 electrons. An *s* subshell has only one orbital, a *p* subshell has 3 orbitals, a *d* subshell has 5 orbitals, and an *f* subshell has 7 orbitals (**Table 3.8**).

Table 3.8

Number of Orbitals in the Different Subshells

Subshell	Number of Electrons	Number of Orbitals
s	2	1
p	6	3
d	10	5
f	14	7

The shape of a particular orbital will depend on the subshell in which it is contained. For example, the orbitals contained in an *s* subshells ($l = 0$) all have spherical shaped regions (**Figure 3.14**). As mentioned previously, each *s* subshell is composed of one orbital that can hold a maximum of two electrons. As we move from 1*s* to 2*s* to 3*s* ...the size of the *s* orbitals increase, but the shapes remain the same.

Figure 3.14

Representation of a spherical shaped *s* orbital.

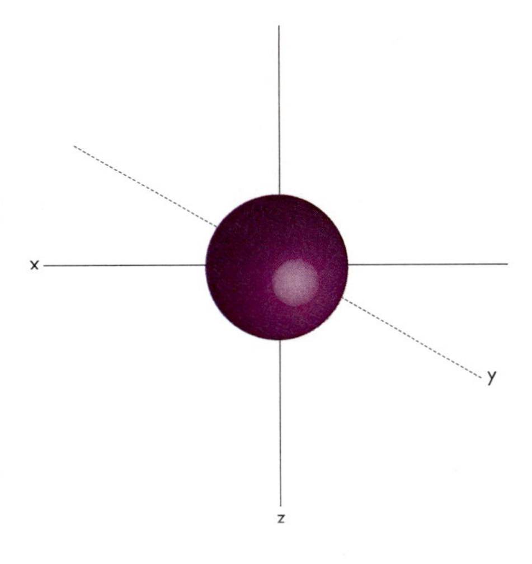

The three orbitals within a *p* subshell have a dumbbell shaped region consisting of two lobes that intersect at the nucleus (**Figure 3.15**). Each of the *p* orbitals is oriented along one of the three major axes. In spectroscopic terms they are named the p_x, p_y, and p_z orbitals. Because there are three separate orbitals in each *p* subshell, the *p* subshell can hold up to six electrons. Electrons can exist in the positive or negative lobes, but not at their point of intersection (i.e., the position of the nucleus). The region where electrons cannot exist is called a *node.* Note that *s* orbitals do not have nodes whereas each *p* orbital has one node. In Figure 3.15, two colors are used to indicate that one lobe is designated a positive phase and the other a negative phase. These phase designations come about from mathematical calculations and are important when considering chemical bonding.

Figure 3.15

Representation of the three *p* orbitals.

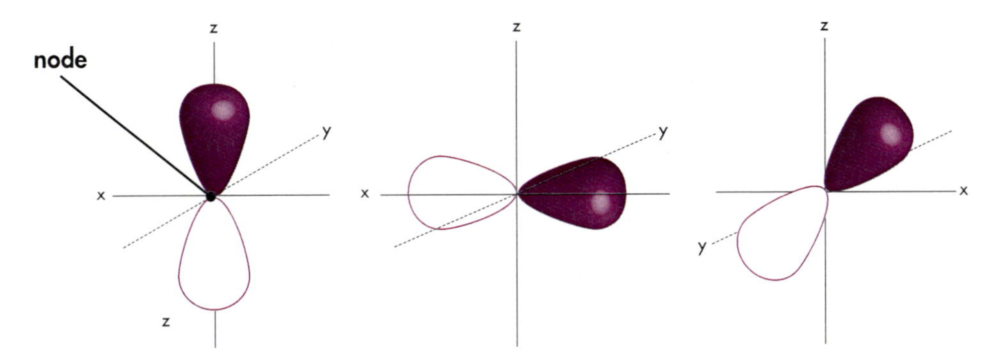

Four of the orbitals in the d subshell have 4 lobes with equivalent phases in opposite lobes. There are two nodal planes per orbital. Two of the orbitals, $d_{x^2-yx^2}$ and d_{xy}, are found in the *xy* plane, one orbital is found in the *xz* plane, d_{xz}, and a fourth, d_{yz}, in the *yz* plane (**Figure 3.16, a–d**). The four orbitals all have different orientations, but geometrically they are the same. Unlike the first four d orbitals, the d_{z^2} orbital has different geometric considerations, the two lobed-shaped regions along the *z*-axis and a donut-shaped region of opposite phase surrounding the origin (**Figure 3.16e**).

Figure 3.16

The five *d* orbitals.

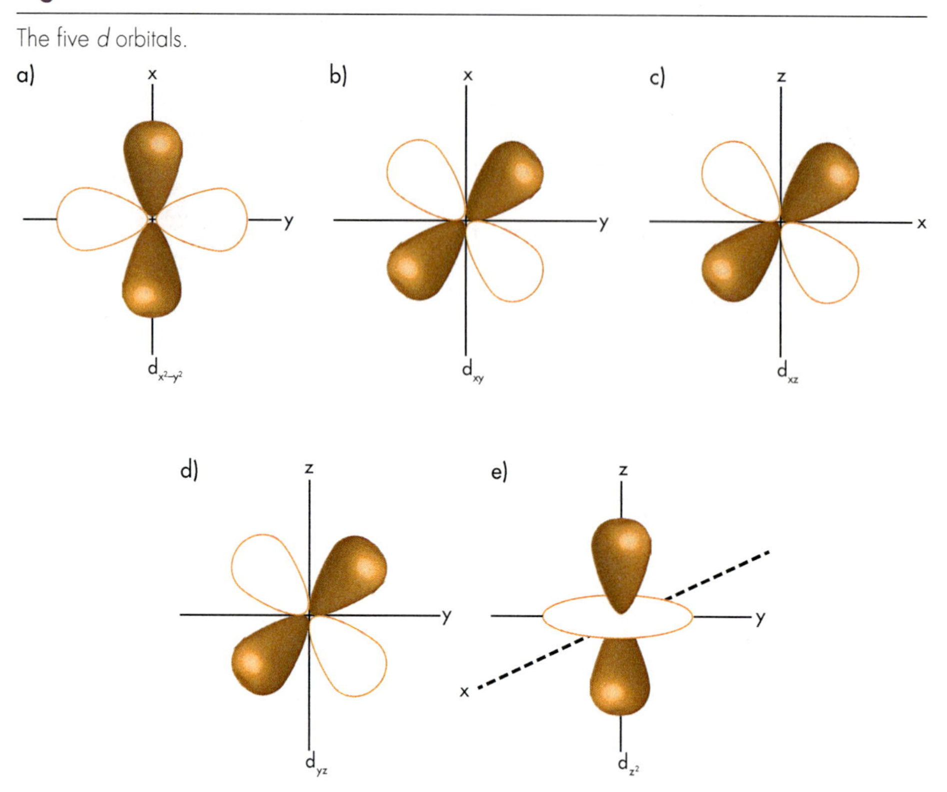

Up to this point we have used quantum numbers to determine what shell and subshell an electron is in. With the use of the *magnetic quantum number*, m_l we can also indicate which particular orbital within a subshell an electron occupies. Take for instance the *p* subshell orbitals, *px*, *py*, and *pz*. We can assign a magnetic quantum number to each of these three orbitals to indicate the location of the electron. The values are −1, 0, and +1 (**Figure 3.17**). The *d* subshell containing 5 orbitals has 5 magnetic quantum numbers, −2, −1, 0, +1, and +2, each corresponding to a particular *d* orbital. In a similar fashion, magnetic quantum numbers can be assigned to the orbitals of the *s* and *f* subshells. The values of m_l range from $-l$ to $+l$ and include zero. By now you see that the magnetic quantum number describes the orientation of a particular orbital.

Figure 3.17

Three p orbitals each with an assigned magnetic quantum number, m_l.

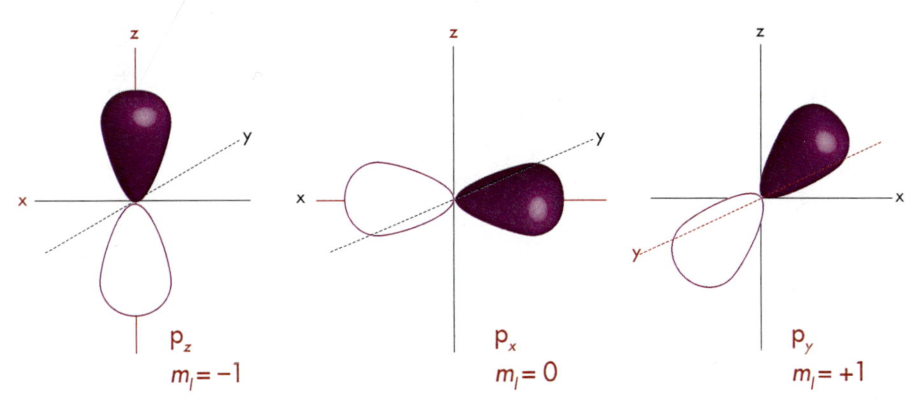

Just as the orbitals of the p and d subshells consist of 2 and 4 lobes respectively, the orbitals of the f subshell consist of 8 lobes. Aside from the complexity of their structures, these orbitals are not considered in normal chemical bonding. For these reasons, a detailed discussion of them will not be necessary at this time. The f subshell contains 7 orbitals and can hold up to 14 electrons with $l = 3$, and $m_l = -3, -2, -1, 0, 1, 2, 3$.

For a complete description of an electron we must consider a fourth quantum, the *electron spin quantum number*, m_s. Recall that a given orbital can contain a maximum of two electrons. When electrons are paired in an orbital, they must spin in opposite directions. An electron spinning in a clockwise direction is assigned $m_s = +1/2$ and an electron spinning in a counterclockwise direction is assigned $m_s = -1/2$.

In summary, there are four unique quantum numbers assigned to each electron within an atom. These numbers are calculated directly from the *Schroedinger equation,* which mathematically describes all the energy levels within an atom. The energy levels of the electrons are divided into a) shells (principal quantum number: $n = 1, 2, 3, \ldots$), which are divided into b) subshells (azimuthal quantum number: $l = n - 1, \ldots 0$) which contain c) orbitals (magnetic quantum number: $m_l = -l, \ldots 0, \ldots +l$), in which electrons possess one of two d) spins (magnetic spin quantum number: $m_s = +1/2$ or $-1/2$). The first three quantum numbers assigned to electrons in the first four shells are given in **Table 3.8**.

Table 3.8

Quantum Numbers for Electrons in the First Four Shells

First Shell	$n = 1$	$l = 0$ (*s*)	$m_l = 0$
Second Shell	$n = 2$	$l = 0$ (*s*) $l = 1$ (*p*)	$m_l = 0$ $m_l = -1, 0, +1$
Third Shell	$n = 3$	$l = 0$ (*s*) $l = 1$ (*p*) $l = 2$	$m_l = 0$ $m_l = -2, -1, 0, +1,$ $m_l = -2, -1, 0, +1, +2$
Fourth Shell	$n = 4$	$l = 0$ (*s*) $l = 1$ (*p*) $l = 2$	$m_l = 0$ $m_l = -1, 0, +1$ $m_l = -2, -1, 0, +1,$

Orbital Box Diagrams

An electron configuration can be represented with an orbital box diagram. Boxes are used to represent the orbitals and an arrow indicates each electron. For example, the hydrogen atom has the electron configuration $1s^2$ and the orbital box diagram is

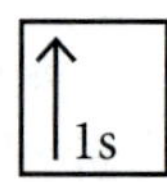

Recall that the two electrons in an orbital must be of opposite spin. In an orbital box diagram, an arrow pointing up indicates that the electron is spinning clockwise ($m_s = +1/2$) and an arrow pointing down indicates a counterclockwise spin ($m_s = -1/2$). In the following example and worksheets, we will fill in orbital box diagrams to guide us in assigning the quantum numbers (i.e., the address) of each electron in the atom or ion. While doing exercises such as these, there are three rules to remember:

1. After determining the total number of electrons for the atom or ion, start filling the lowest energy orbitals first.
2. Each box represents one orbital that can hold up to two electrons.
3. When more than one orbital box occurs in a given subshell level, do not pair electrons until all orbitals in that subshell contain one electron. For example, the *p* subshell contains three orbitals and each *p* orbital must contain one electron before electrons can be paired.

Example 3.16 Orbital Box Diagrams

Problem

Fill in the following orbital box for the neutral carbon atom.

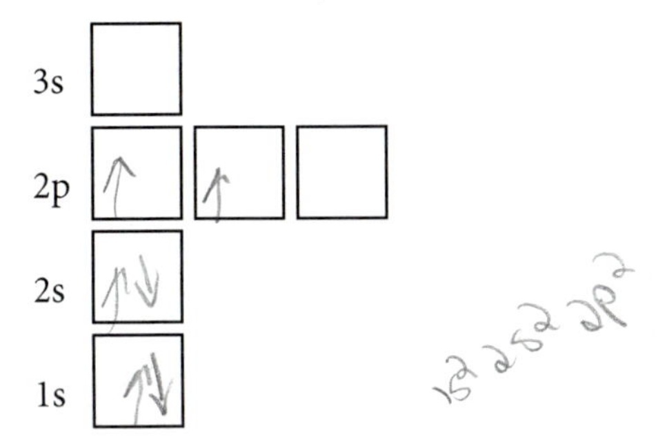

Solution

A neutral carbon atom has $Z = 6$. There are 6 electrons to distribute. We start filling the lower energy orbitals. Two electrons go into the 1*s* orbital, 2 electrons into the 2*s* orbital, and 2 electrons into 2 of the 3 *p* orbitals.

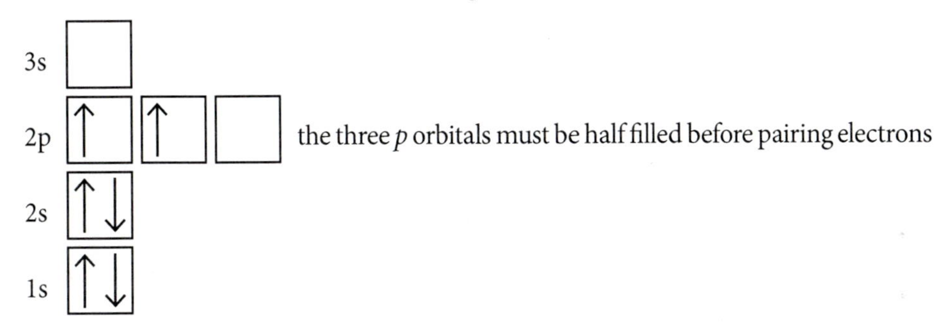

The orbital box diagram corresponds to the electron configuration $1s^2 2s^2 2p^2$.

Example 3.17 Orbital Box Diagrams and Assignment of Quantum Numbers

Problem

a) Fill in the following orbital box diagram for a neutral sodium atom and **b)** use the orbital box diagram to assign quantum numbers to each electron in the neutral sodium atom.

Solution

a) Sodium has $Z = 11$. Two electrons are assigned to the 1*s* orbital, two electrons to the 2*s* orbital, six electrons to the 2*p* orbital, and 1 electron is assigned to the 3*s* orbital.

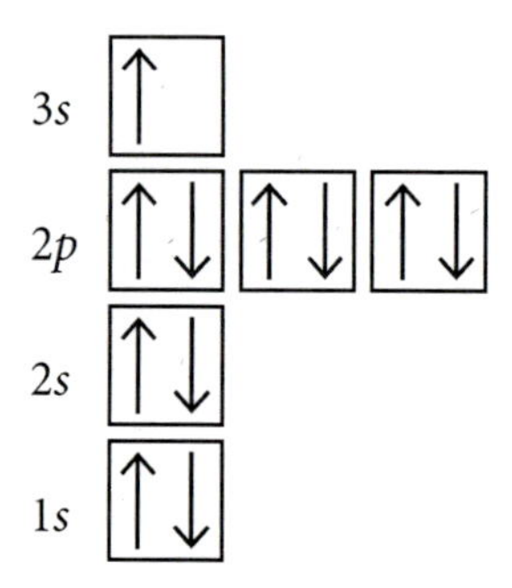

From the orbital box diagram we can write the electron configuration for Na.

$1s^2 2s^2 2p^6 3s^1$

b) It is helpful to set up a table and assign a number to each of the 11 electrons. Refer to the orbital box diagram in a) and start with the first electron. Electron 1 is in a 1*s* orbital so we assign $n = 1$, $l = 0$, $m_l = 0$, and $ms = +1/2$. For the second electron, $n = 1$, $l = 0$, $m_l = 0$, and $m_s = -1/2$. Note that the two electrons in the 1*s* orbital have the same three quantum numbers, n, l, m_l, but the electron spin quantum numbers have different signs. Two electrons in an orbital will always have the same three quantum numbers, and the fourth quantum number, which represents spin, is either a +1/2 or −1/2.

Electron	n	l	m_l	m_s
11	3	0	0	+ ½
10	2	1	1	−½
9	2	1	1	+ ½
8	2	1	0	−½
7	2	1	0	+ ½
6	2	1	−1	−½
5	2	1	−1	+ ½
4	2	0	0	−½
3	2	0	0	+ ½
2	1	0	0	−½
1	1	0	0	+ ½

Summary for Electron Configuration

- When writing an electron configuration, the shells and subshells fill in order of increasing energy.
- Electron configuration notation consists of shell numbers, the subshells within the shells, and the number of electrons within each subshell.
- The orbitals within each type of subshell (*s, p, d, f*) have their own unique shape and orientation.
- Electrons are arranged in orbitals within subshells. An orbital can hold up to two electrons.
- As a result of energy crossovers between subshells, the 4*s* subshell fills before the 3*d* subshell, the 5*s* subshell fills before the 4*d* subshell, and so on.
- No two electrons can have the same four quantum numbers, *n, l, ml,* and *ms.*
- In orbital box diagrams each box represents an orbital and arrows the electrons. Two electrons in an orbital must have opposite spins.

Worksheet 3.6 Orbital Box Diagrams, Electron Configurations, Quantum Numbers, and Valence Electrons (Sections 3.5 and 3.6)

Determine the electron configuration in both an orbital box diagram and spectroscopic notation for each species. Indicate the number of valence electrons in each species, and assign quantum numbers to each valence electron. Use the information from this worksheet to fill in the tables in Worksheet 3.7.

a) Beryllium (Be)

Z = 4

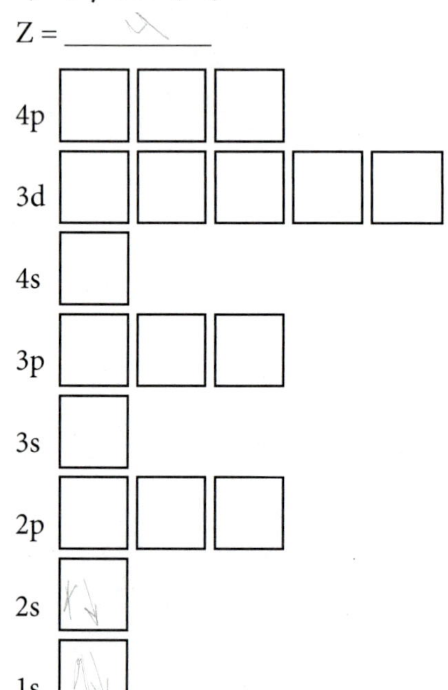

Electron Configuration (spectroscopic notation) $1s^2 2s^2$

Number of valence electrons 2

Provide the quantum numbers for each valence electron (a species can have up to 8 valence electrons)

Valence electron

8 n = ______ l = ______ m_l = ______ m_s = ______

7 n = ______ l = ______ m_l = ______ m_s = ______

6 n = ______ l = ______ m_l = ______ m_s = ______

5 n = ______ l = ______ m_l = ______ m_s = ______

4 n = ______ l = ______ m_l = ______ m_s = ______

3 n = ______ l = ______ m_l = ______ m_s = ______

2 n = ______ l = ______ m_l = ______ m_s = ______

1 n = ______ l = ______ m_l = ______ m_s = ______

b) Sulfur (S)

Z = 16

4p					
3d					
4s					
3p	↑↓	↑	↑		
3s	↑↓				
2p	↑↓	↑↓	↑↓		
2s	↑↓				
1s	↑↓				

$1s^2 2s^2 2p^6 3s^2 3p^4$

Electron Configuration (spectroscopic notation) ____________________

Number of valence electrons____________________

Provide the quantum numbers for each valence electron (a species can have up to 8 valence electrons)

Valence electron

8	n = ______	l = ______	m_l = ______	m_s = ______
7	n = ______	l = ______	m_l = ______	m_s = ______
6	n = ______	l = ______	m_l = ______	m_s = ______
5	n = ______	l = ______	m_l = ______	m_s = ______
4	n = ______	l = ______	m_l = ______	m_s = ______
3	n = ______	l = ______	m_l = ______	m_s = ______
2	n = ______	l = ______	m_l = ______	m_s = ______
1	n = ______	l = ______	m_l = ______	m_s = ______

c) Oxygen (O)

Z = __________

4p	☐	☐	☐		
3d	☐	☐	☐	☐	☐
4s	☐				
3p	☐	☐	☐		
3s	☐				
2p	☐	☐	☐		
2s	☐				
1s	☐				

Electron Configuration (spectroscopic notation) ____________________________

Number of valence electrons____

Provide the quantum numbers for each valence electron (a species can have up to 8 valence electrons)

Valence electron

8 n =__________ l =__________ m_l =__________ m_s =__________

7 n =__________ l =__________ m_l =__________ m_s =__________

6 n =__________ l =__________ m_l =__________ m_s =__________

5 n =__________ l =__________ m_l =__________ m_s =__________

4 n =__________ l =__________ m_l =__________ m_s =__________

3 n =__________ l =__________ m_l =__________ m_s =__________

2 n =__________ l =__________ m_l =__________ m_s =__________

1 n =__________ l =__________ m_l =__________ m_s =__________

d) Boron (B)

Z = __________

4p

3d

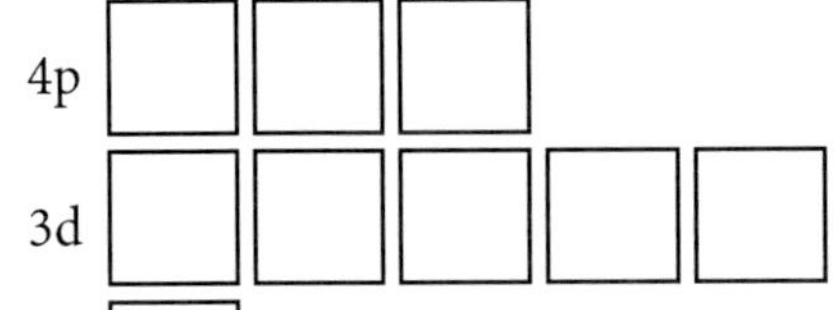

4s

3p

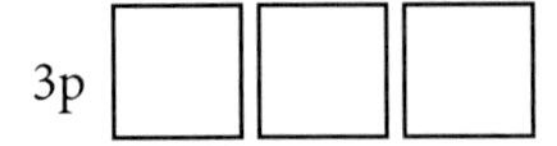

3s

2p

2s

1s

Electron Configuration (spectroscopic notation) __________________________

Number of valence electrons____

Provide the quantum numbers for each valence electron (a species can have up to 8 valence electrons)

Valence electron

8 n =__________ l =__________ m_l =__________ m_s =__________

7 n =__________ l =__________ m_l =__________ m_s =__________

6 n =__________ l =__________ m_l =__________ m_s =__________

5 n =__________ l =__________ m_l =__________ m_s =__________

4 n =__________ l =__________ m_l =__________ m_s =__________

3 n =__________ l =__________ m_l =__________ m_s =__________

2 n =__________ l =__________ m_l =__________ m_s =__________

1 n =__________ l =__________ m_l =__________ m_s =__________

e) Argon (Ar)

Z = __________

4p ☐☐☐

3d ☐☐☐☐☐

4s ☐

3p ☐☐☐

3s ☐

2p ☐☐☐

2s ☐

1s ☐

Electron Configuration (spectroscopic notation) ______________________________

Number of valence electrons____

Provide the quantum numbers for each valence electron (a species can have up to 8 valence electrons)

Valence electron

8 n =__________ l =__________ m_l =__________ m_s =__________

7 n =__________ l =__________ m_l =__________ m_s =__________

6 n =__________ l =__________ m_l =__________ m_s =__________

5 n =__________ l =__________ m_l =__________ m_s =__________

4 n =__________ l =__________ m_l =__________ m_s =__________

3 n =__________ l =__________ m_l =__________ m_s =__________

2 n =__________ l =__________ m_l =__________ m_s =__________

1 n =__________ l =__________ m_l =__________ m_s =__________

f) Vanadium (V)

Z = __________

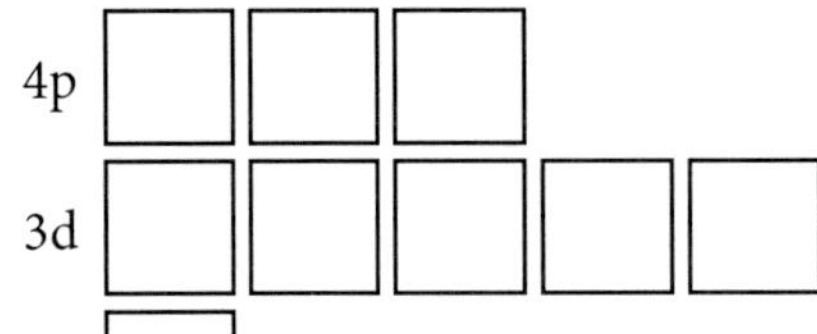

4s

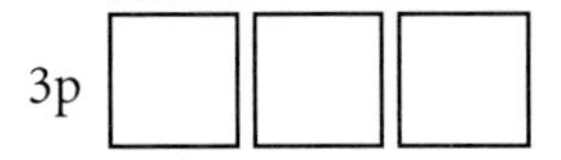

3s

2s

1s

Electron Configuration (spectroscopic notation) ______________________________

Number of valence electrons____

Provide the quantum numbers for each valence electron (a species can have up to 8 valence electrons)

Valence electron

8 n =__________ l =__________ m_l =__________ m_s =__________

7 n =__________ l =__________ m_l =__________ m_s =__________

6 n =__________ l =__________ m_l =__________ m_s =__________

5 n =__________ l =__________ m_l =__________ m_s =__________

4 n =__________ l =__________ m_l =__________ m_s =__________

3 n =__________ l =__________ m_l =__________ m_s =__________

2 n =__________ l =__________ m_l =__________ m_s =__________

1 n =__________ l =__________ m_l =__________ m_s =__________

g) Zirconium (Zr)

Z = __________

4p ☐☐☐

3d ☐☐☐☐☐

4s ☐

3p ☐☐☐

3s ☐

2p ☐☐☐

2s ☐

1s ☐

Electron Configuration (spectroscopic notation) ______________________________

Number of valence electrons______________________________

Provide the quantum numbers for each valence electron (a species can have up to 8 valence electrons)

Valence electron

8	n = ________	l = ________	m_l = ________	m_s = ________
7	n = ________	l = ________	m_l = ________	m_s = ________
6	n = ________	l = ________	m_l = ________	m_s = ________
5	n = ________	l = ________	m_l = ________	m_s = ________
4	n = ________	l = ________	m_l = ________	m_s = ________
3	n = ________	l = ________	m_l = ________	m_s = ________
2	n = ________	l = ________	m_l = ________	m_s = ________
1	n = ________	l = ________	m_l = ________	m_s = ________

h) Bromine (Br)

Z = ________

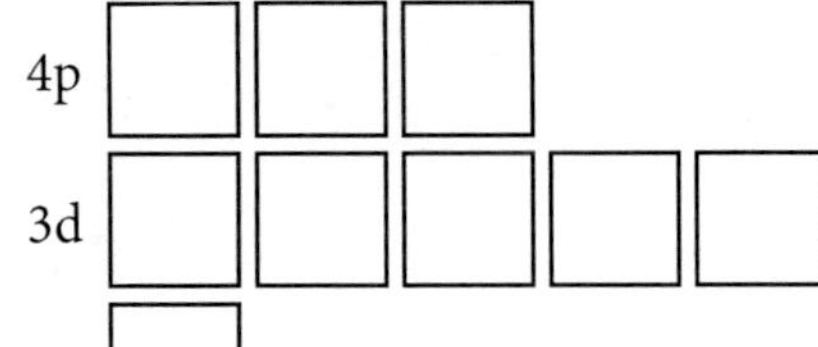

4s

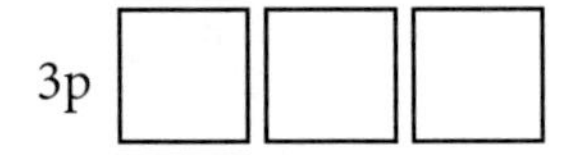

3s

2s

1s

Electron Configuration (spectroscopic notation) ________________________

Number of valence electrons ________________________

Provide the quantum numbers for each valence electron (a species can have up to 8 valence electrons)

Valence electron

8 n = ________ l = ________ m_l = ________ m_s = ________

7 n = ________ l = ________ m_l = ________ m_s = ________

6 n = ________ l = ________ m_l = ________ m_s = ________

5 n = ________ l = ________ m_l = ________ m_s = ________

4 n = ________ l = ________ m_l = ________ m_s = ________

3 n = ________ l = ________ m_l = ________ m_s = ________

2 n = ________ l = ________ m_l = ________ m_s = ________

1 n = ________ l = ________ m_l = ________ m_s = ________

i) Potassium ion (K^+)

Z = __________

4p ☐ ☐ ☐

3d ☐ ☐ ☐ ☐ ☐

4s ☐

3p ☐ ☐ ☐

3s ☐

2p ☐ ☐ ☐

2s ☐

1s ☐

Electron Configuration (spectroscopic notation) ______________________________

Number of valence electrons____

Provide the quantum numbers for each valence electron (a species can have up to 8 valence electrons)

Valence electron

8 n = __________ l = __________ m_l = __________ m_s = __________

7 n = __________ l = __________ m_l = __________ m_s = __________

6 n = __________ l = __________ m_l = __________ m_s = __________

5 n = __________ l = __________ m_l = __________ m_s = __________

4 n = __________ l = __________ m_l = __________ m_s = __________

3 n = __________ l = __________ m_l = __________ m_s = __________

2 n = __________ l = __________ m_l = __________ m_s = __________

1 n = __________ l = __________ m_l = __________ m_s = __________

j) Chloride ion (Cl^-)

Z = __________

4p ☐ ☐ ☐

3d ☐ ☐ ☐ ☐ ☐

4s ☐

3p ☐ ☐ ☐

3s ☐

2p ☐ ☐ ☐

2s ☐

1s ☐

Electron Configuration (spectroscopic notation) ______________________

Number of valence electrons______________________

Provide the quantum numbers for each valence electron (a species can have up to 8 valence electrons)

Valence electron

8 n = ________ l = ________ m_l = ________ m_s = ________

7 n = ________ l = ________ m_l = ________ m_s = ________

6 n = ________ l = ________ m_l = ________ m_s = ________

5 n = ________ l = ________ m_l = ________ m_s = ________

4 n = ________ l = ________ m_l = ________ m_s = ________

3 n = ________ l = ________ m_l = ________ m_s = ________

2 n = ________ l = ________ m_l = ________ m_s = ________

1 n = ________ l = ________ m_l = ________ m_s = ________

k) Iron (II) ion (Fe^{2+})

Z = __________

4p
3d
4s
3p
3s
2p
2s
1s

Electron Configuration (spectroscopic notation) ____________________________

Number of valence electrons____________________________

Provide the quantum numbers for each valence electron (a species can have up to 8 valence electrons)

Valence electron

8 n =__________ l =__________ m_l =__________ m_s = __________

7 n =__________ l =__________ m_l =__________ m_s = __________

6 n =__________ l =__________ m_l =__________ m_s = __________

5 n =__________ l =__________ m_l =__________ m_s = __________

4 n =__________ l =__________ m_l =__________ m_s = __________

3 n =__________ l =__________ m_l =__________ m_s = __________

2 n =__________ l =__________ m_l =__________ m_s = __________

1 n =__________ l =__________ m_l =__________ m_s = __________

End of Chapter 3 Questions

Section 3.1 The Nuclear Region

1. Which of the following statements is correct?
   a) All subatomic particles are the same size.
   b) Protons and electrons are located in the nuclear region.
   c) Protons have a positive charge.
   d) The extranuclear region contains electrons and neutrons.

2. Define the following
   Atom

   Proton

   Neutron

   Electron

3. What is the name of the instrument that measures the relative mass of atoms?

4. The nuclear region includes which particles?
   a) electrons b) protons c) atoms d) neutrons

5. If a magnesium atom is 2.0002 times the mass of a carbon atom that contains 6 protons and 6 electrons, what is the relative mass of the magnesium atom in amu?

6. If a lithium atom has a mass that is 0.5784 times the mass of a carbon atom that contains 6 protons and 6 neutrons, what is the relative mass of the lithium atom in amu?

7. Which subatomic particles are called nucleons?

8. What part of an atom contains most of the atom's mass?

9. An element has an atomic number of 14. What is the name of this element?

10. What is the atomic number of the element bromine, Br?

11. An element has 17 protons, 18 neutrons, and 17 electrons. What is its atomic number? What is its mass number?

12. An element has 8 protons, 10 neutrons, and 8 electrons.
    a) What is the name of the element?

    b) What is the atomic number of this element?

    c) What is the mass number of this element?

    d) Does this atom have a charge?

13. An element can have different atoms? How do these atoms differ from one another?

Section 3.2 Isotopes

14. Define isotope.

15. Isotopes of the same element have
    a) The same mass number but a different atomic number.
    b) The same atomic number but a different number of electrons.
    c) The same atomic number but a different mass number
    d) The same mass number but a different number of electrons.

16. The uranium-238 isotope has ___________ protons and ___________ neutrons.

17. The isotope ^{40}Ca has _______ protons, _______ neutrons, and ______ electrons.

18. An isotope has 47 protons, 60 neutrons, and 47 electrons.
a) What is the name of this isotope?
b) Write the symbol for this isotope using the isotope's symbol, mass number, and atomic number?

19. What is percent abundance?

20. Two of the naturally occurring isotopes of carbon have mass numbers of 12 and 13. How many neutrons does each isotope have? Write the isotopic symbols for both isotopes.

21. Fill in the blanks for the following:

a) $^{6}_{\_}O$ b) $^{0}_{5}\_\_\_\_$ c) $^{55}_{\_}Mn$

22. An element has three isotopes. One isotope has a percent abundance of 35.4%, and another isotope has an abundance of 50.5%, what is the percent abundance of the third isotope?

23. An element has three isotopes. The first isotope has a mass of 27.9769 amu and a percent abundance of 92.2%. The second isotope has a mass of 28.9765 amu and a percent abundance of 4.7%, and the third isotope has a mass of 29.9734 and a percent abundance of 3.1%. Calculate the atomic mass of the element.

24. Given the isotopic masses and their percent abundances, a student calculates the atomic mass of an element by adding the isotopic masses and dividing by the number of isotopes. Is there a chance that his answer will be correct? Explain.

25. Define atomic mass.

26. An element has $Z = 31$ and 38 neutrons
a) What is the identity of the element?

b) What is the mass number?

c) What is the atomic mass?

Section 3.3 Charge Balance

27. The isotope $^{138}Ba^{2+}$ contains __________ protons, _________ neutrons, and __________ electrons.

28. Write the symbol for a nitrogen atom that loses 3 electrons. Is this a cation or an anion?

29. How many electrons does the Al^{3+} ion have?

30. The selenium ion has a charge of −2. Is this a loss or gain of electrons? Write the symbol of the ion.

31. An ion has 24 protons and 19 electrons. Write the symbol for this ion.

32. Why is it important not to state that the atomic number is always the same as the number of electrons?

Section 3.4 Extranuclear Region

33. How does the energy that we experience in everyday life differ from the energy that an electron in an atom experiences?

34. What does it mean to say that the energy of an electron is quantized?

35. In the model of the hydrogen atom that Bohr proposed, the electron circles the nucleus in ____________ around the nucleus.

36. For an electron to move from $n = 1$ to $n = 2$, the electron would the electron need to absorb or release energy?

37. If an electron in a hydrogen atom moves from $n = 3$ to $n = 2$, is energy absorbed or released?

38. As the values of n increase, the energies of the orbits
    a) increase b) decrease c) stay the same

39. List three points that Bohr stated about the electron in the hydrogen atom.

Section 3.5 Electron Shells

40. The letter n is called ___________________, and it is used to indicate the ________________________.

41. As the value of n increases, so does ______________________________ and ________________________.

42. The maximum number of electrons in the 4th shell is ______________.

43. What is the maximum number of electrons that can be contained in $n = 4$?

44. In the sodium atom, the electrons are arranged as ___________electrons in the first shell, ___________ electrons in the second shell, and ______________ electrons in the third shell.

45. A shell can hold a maximum of 2 electrons. What is the principle quantum number, n?

46. Define valence shell.

47. How many valence electrons does a neutral potassium atom have?

48. Which electrons are the most important for chemical bonding and chemical reactivity?

49. A neutral atom has an atomic number of 15.
a) What is its total number of electrons?

b) How many electrons are in its second shell?

c) How many valence electrons does it have?

d) What is its valence shell number?

50. A main group atom has 6 valence electrons. What group is it in?

51. An atom has 5 valence electrons in $n = 1$. What is the identity of this element?

52. How many subshells are contained in $n = 3$? In $n = 2$? In $n = 5$?

3.6 Electron Configuration

53. What is the name of the quantum number that indicates a subshell?

54. Which subshell is higher in energy? A p subshell or the d subshell?

55. The quantum number $l = 2$ corresponds to what principle energy level?

56. What is the quantum number that indicates a *d* subshell?

57. How many electrons can fill a *p* subshell?

58. An electron has the quantum numbers $n = 2$ and $l = 0$. What is the spectroscopic notation for this electron?

59. An electron has the quantum numbers $n = 3$ and $l = 1$. What is the spectroscopic notation for this electron?

60. A *d* subshell can hold a maximum of how many electrons?

61. In the 4th period, the energies of the 4*s* and the 3*d* interchange. Why does this happen?

62. What is the electron configuration of a neutral iron, Fe, atom?

63. Each orbital can hold a maximum of ______ electrons.

64. How many orbitals does the d subshell contain?

65. The orbits within the s subshell have a ____________________ shape.

66. The three *p* subshell orbitals lie along the ____________________________ axes.

67. The *p* subshell orbitals all have a point where the electrons cannot be found. This place is called a ________________.

68. The d_{xz} orbital lies within the ______ plane and has ______ lobes.

69. The quantum number that indicates the orientation of an orbital is called the ______________ quantum number.

70. What are the two values for the spin magnetic quantum number m_s? What do these values mean?

71. Write an electron configuration for each of the following:
 a) Ca b) Br c) Al d) F e) As f) S^2

72. State the number of valence electrons for each atom/ion in problem 70.

73. Construct an orbital box diagram for the selenium atom. $Z = 34$.
 a) What is the electron configuration for selenium?

 b) What is the electron configuration for the valence shell of the selenium atom?

 c) How many valence electrons does the selenium atom have?

 d) What are the four quantum numbers for the valence electrons of the selenium atom?

General Questions

74. Given the following atoms, indicate the highest energy subshell that holds electrons.
 a) carbon b) sodium c) Ar d) Zn

75. Gallium has two stable isotopes, ^{69}X and ^{71}X, and has an average weighted atomic mass of 69.72 amu. What are the relative percent abundances of these isotopes?

76. Calculate the mass of the nucleus of ^{39}K. Include the electrons in the mass calculation.

77. An atom has the following quantum numbers. Identify the element.

78. Write the electron configuration for Boron. Assign quantum numbers to each electron.

79. Consider $^{26}Mg^{2+}$ to answer the following questions.
 a) What is the mass number?

 b) How many electrons, protons, and neutrons?

c) What is the atomic mass?

d) Is this a cation or anion?

e) Are electrons lost or gained? How many electrons lost or gained?

d) How many valence electrons?

80. Which of the following sets of quantum numbers are not possible? Explain
a) $n = 3, l = 3, ml = 0$ b) $n = 6, l = 5, ml = 1$
c) $n = 2, l = 1, ml = 0$ d) $n = 4, l = 3, ml = 4$

81. Write a set of quantum numbers for the 3*d* orbital.

82. Which would be the most stable configuration and why?

a) [↑↓] [ ] [ ]

b) [↑] [↑] [ ]

c) [↓] [↑] [ ]

83. Consider the electron configuration to answer the questions below.
$1s^2 2s^2 2p^6 3s^2 3p^6 4s^2 3d^{10} 4p^4$
a) What is the identity of the element?

b) What is the valence shell?

c) How many valence electrons?

d) Assign quantum numbers to the valence electrons

84. How does the relative mass of lithium compare to the mass of carbon-12.

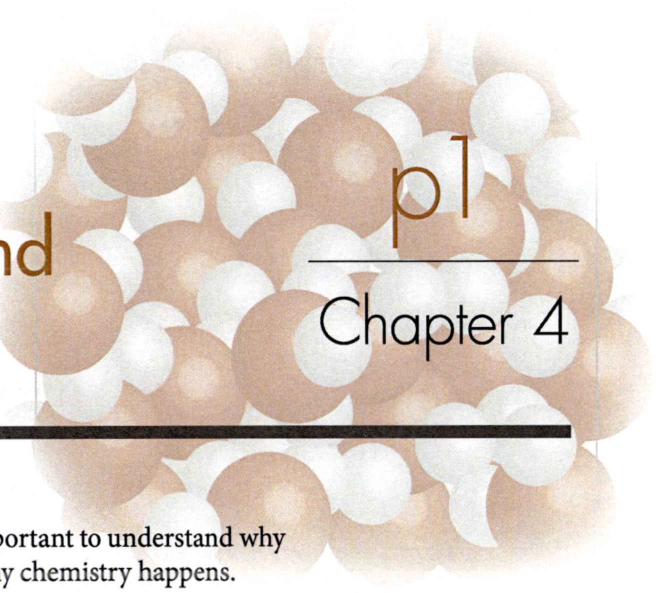

Chemical Bonding and Structure

p1

Chapter 4

Before talking about how atoms bond together, it's important to understand why atoms bond together, and in more general terms; why chemistry happens.

Why does a chlorine atom (Cl) become a chloride ion (Cl^-)?

Why do two hydrogen atoms (H) come together to form a molecule of hydrogen (H_2)?

Why do copper atoms (Cu) conduct electricity and nitrogen atoms (N) do not?

By the same token, chemical elements and compounds will normally react to form new products which are lower in energy than the starting materials. A thorough discussion on this matter is part of the focus of an advanced chemistry course, i.e. physical chemistry, but at this level of chemistry, we will introduce some of the basic concepts.

Gaining Stability

The first rule to remember is that normally, most chemical phenomena occur for the sake of stability. Atoms may gain or lose electrons to become more stable, i.e. they become lower energy species.

As previously stated regarding main group elements, the Noble Gases (Group 8A), have little if any reactivity since these atoms have a very stable valence electron configuration.

In this section, we will see two ways atoms, other than noble gases, can obtain this configuration.

Figure 4.1

The Periodic Table

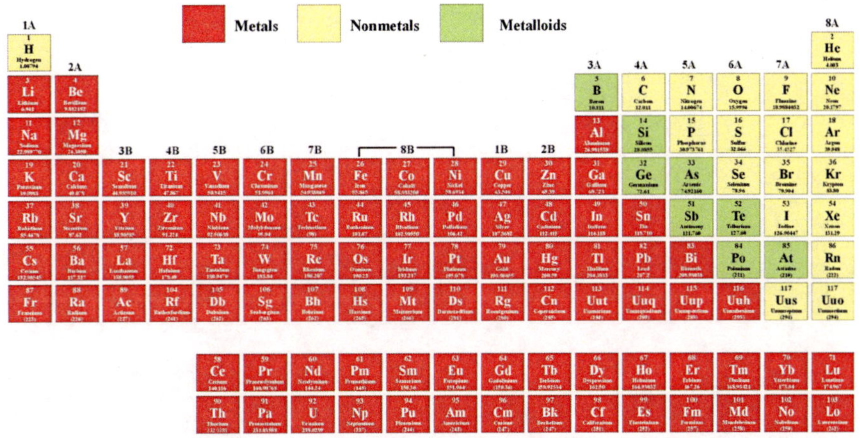

Review of Main Group Valence Structure

Recall how the **valence s and p subshells** fill as we move across the second period:

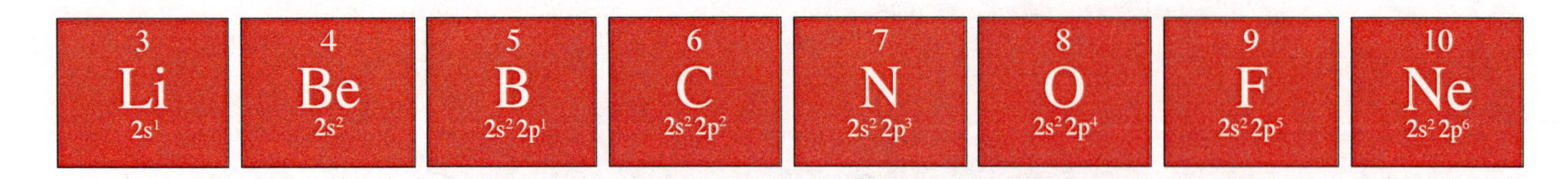

Notice how the valence s and p subshells fill moving from left to right. Arriving at the Noble Gas neon, (Ne), both the s and p subshells are filled:

$$\mathbf{2s^2\ 2p^6}$$

Not only does neon have this valence electron configuration, so do the remaining Noble Gases.

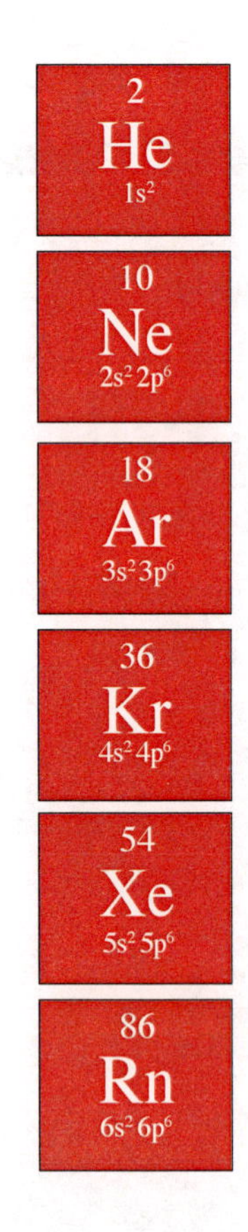

All the noble gases have filled valence s and p subshells.

In more general terms, this configuration can be stated as

$ns^2\ np^6$
(n = valence shell number)

From your knowledge of atomic structure, explain: why is helium different?

Helium's valence shell contains 2 electrons (a doublet).

The other noble gases have valence shells containing 8 electrons (an octet).

Ion Formation in Main Group Elements

One of the most fundamental chemical phenomena is the formation of ions from atoms. Recall that we refer to atoms containing an equal number of protons and electrons as being neutral (in net charge).

Sometimes an atom of the main group elements will take on extra electrons to obtain a stabilizing octet, that is a completely filled valence p subshell:

$\mathbf{F}$	+ $\mathbf{1\ e^-}$	$\longrightarrow$ $\mathbf{F^-}$
fluorine atom	one electron	fluoride ion (negative one charge)
$1s^2\ 2s^2\ 2p^5$		$\mathbf{1s^2\ 2s^2\ 2p^6}$

By accepting an additional electron, the fluorine atom obtains the same electron configuration as the noble gas neon:

Ne

$\mathbf{1s^2\ 2s^2\ 2p^6}$

And sometimes an atom in the main group takes on more than one electron:

O	+	**2 e⁻**	**O²⁻**
oxygen atom		two electrons	oxide ion (negative 2 charge)
$1s^2\ 2s^2\ 2p^4$			$\mathbf{1s^2\ 2s^2\ 2p^6}$

By accepting two electrons in its valence shell, oxygen has also obtained the same electron configuration as the noble gas neon.

Assignment 4.1

In the following exercise, show the change of **valence electron** configuration upon ion formation and the resulting charge when noble gas configuration is achieved. Name the Noble Gas that has the same configuration.

O $2s^2\ 2p^4$	O^{2-} $2s^2\ 2p^6$	Noble Gas Neon
N	N^{3-}	Noble Gas
Cl	Cl^-	Noble Gas
S	S^{2-}	Noble Gas
Br	Br^-	Noble Gas
Se	Se^{2-}	Noble Gas

Assignment 4.2

From the preceding exercise, what trend can you see in the ion charge and the group number?

	Ion Charge	Group Number
O^{2-}	– 2	6A
N^{3-}		
Cl^{-}		
S^{2-}		
Br^{-}		
Se^{2-}		

Observations

Assignment 4.3

Ion	Charge	Group Number
Li^{+}	+1	1A
Na^{+}		
Mg^{2+}		
K^{+}		
Rb^{+}		
Sr^{2+}		

Observations

Just as some atoms of main group elements gain valence electrons, others will lose valence electrons to obtain Noble Gas electron configuration:

Li →	Li^{+}	+	$1e^{-}$
lithium atom	lithium ion (one positive charge)		one electron (lost by lithium atom)
$1s^2\ 2s^1$	$1s^2$		

By losing one valence electron, the lithium atom obtains the same electron configuration as the noble gas helium:

He

$1s^2$

Another Noble Gas configuration can be obtained by a magnesium atom as it loses two valence electrons:

Mg →	Mg^{2+}	+	$2e^{-}$
magnesium atom	magnesium ion (two positive charges)		two electrons (lost by magnesium atom)
$1s^2\ 2s^2\ 2p^6\ 3s^2$	$1s^2\ 2s^2\ 2p^6$		

The result is the same ten electron configuration as the noble gas neon.

Shorthand Notation

A shorthand notation can be used in the Electron Configuration: Instead of using full notation for both core and valence electrons, the core electrons can be conveniently represented by the bracketed symbol of the Noble Gas that represents.

For example,

instead of stating the electron configuration of chlorine as:

$1s^2\ 2s^2\ 2p^6\ 3s^2\ 3p^5$,

$\underbrace{1s^2\ 2s^2\ 2p^6}_{\text{core electrons}}\quad \underbrace{3s^2\ 3p^5}_{\text{valence electrons}}$,

we can substitute the bracketed symbol for neon [Ne] to represent the core electrons.

Note: the notation for neon is $1s^2\ 2s^2\ 2p^6$.

$\underbrace{1s^2\ 2s^2\ 2p^6}_{[Ne]}\quad \underbrace{3s^2\ 3p^5,}_{3s^2\ 3p^5,}$

Shorthand notation for the electron configuration for chlorine is

$[Ne]3s^2\ 3p^5$,

By proper selection of the Noble Gas, a long electron configuration such strontium, Sr

$1s^2\ 2s^2\ 2p^6\ 3s^2\ 3p^6\ 4s^2\ 3d^{10}\ 4p^6\ 5s^2$

can be stated in shorthand notation as

$[Kr]\ 5s^2$

In the following exercise write out the full electron configuration for the element, determine the proper Noble Gas to represent the core electrons, and write the short hand notation of the configuration.

Assignment 4.4

Full Electron Configuration	Shorthand Notation
Na $1s^2\ 2s^2\ 2p^6\ 3s^1$	[Ne] $3s^1$
O	
Ca	
Fe	
P	
Br	
De	
Rh	
C	
Al	

Assignment 4.5

Before proceeding to the next discussion, work the following review exercise on ionic charge.

Symbol	Name	Total Electrons in Ion	Total Electrons in Noble Gas
Li^{+}	lithium ion	2	helium
Sr^{2+}	________	________	________
I^{-}	________	________	________
S^{2-}	________	________	________
P^{3-}	________	________	________
K^{+}	________	________	________
Ca^{2+}	________	________	________
Cs^{+}	________	________	________
O^{2-}	________	________	________
Al^{3+}	________	________	________

Lewis Dot Symbols and Summation of Valence Electrons

It is the outermost electrons of an atom, i.e. the valence electrons, which are responsible for chemical activity between atoms, ions, and molecules. For this reason it is very important that we know the electron configuration of the individual atoms and ions.

Recall the arrangement of electrons in an atom of nitrogen:

$$\mathbf{N} \quad 1s^2\, 2s^2\, 2p^3$$

The valence shell of the nitrogen atom is the second shell ($2s^2\ 2p^3$), which contains a total of five electrons available for forming covalent bonds or lone pairs.

As part of the chemistry language, we learn to represent both chemicals and chemical reactions symbolically.

Since, chemists are, for the most part, only interested in the outermost electrons a notation called **Lewis Dot Symbols** is used. It consists of the atom's symbol from the periodic table, and the valence electrons surrounding the symbol.

Looking at the electron configuration for nitrogen, we see two shells containing electrons:

$$1s^2\ 2s^2\ 2p^3$$

The first shell, represented as $1s^2$, contains two electrons in the s subshell.

The second shell (the outer most shell), represented as $2s^2\ 2p^3$, contains five electrons: **two electrons in the s-subshell** and **three electrons in the p-subshell** (resulting in a total of five electrons in the outermost shell, i.e. the valence shell).

These five electrons are referred to as **valence electrons.**

Using Lewis dot structure, we place the five valence electrons around the nitrogen symbol:

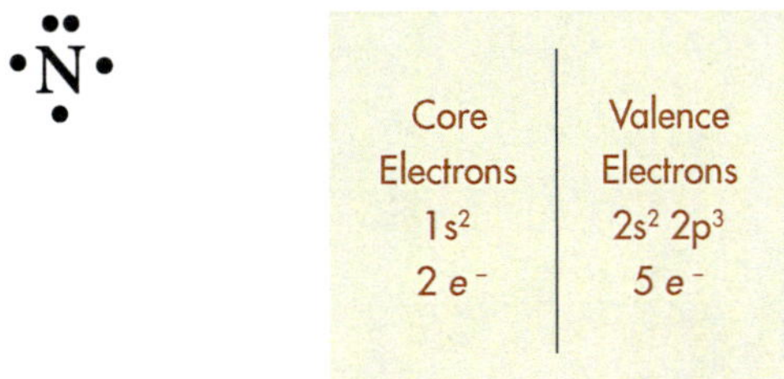

When placing the valence electrons around a main group atom, follow the method shown for the second period atoms:

Li•　　•Be•

•B•　•C•　•N•　•O•　:F•　:Ne:

Recall that the elements within a group

(i.e. 1A, 2A, 3A, 4A, 5A, 6A, 7A, or 8A)

will have the same number of valence electrons as the group number:

e.g.

IA:　Li, Na, K, Rb, Cs, Fr

will all have a single electron in the outermost shell, or in terms of Lewis Dot structures:

Li•　Na•　K•　Rb•　Cs•　Fr•

Figure 4.2

The Periodic Table

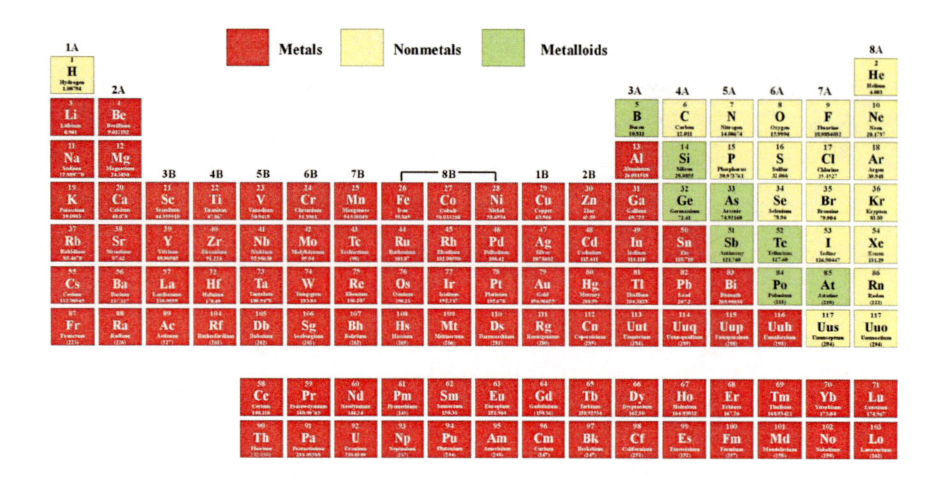

Assignment 4.6

Referring to the Periodic Table, fill in the following worksheet:

Atom or Ion	Symbol	Group Number	# Valence Electrons	Lewis Dot Structure
carbon	C	4A	4	$\cdot\dot{\underset{\cdot}{C}}\cdot$
chlorine	Cl	7A	7	:Cl:
iodide ion				
phosphorus				
arsenide ion				
krypton				
xenon				
potassium				

rubidium ion				
magnesium				
strontium ion				

Obtaining Noble Gas Configurations in Main Group Elements by Sharing of Electrons

Lewis Structures

In the previous section, we discussed how atoms obtain Noble gas configurations by either accepting or losing electrons.

Octets can also be completed by the **sharing of electrons between atoms**.

Consider the fluorine atom's valence shell:

One way the fluorine atom can complete the octet of its valence shell is by accepting an additional electron and forming the negatively charged fluoride ion:

Besides providing a shorthand notation of an atom's valence electron count and its configuration, Lewis Dot Symbols also set down the foundation for connecting the atoms in a molecule. This results in the Lewis Structure.

$$:\ddot{\underset{..}{F}}: \;+\; 1\,e^- \longrightarrow \;:\ddot{\underset{..}{F}}:$$

$1s^2\,2s^2\,2p^5$ $\qquad\qquad$ $1s^2\,2s^2\,2p^6$

The second way the fluorine atom can obtain noble gas configuration is by sharing electrons.

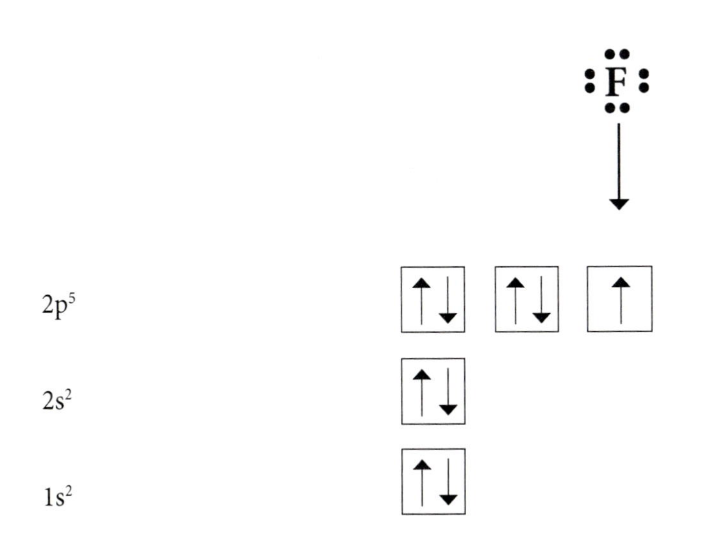

The unshared electron on the right hand side of the Lewis Dot structure corresponds to the unshared electron in the orbital box diagram directly below it.

As you can see this half-filled orbital is a p-orbital, and in the following discussion the unshared electrons, like in fluorine, will be represented as

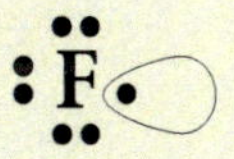

The process by which electrons are shared between two atoms is possible through the overlapping of the orbitals containing the unshared electrons:

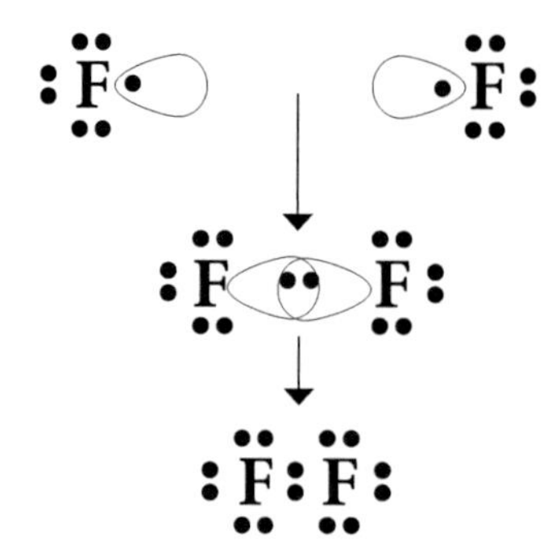

Notice that in the formation of the diatomic molecular fluorine, F_2 , each of the fluorine atoms has an octet of electrons in its valence shell:

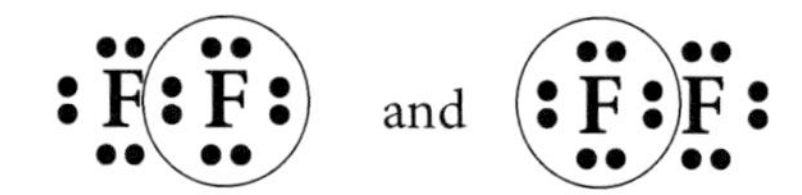

Octets and doublets can also be completed by two dissimilar atoms sharing valence electrons.

For example, consider the molecule hydrogen chloride, **HCl**.

The Lewis Dot Structures show hydrogen with one valence electron, and chlorine with seven valence electrons:

H •• Cl:

By overlapping the orbitals of the single unshared electron in the hydrogen's 1s orbital with the single unshared electron in a 3p orbital of the chlorine, hydrogen and the chlorine can each share an extra electron. This will complete a doublet for the hydrogen,

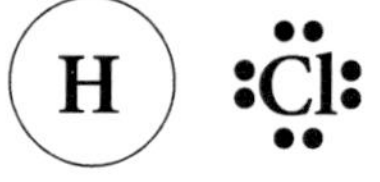

and complete the octet for the chlorine atom.

Note:

the two shared electrons are now represented by a straight line while the unshared electrons on the chlorine atom (known as lone pairs) remain in dot notation.

H — :Cl:

There are two approaches to determining Lewis Structures.

Part I

The first approach is quite simple:

1. Consider the Lewis Dot Symbols and complete octets and doublets by sharing electrons between atoms:

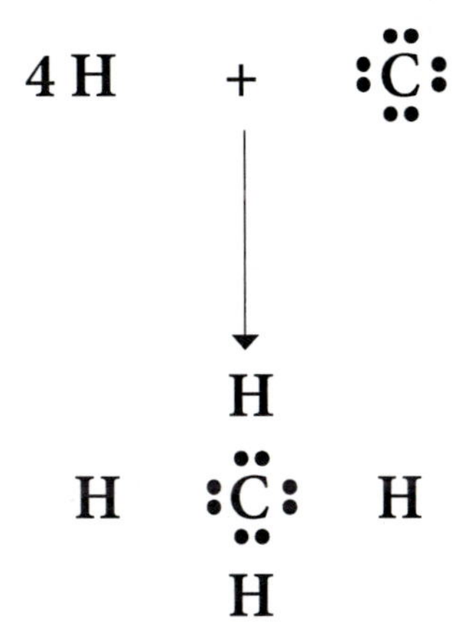

Each hydrogen atom now has a doublet in its valence shell.

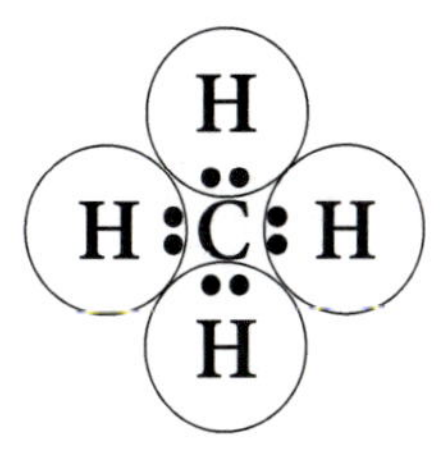

The carbon atom now has a completed octet in its valence shell.

H

H :C: H

H

Assignment 4.7

Using the Lewis Dot Symbols below, determine the Lewis Structures for the following

H• •B• •C• •N• •O• :F•

1C	4H	H H :C: H H H H—C—H H
1N	3F	
1B	3H	
2C	6H	

Notice that there is a relationship between the number of unpaired valence electrons, and the number of bonds that the atom can form:

C (four bonds) N (three bonds) O (two bonds) F (one bond)

These are called the **common bonding patterns** for these particular atoms. We shall see that there are exceptions to these bonding patterns.

Additional Exercises

Assignment 4.8

Using the Common Bonding patterns, determine the structures of the following compounds.

Chemical Formula	Structure
C_3H_6	
BCl_3	
CF_4	
CH_4O	
C_2H_6O	

Scratch Sheet

In the preceding exercise, the last formula contained two carbon atoms, six hydrogen atoms, and one oxygen atom:

$$C_2H_6O$$

Given the common bonding patterns of carbon, hydrogen, and oxygen, two different structures can be obtained:

Ethyl alcohol

```
    H   H
    |   |
H — C — C — O — H
    |   |
    H   H
```

Dimethyl ether

```
    H       H
    |       |
H — C — O — C — H
    |       |
    H       H
```

When more than one structure can be drawn from one chemical formula, the resulting structures are called isomers. In this example the formula, C_2H_6O yields two isomers:

Ethyl alcohol and dimethyl ether

Assignment 4.9 How many isomers can be drawn for the chemical formula, $C_6H_{14}O$?

Before proceeding, we need to understand that within a molecule or polyatomic ion all the electrons shared between atoms (covalent bonds) and those not shared (lone pairs), are from the total number of available valence electrons.

Part 2

The general approach in determining Lewis Structures requires a few simple rules, but this method is much more versatile than the previous one.

A very simple example of this is the methane molecule, CH_4. As previously discussed, the four hydrogen atoms are directly bonded to the carbon atom,

```
     H                 H
     ••                |
H  :C:  H    =    H — C — H
     ••                |
     H                 H
```

resulting in four single bonds requiring a total of eight valence electrons: each of the four hydrogen atoms contribute one electron and the carbon contributes four electrons.

For a more complicated structure, exchange the hydrogens in methane for chlorines:

carbon tetrachloride
or
tetrachloromethane

```
           ••
          :Cl:
           |
  ••       |       ••
:Cl  —  C  —  Cl:
  ••       |       ••
           |
          :Cl:
           ••
```

Aside from the eight valence electrons needed for the four single bonds, another twenty-four electrons are required for the twelve lone pairs on the four chlorine atoms resulting in a total of thirty-two electrons needed in the carbon tetrachloride molecule.

Problem 4.1
Given four chlorine atoms and one carbon atom, what is the total number of valence electrons available?

* *Each of the chlorines contributes seven electrons and the carbon contributes four. The total number of valence electrons available is thirty-two.*

Up to this point all we have established is the relationship between total number of available valence electrons and the number of electrons needed for the molecule's electronic structure (i.e. the bonding and non bonding electrons).

Now we need to look at some rules for distributing the valence electrons in the molecule. Sometimes it can be as easy as the last two examples, other times it's not.

1. Given a molecular formula, count the total number of valence electrons that are available.

e.g. ammonia, NH_3

1 N atom	5 valence electrons
3 H atoms	+3 valence electrons
	8 valence electrons (available)

2. Write the atoms in some symmetrical arrangement.

H N H
H

3. Connect the hydrogen atoms to the nitrogen atom using single bonds.

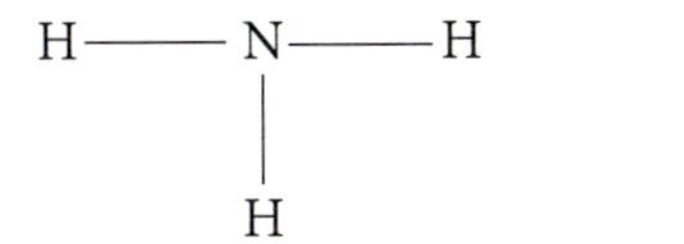

Note: Each single bond requires two electrons

With the ammonia molecule there are eight available valence electrons. It has taken six of these electrons to form the three single bonds.

8	available valence electrons
– 6	valence electrons needed for 3 single bonds
2	valence electrons remaining

4. With the initial bonding done, begin completing octets or doublets for the atoms in the molecule, starting with the **peripheral atoms**, i.e., the hydrogen atoms.

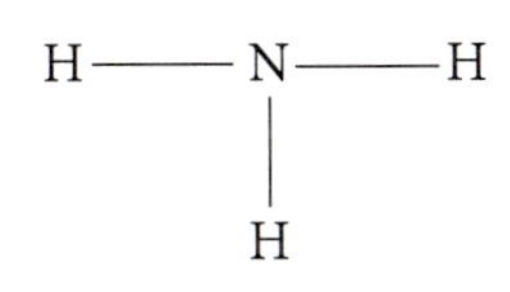

Each of the hydrogen atoms has a doublet.

The nitrogen atom has only six electrons, and needs two more.

The remaining two valence electrons are placed on the nitrogen atom.

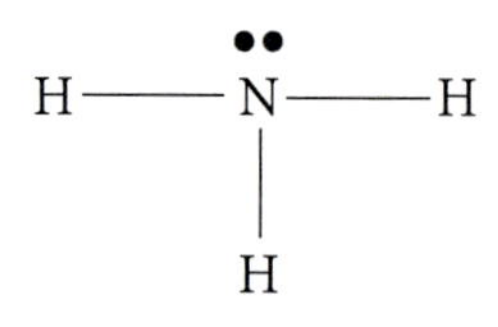

The octet of the nitrogen is now complete and in the process all available valence electrons have been used.

But what happens when there are not enough valence electrons to complete the octet on the central atom?

For example, formaldehyde

$$H_2CO$$

Add available valence electrons

2 H atoms	2	valence electrons
1 C atom	4	valence electrons
1 O atom	+ 6	valence electrons
	12	available valence electrons

Arrange the atoms

(*hint: since the carbon atom has the highest common bonding pattern, place it in the center of the arrangement)*

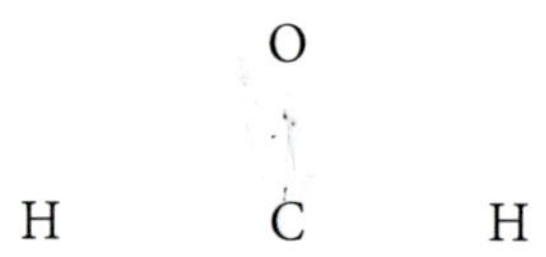

Connect the peripheral atom to the central atom

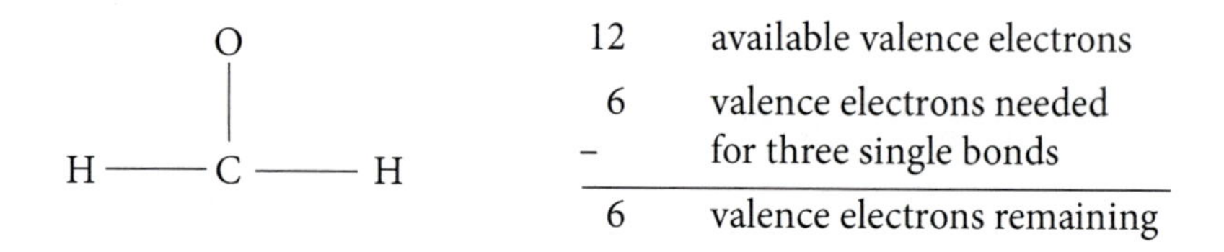

	12	available valence electrons
–	6	valence electrons needed for three single bonds
	6	valence electrons remaining

Complete doublets and octets on the peripheral atoms.

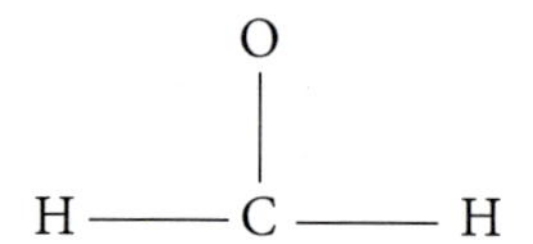

Oxygen has only 2 valence electrons associated with it. It needs 6 more electrons to complete its octet

Both hydrogen atoms have 2 valence electrons associated with them. Their doublets are complete

Complete the octet on the central atom

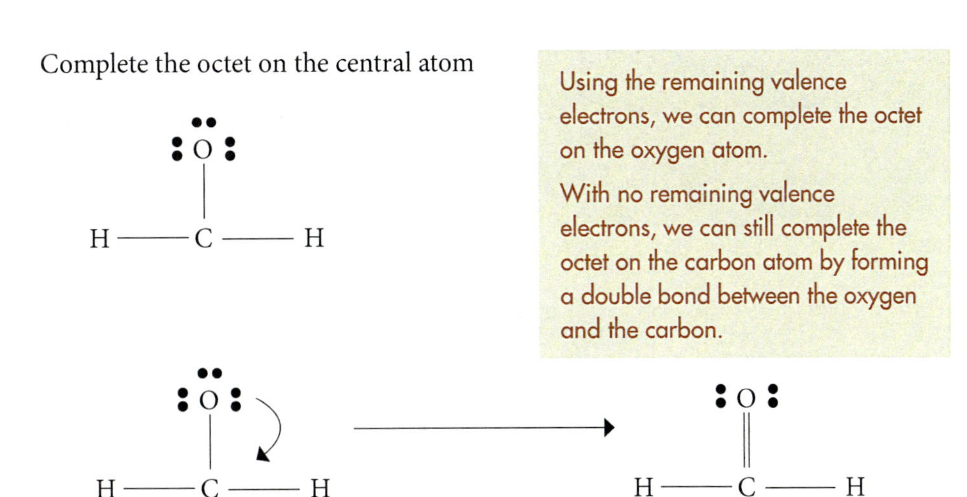

Using the remaining valence electrons, we can complete the octet on the oxygen atom.

With no remaining valence electrons, we can still complete the octet on the carbon atom by forming a double bond between the oxygen and the carbon.

By shifting one of the lone pairs on the oxygen atom to form a double bond between the oxygen and the carbon, both atoms now have complete octets.

Assignment 4.10

Draw the Lewis Structure for phosgene, $COCl_2$.

(Note: again, the carbon atom is in the center.)

1. Add available valence electrons

2. Arrange the atoms

3. Connect the peripheral atoms to the central atom

4. Complete doublets and octets on the peripheral atoms.

5. Complete the octet on the central atom.

Scratch Sheet

Assignment 4.11

Determine the Lewis Structures for the following molecules:

SO_2

CBr_2F_2

F_2O

Scratch Sheet

Lewis Structures of Polyatomic Ions

Determination of Lewis Structures also extends to polyatomic ions. Structurally, the polyatomic ions are the same as any covalent compound with the one exception that they carry a net charge.

If the polyatomic ion has a net negative charge, add in additional electrons when counting up the total available valence electrons in step one of determining the Lewis Structure.

Add in one additional electron for each negative charge.

For example, the nitrate ion, NO_3^-, has one negative charge. Add up the total number of valence electrons as,

1	N	=	5	valence electrons
3	O	=	18	valence electrons
1	negative charge	=	1	valence electron
Total available valence electrons		=	24	valence electrons

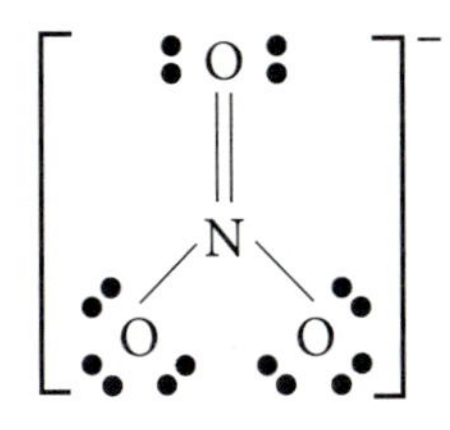

Following the procedure for determining the Lewis Structure of the nitrate ion, containing 24 valence electrons, results in the figure to the left.

Note that when drawing the Lewis Structure for a polyatomic ion, the structure is always enclosed in a bracket and the charge outside.

Assignment 4.12

Compare the Lewis Structures of sulfur trioxide, SO_3, and the sulfite ion, SO_3^{2-}.

SO_3

SO_3^{-2}

Scratch Sheet

Additional Assignments

Assignment 4.13

Draw the Lewis Structures for the following compounds:

Formula	Lewis Structure
bromine trifluoride BrF_3	
hydrogen cyanide HCN	
germane GeH_4	
carbon dioxide CO_2	
xenon tetrafluoride XeF_4	
nitrous oxide N_2O	
ozone O_3	

Scratch Sheet

Scratch Sheet

Assignment 4.14

Draw the Lewis Structures for the following polyatomic ions:

Formula	Lewis Structure
carbonate ion CO_3^{2-}	
cyanide ion CN^-	
nitrite ion NO_2^-	
sulfate ion SO_4^{2-}	
azide ion N_3^-	
nitrous oxide N_2O	
hydrogen carbonate ion HCO_3^-	

Scratch Sheet

Departures from the Octet Rule

Less is More (Reactive): Boron

Placement of the three valence electrons in Lewis Dot Structure of atomic boron indicate the possible formation of three covalent bonds.

$1s^2 2s^2 2p^1$ • B • — B —

In the formation of three covalent bonds, boron's valence shell shares only six electrons.

Unlike nitrogen, which also forms three covalent bonds, boron does not achieve an octet as a trivalent compound (e.g. BCl_3, BF_3, $B(OH)_3$). Lacking an octet in the valence shell, trivalent boron is electron deficient and reactive. That is, it will readily accept an electron pair from the lone pair of another atom to form a fourth covalent bond, completing its octet. Note that when an atom, in case fluorine, contributes the two electrons that form the single bond, it is called a **coordinate covalent bond.**

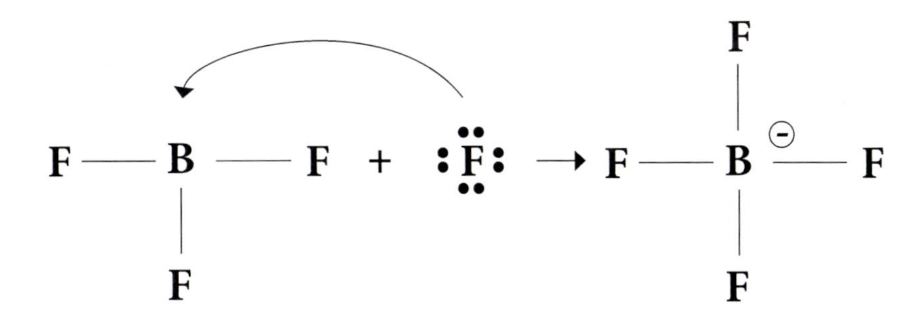

Aside from completing its octet, by taking on an additional electron pair, the boron atom now has a negative charge.

Beyond the Second Period: When Eight is Not Enough

Having done our best to account for noble gas configurations in the Lewis structure, let's now consider phosphorous pentafluoride, PF_5, and sulfur hexafluoride, SF_6.

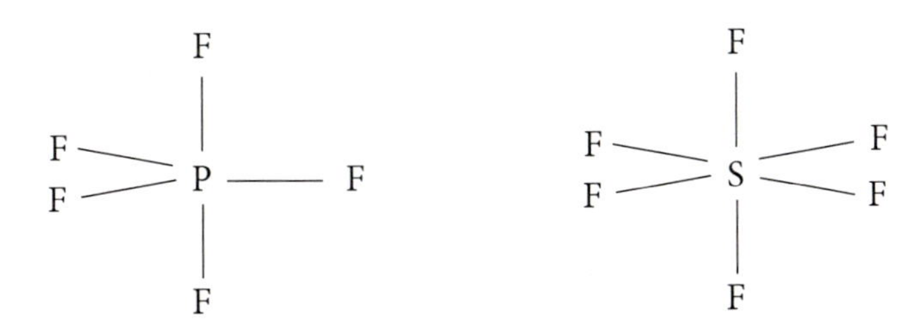

Since such compounds do exist, we can conclude that the octet can be exceeded. The question is, why?

Consider nitrogen and phosphorous: both are in the same group and have five valence electrons,

$$\mathbf{ns^2\ np^3.}$$

Both form three covalent bonds,

In both compounds the central atoms achieve octets, and we have seen that phosphorous can accommodate two more fluorine atoms.

Can nitrogen do the same?

The answer is no, and let's consider why.

It's all a matter of available valence orbitals. Recall that nitrogen is a second period element, which means that its valence shell is n = 2. And within the second shell only two subshells (s and p) are available. The s-subshell has a capacity of 2 electrons and the p-subshell has a capacity of 6 electrons. Therefore, the nitrogen atom can only support up to eight valence electrons.

It is the available valence subshells that determine how many bonding and non bonding electrons an atom can support in its outermost shell.

Phosphorous, on the other hand, is a third period element, containing three valence subshells (s, p, and d). For phosphorous to achieve noble gas configuration, it uses only its s and p subshells. But there still remains an available empty d subshell which can take on additional electrons when further bonding occurs. Think of this empty d-subshell as an attic on a house (i.e. storage space).

To sum up the correlation between Noble Gas configuration and the periods on the Periodic Table:

Period	Subshell	Electron Capacity	Noble Gas Configuration
1st	1 subshell (s)	2 electrons	doublet
2nd	2 subshells (s and p)	8 electrons	octet
3rd	3 subshells (s, p, and d)	18 eletrons	octet and greater
4th and beyond	4 subshells (s,p,d, and f)	32 and greater	octet and greater

Assignment 4.15

In the following exercise, based on the number of valence subshells, state whether the atom can exceed an octet when forming covalent bonds.

Element	exceed octet (Yes or No)
As	
N	
I	
C	
Si	
He	
Ne	
O	
I	
B	

Assignment 4.16

For each of the following compounds, draw a Lewis structure and using the above information. Based on the number of valence subshells on the central atom, determine if the compound can actually exist.

BCl_3 yes __ no __

CCl_5 yes __ no __

FO_4^- yes __ no __

ClO_4^- yes __ no __

BCl_5^{2-} yes __ no __

CO_3^{2-} yes __ no __

CO_4^{3-} yes __ no __

Scratch Sheet

Scratch Sheet

Other Anomalies

Odd Number of Electrons: Radical Thinking

Up to now, we have only discussed compounds which contained even numbers of electrons in the valence shells, i.e. electrons occurring in pairs. There are covalent compounds which contain odd numbers of valence electrons, one or more unpaired electrons.

Compounds containing unpaired electrons are called **radicals**.

Nitrogen dioxide is an example of a radical compound which has an odd number of valence electrons.

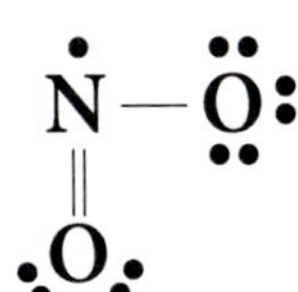

The total number of valence electrons is 17. After completing the octets on the peripheral atoms (the oxygens), the remaining electron is placed on the nitrogen.

Instead of an octet, the nitrogen has only 7 valence electrons.

Resonance Structures

The Lewis structure for the carbonate ion, $CO_3^{\,2-}$, contains one double bond and two single bonds between the carbon atom and the three oxygen atoms. There are also two negative charges on two of the oxygens

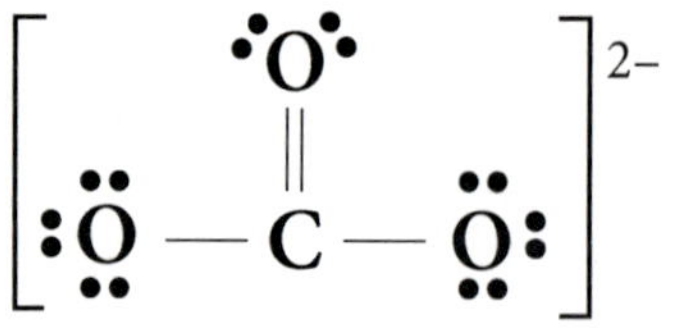

Keeping the positions of all atoms the same, we can reposition the single and double bonds (as well as the lone pairs) as such:

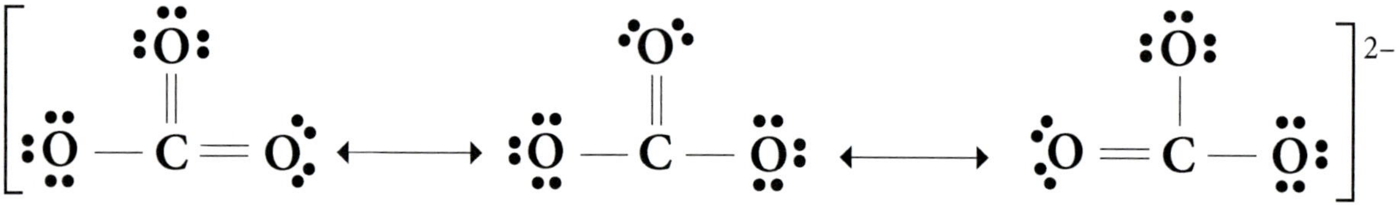

These three structures of the carbonate ion are called **resonance structures**.

Note that resonance structures are drawn within parentheses with a double headed arrow separating the structures.

Exercise 4.5 Draw the resonance structures of the following compounds and ions.

O_3

NO_3^-

SCN^-

Additional Assignments

4.17 Draw three resonance structures for nitric acid, HNO_3.

4.18 Draw two resonance structures for carbon dioxide, CO_2.

4.19 Draw two resonance structures for nitrite ion, NO_2^-

4.20 Draw all resonance structures for dinitrogen tetroxide, N_2O_4.

($O_2N–NO_2$)

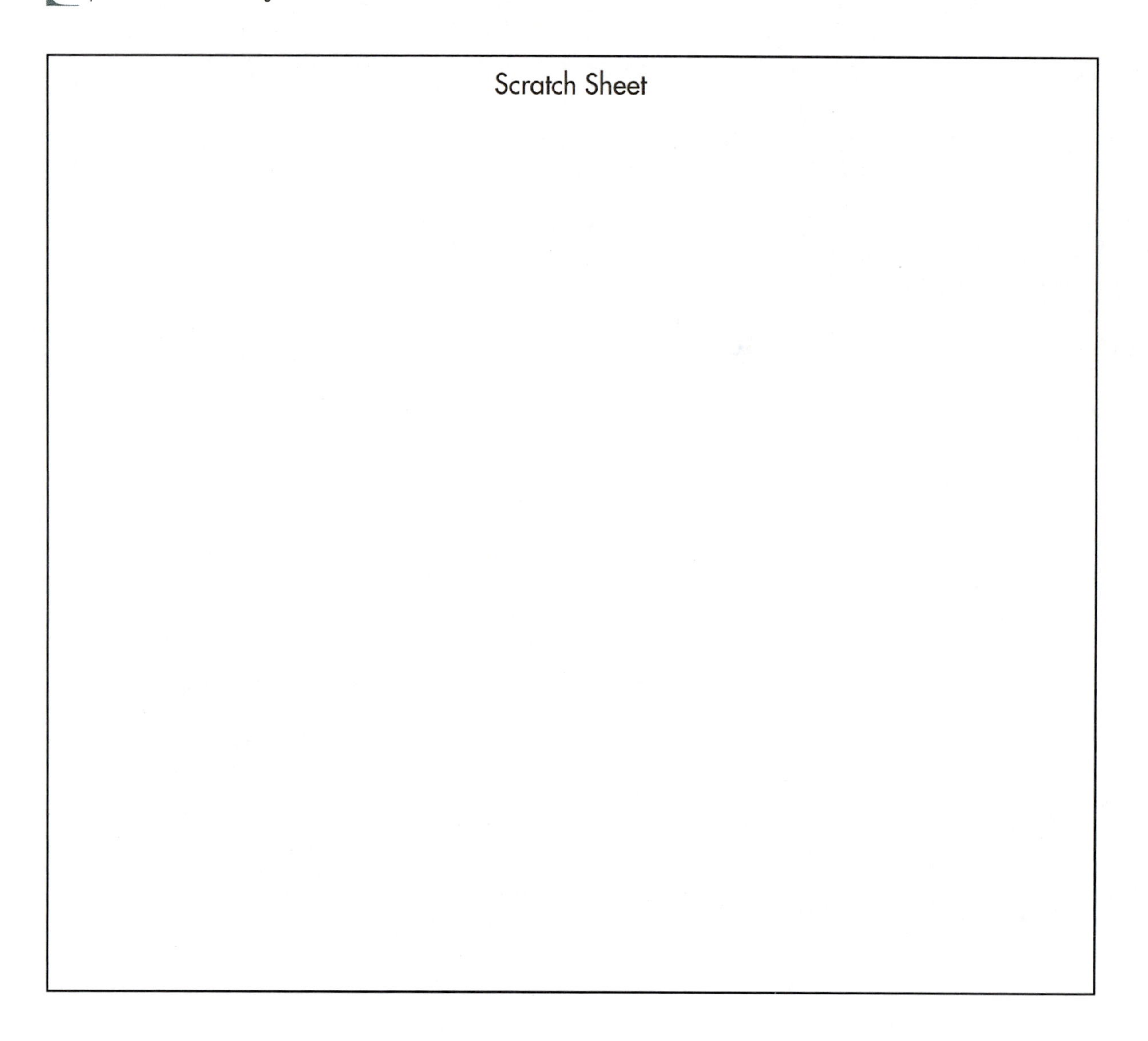

Formal Charge

Up to now, we have been keeping track of the valence electrons of covalently bonded atoms in terms of completion of Noble Gas configurations, i.e. doublets and octets.

In this process, we have been counting the total number of electrons that an atom "sees" in its valence shell as a result of covalent sharing. In the case of the following two examples of the cyanide ion and the carbon monoxide molecule, we will see that the sharing of electrons affects the charges on the both of the bonded atoms.

Consider how a neutral atom, containing a balanced number of electrons and protons, can undergo ionization by either the gain or loss of electrons within its valence shell.

Na	Na^+ + e^-
11 protons	11 protons
11 electrons	10 electrons

In this example, it is easy to account for the development of the positive charge on the sodium atom by the loss of a single valence electron.

We can also observe the positive charge formation in the reaction of the ammonia molecule and a hydrogen ion,

$$H-\ddot{N}H-H \quad + \quad H^{+} \longrightarrow \left[H-\underset{H}{\overset{H}{N}}-H \right]^{+}$$

ammonia | hydrogen ion | ammonium

Taking a closer look, we see that the new N–H bond is formed from the lone pair of electrons on the nitrogen:

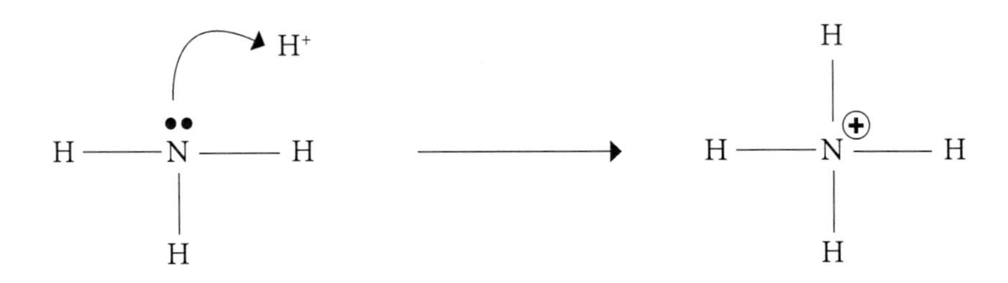

When each atom contributes one electron to the formation of a covalent bond, it can regarded as electron – **electron pairing.**

In the case of the ammonium ion, the new covalent bond results from the nitrogen atom donating both of the bonding electrons.

This is called **coordinate covalent bond formation**.

Since the nitrogen atom is "giving up" one of its valence electrons to the hydrogen atom, a positive charge develops on the nitrogen atom.

This charge is called a **formal charge**.

Actually, all of the atoms within covalent compounds and polyatomic ions have formal charges, which can be positive, negative, or zero.

With this in mind, we can see that the formal charge on the nitrogen atom in the ammonia molecule is zero, and after the reaction the formal charge of the same atom has changed to +1.

In this section, we will develop a method for determining the formal charges of covalently bonded atoms.

To better understand the nature of this charge development let's first consider the structural considerations of the cyanide ion and the carbon monoxide molecule.

CN^-	CO
cyanide ion	carbon monoxide

Although the net charge on the cyanide ion is easily accounted for by counting the number of valence electrons, designating on which atom the charge resides isn't so obvious.

In the Lewis structure of the cyanide ion,

$$[:C\equiv N:]^-$$

both carbon and nitrogen contain complete octets within their valence shells.

For the next step, just consider the three pairs of electrons within the triple bond as being equally shared.

As indicated in the figure below, both the carbon atom and the nitrogen atom have five electrons in their valence shells.

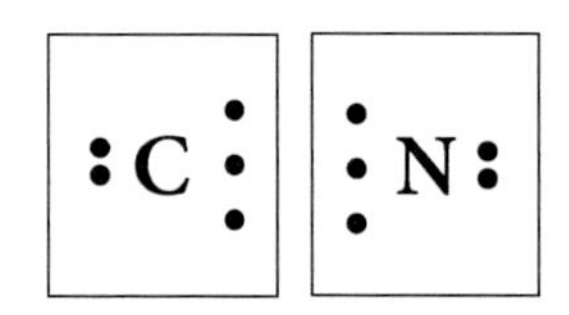

Recall that the neutral carbon atom requires four valence electrons and the neutral nitrogen requires five.

By the distribution of the valence electrons in the cyanide ion, the nitrogen atom has the five electrons it needs to remain neutral, but the carbon atom contains an extra valence electron.

This results in the carbon atom developing a **formal charge** of negative one, – 1. Since there are no other formal charges in this structure, the net charge is also negative one.

$$[:\overset{\ominus}{C}\equiv N:]^-$$

Knowing the position of the formal charge(s) within a structure is very important when considering reaction mechanisms.

The carbon monoxide molecule is interesting from the viewpoint that it is considered a neutral compound, but actually contains two formal charges.

As with the cyanide ion, draw the Lewis structure

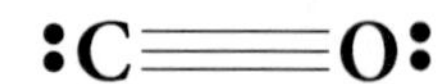

and equally divide up the covalently bonded electrons. Again, each atom will account for five electrons in its valence shell.

As in the cyanide ion, the carbon supports one additional electron resulting in a formal charge of −1, while the oxygen atom contains one less electron than its valence of six requires for neutrality. This will result in a formal charge of positive one, +1, on the oxygen atom.

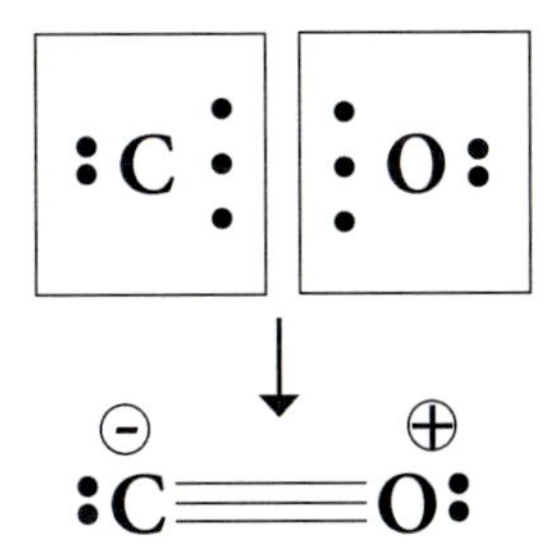

The combination of the two formal charges results in an overall net charge of zero, for the neutral carbon monoxide molecule.

Charged Species

Having worked several examples of covalent compounds in which all valence requirements are satisfied, let's consider covalently bonded species which contain formal charges and may or may not result in overall net charges.

Consider the Lewis structure of the ammonium ion, NH_4^+,

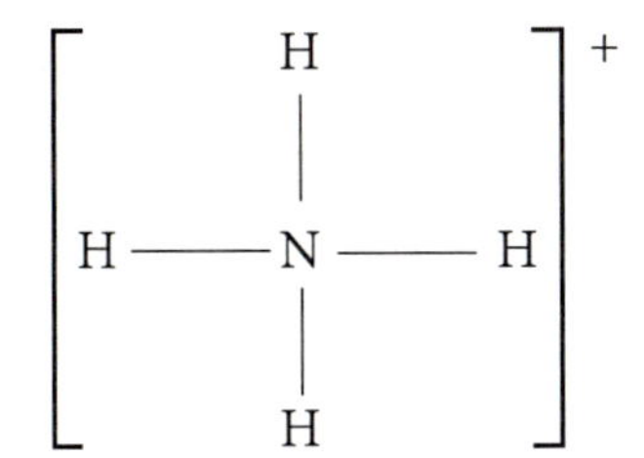

Although a polyatomic ion, such as the ammonium ion, contains a net charge, it may not be obvious on which atom or atoms the charge resides. As a visual exercise let's consider the ammonium ion and divide the electrons by rotating the bonding pairs.

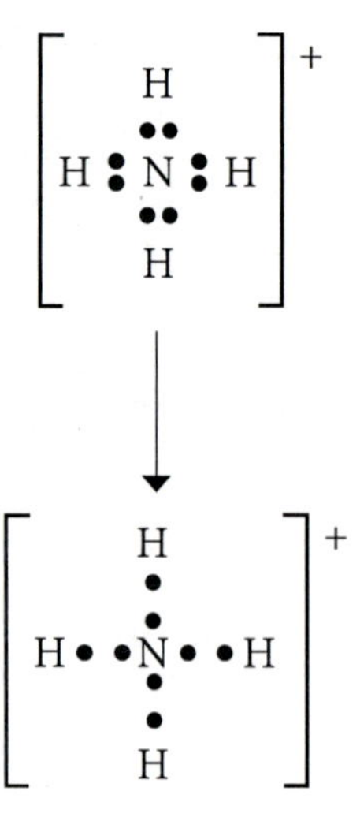

As an alternative approach, rotate bonding electrons 90°.

Divide up electrons between atoms.

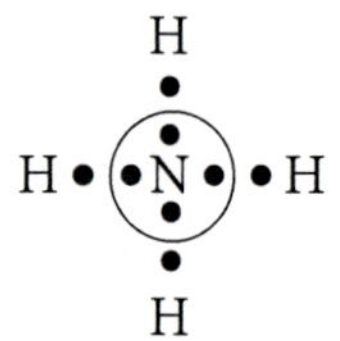

Since nitrogen in ammonium has only 4 valence electrons, it will have a formal charge of +1.

Valence electrons on nitrogen in ammonium ion.

Valence electrons on neutral nitrogen atom.

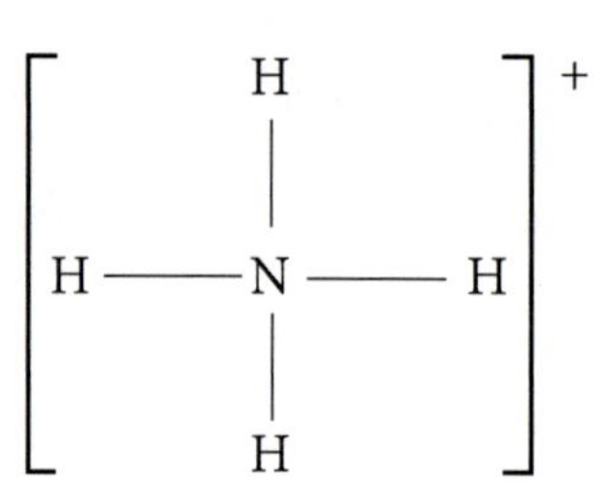

By determining the formal charge in the ammonium ion, we can indicate the source of the single positive charge as the nitrogen atom.

A circle drawn around the charge, indicates a formal charge.

Polar molecules, such as carbon monoxide and hydrogen fluoride may or may not contain formal charges.

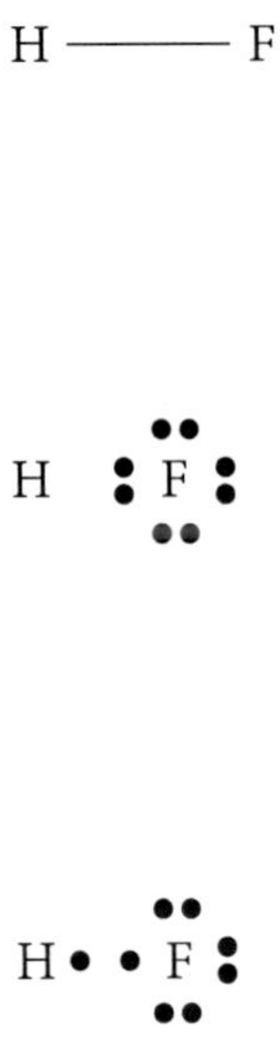

H —— F

H :F:

H• •F:

Applying the formal charge method not only helps to determine charge distribution within a structure, it is also very useful in selecting the most stable Lewis structure, as demonstrated in the following discussion.

When drawing the Lewis structure of the sulfate ion, SO_4^{2-}, the following structures are possible:

1 2 3

Which structure is the most stable?

> **Method**
>
> 1. Determine the formal charges on each atom within each structure.
> 2. The structure in which the individual formal charges are the smallest, i.e. closest to zero, is the most stable.

Structure 1

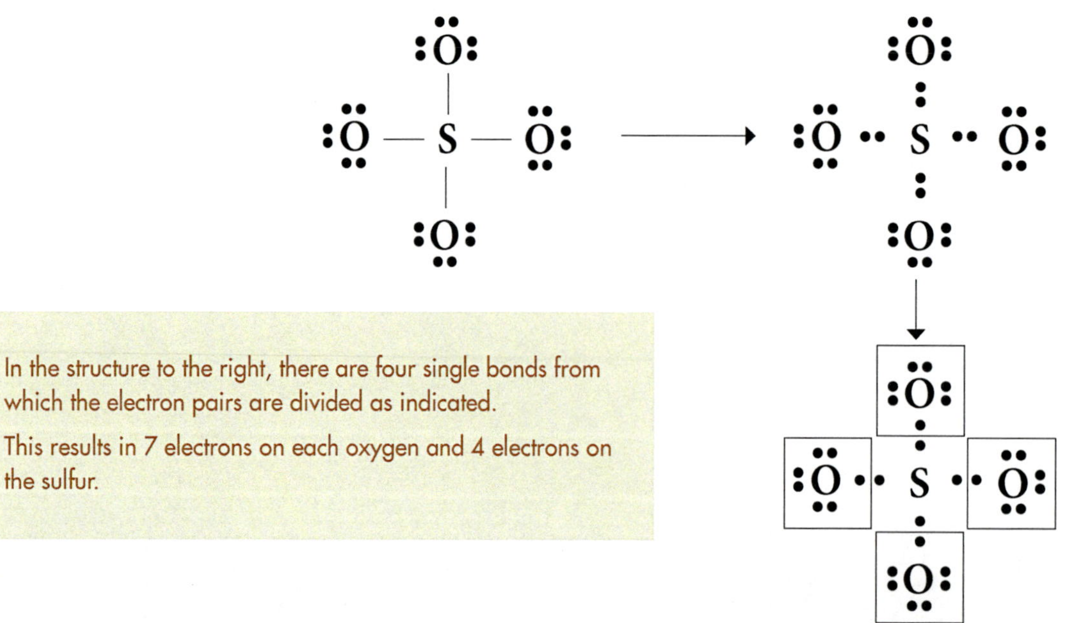

In the structure to the right, there are four single bonds from which the electron pairs are divided as indicated.

This results in 7 electrons on each oxygen and 4 electrons on the sulfur.

Since neutral oxygen only requires 6 electrons the extra electron will impose a single negative charge on each oxygen atom.

The sulfur atom, with only 4 valence electrons, has a +2 charge on it.

Note that the net charge for all structures will remain at −2.

Structure 2

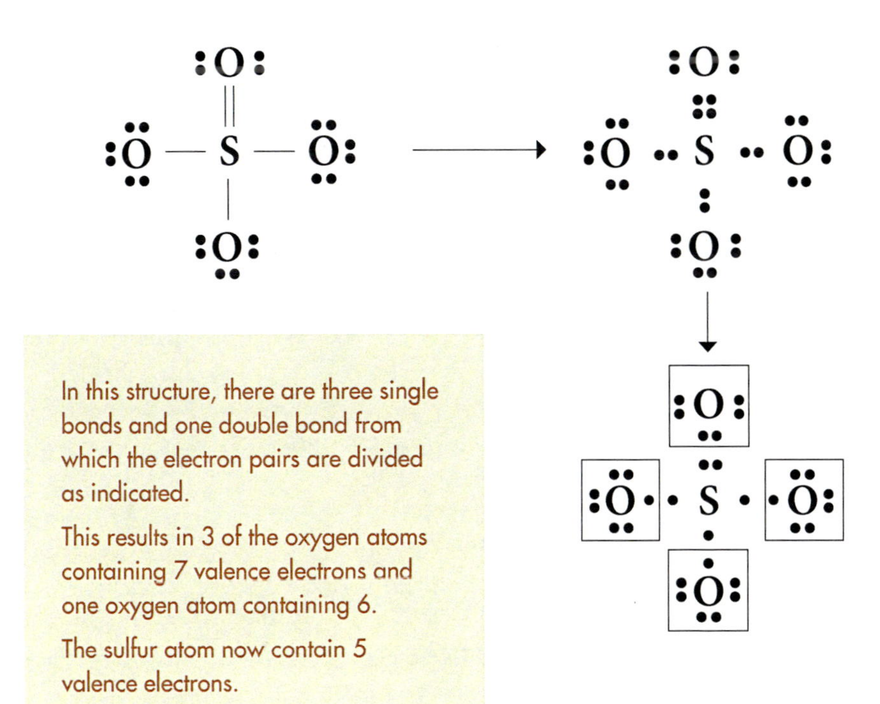

In this structure, there are three single bonds and one double bond from which the electron pairs are divided as indicated.

This results in 3 of the oxygen atoms containing 7 valence electrons and one oxygen atom containing 6.

The sulfur atom now contain 5 valence electrons.

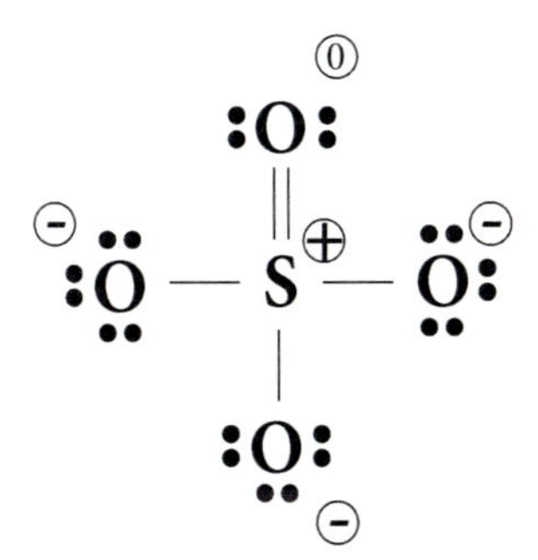

Since the oxygen atom directly above the sulfur contains 6 valence electrons, its formal charge is zero.

The 3 remaining oxygen atoms still containing 7 valence electrons each still possesses a single negative formal charge.

The sulfur atom, with 5 valence electrons, has a +1 charge on it.

Due to the reduction of formal charges on the one oxygen atom and the sulfur atom **structure 2 is more stable than structure 1.**

Structure 3

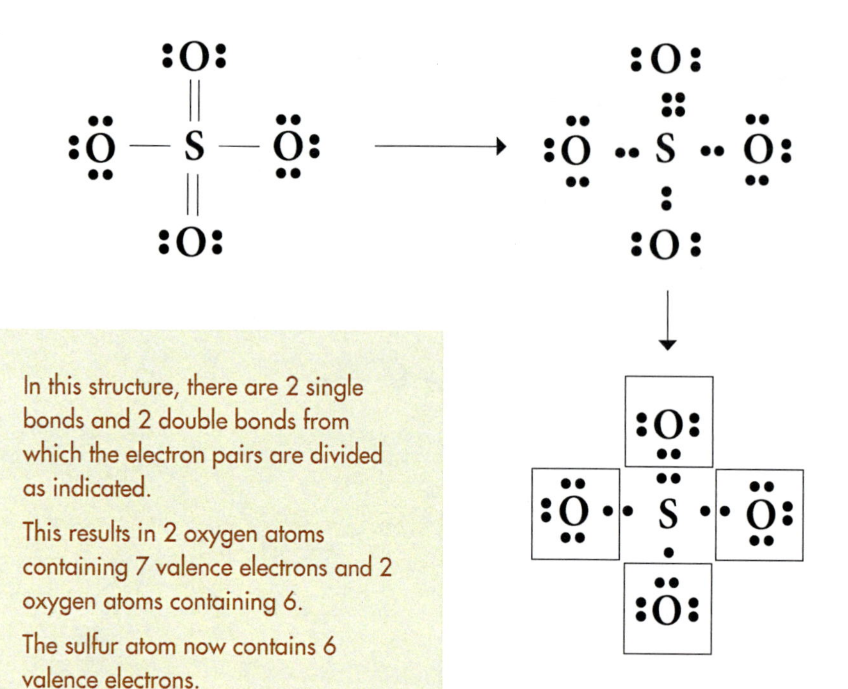

In this structure, there are 2 single bonds and 2 double bonds from which the electron pairs are divided as indicated.

This results in 2 oxygen atoms containing 7 valence electrons and 2 oxygen atoms containing 6.

The sulfur atom now contains 6 valence electrons.

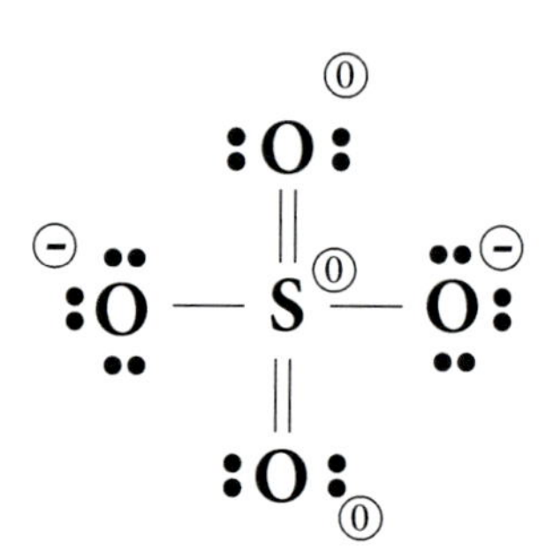

There are now 3 atoms with zero formal charges: the 2 oxygen atoms and the sulfur atom

The 2 remaining oxygen atoms still containing 7 valence electrons each still possesses a single negative formal charge.

Of the three sulfate ion structures, the third structure is the most stable.

Assignment 4.21

1. Draw the Lewis structures of the following compounds, and determine the formal charges on all the atoms within the following compounds:

a. CCl_4

b. NO

c. PF_5

d. O_3

2. Draw the resonance structures of the carbonate ion, and assign the formal charges to all the atoms within the structures.

3. Draw three possible Lewis structures for the phosphate ion, and using formal charges, determine which of the three structures is the most stable.

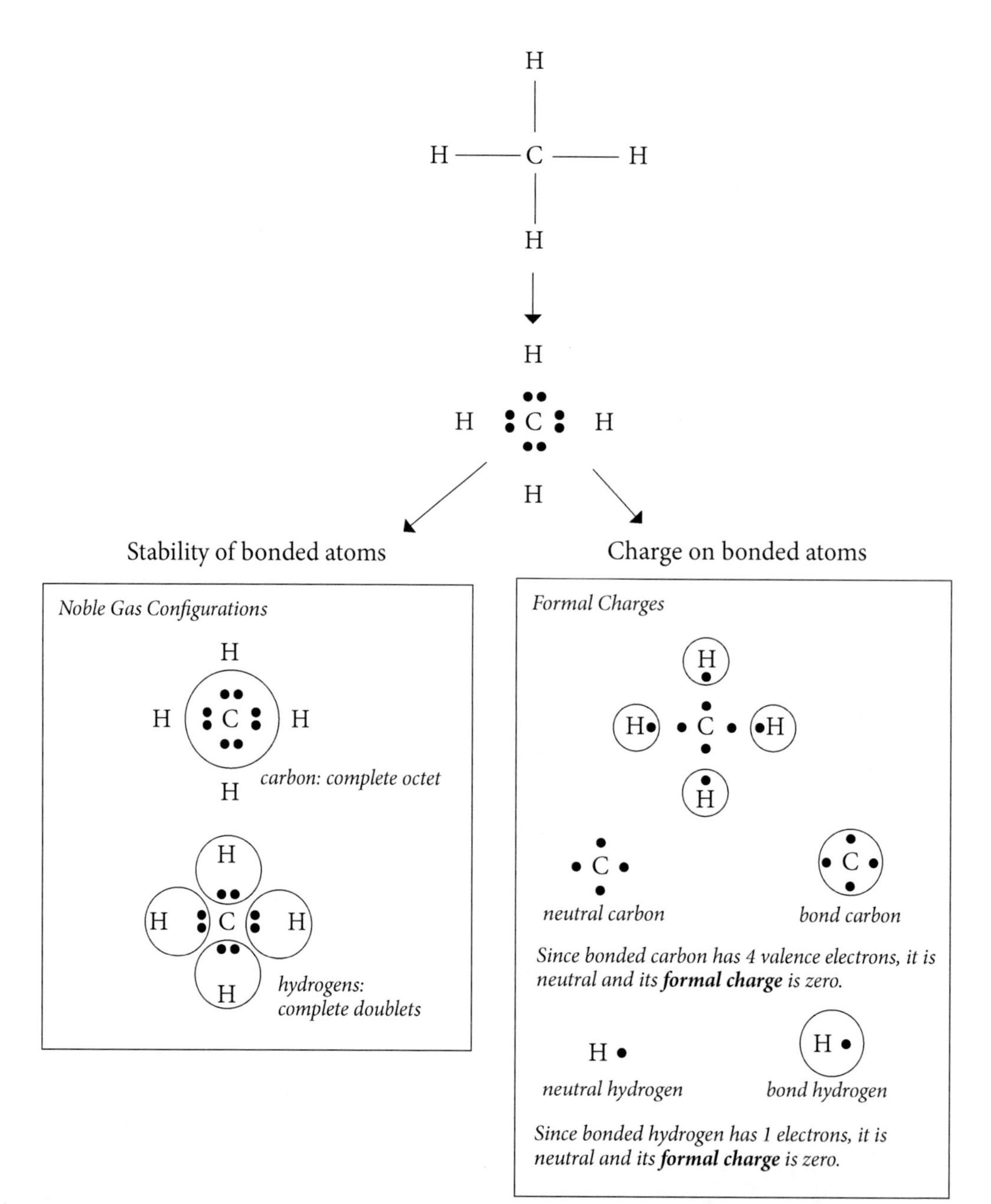

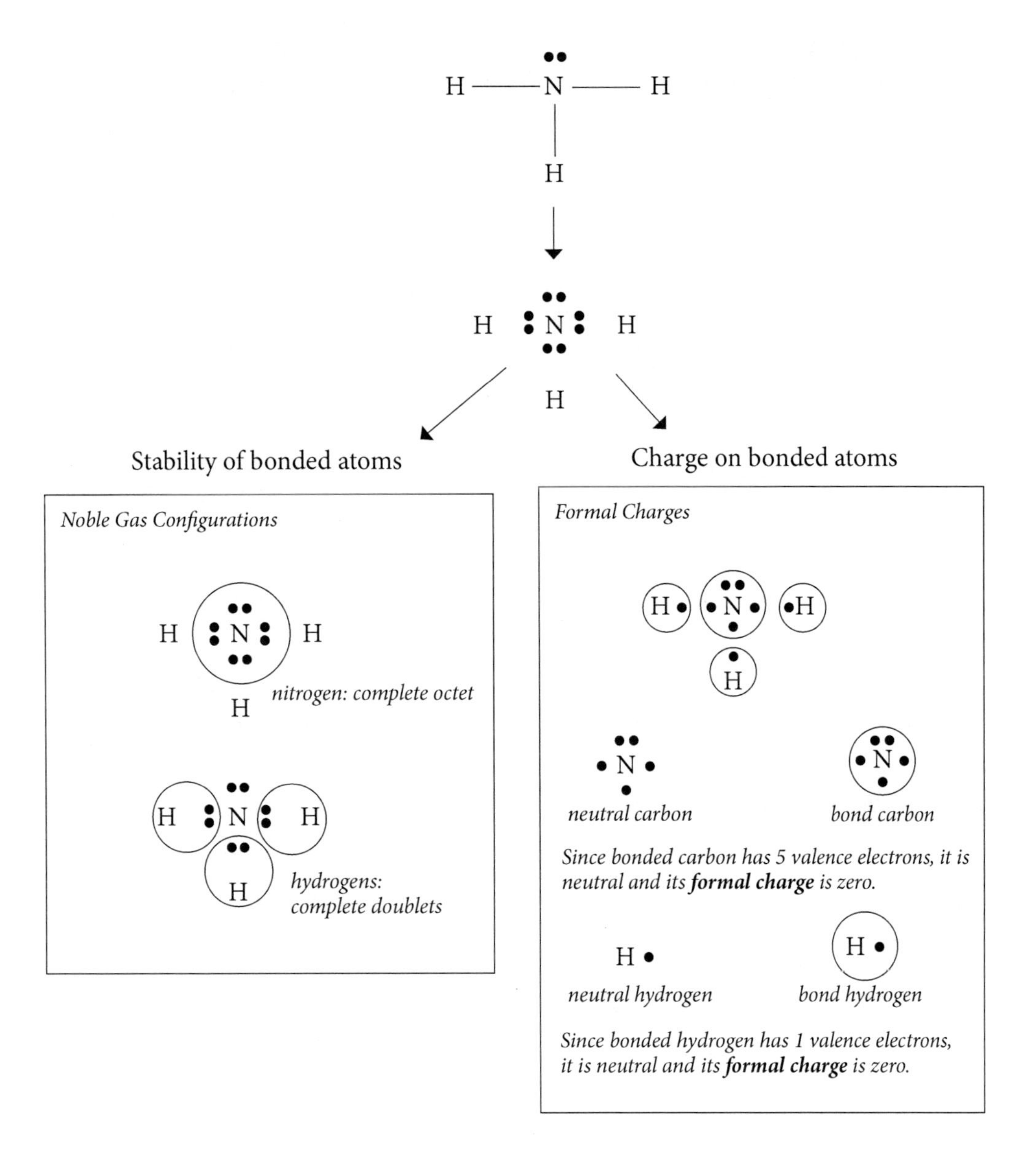

In the following exercise, determine that the Noble Gas configurations are satisfied and the formal charge of each atom is neutral.

C_3H_6O

Lewis Structure

Noble Gas Configurations

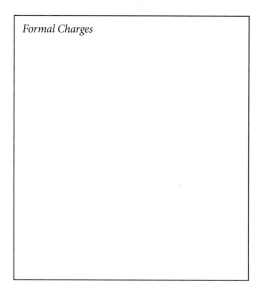

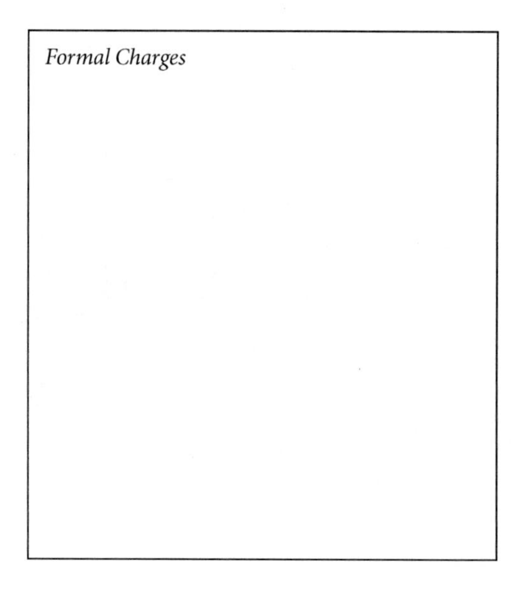

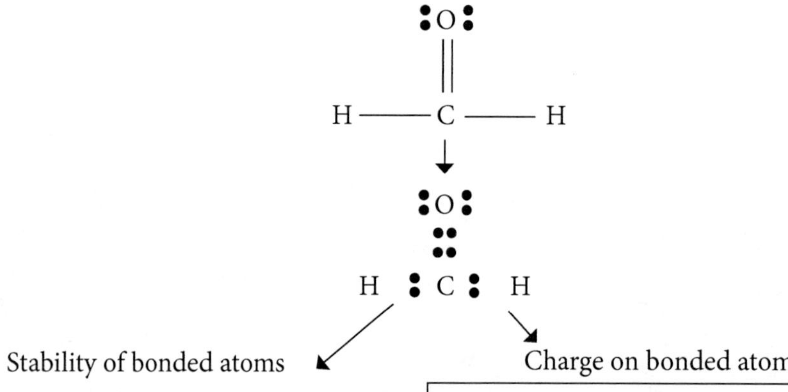

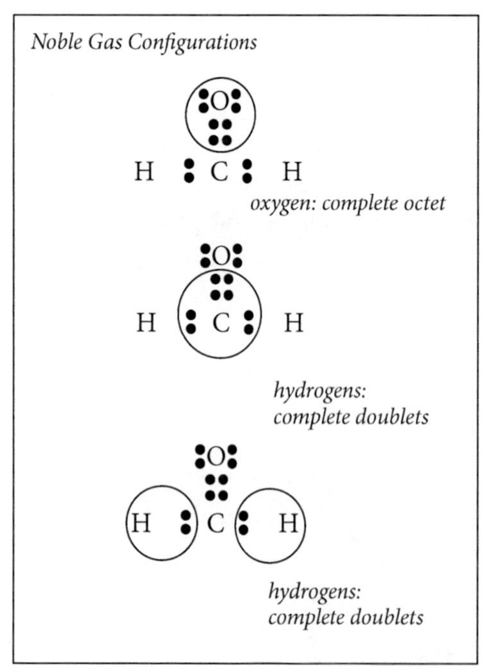

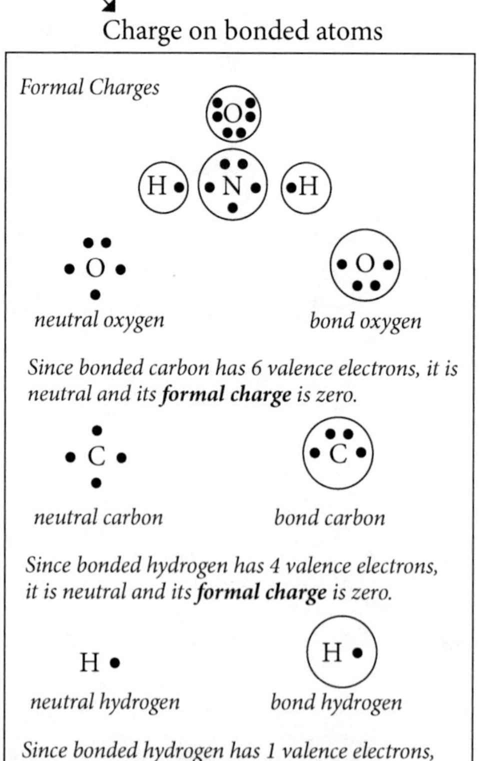

CH_4O
Lewis Structure

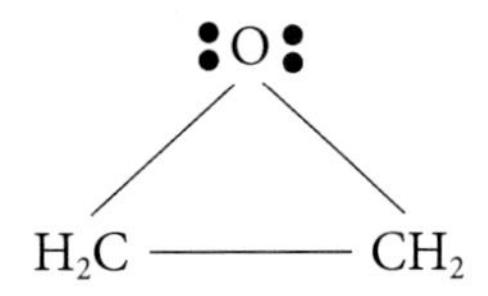

Noble Gas Configurations	*Formal Charges*

CH_4O

Lewis Structure

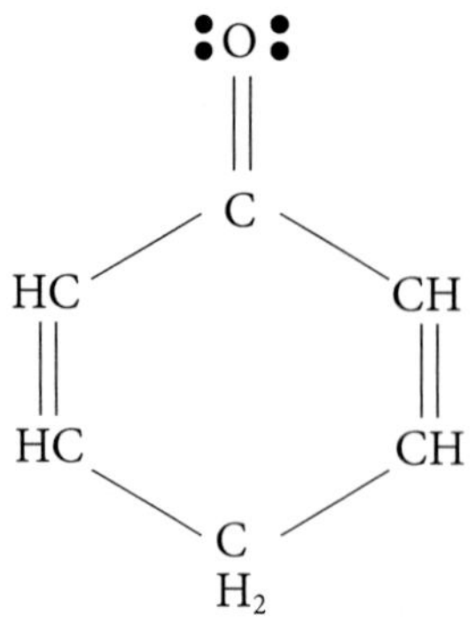

:O:
||
C
HC   CH
|| ||
HC   CH
C
H_2

Noble Gas Configurations

Formal Charges

Assignment 4.22

For each of the following chemical formulas,

a. Draw the Lewis Structure.
b. Draw any possible resonance structures.
c. Assign formal charges to all atoms.
d. If resonance structures exist, determine which is the most stable.

hydrazine, N_2H_4

thionyl chloride, $SOCl_2$ (S is the central atom)

sulfuryl chloride, SO_2Cl_2 (S is the central atom)

Acetate ion, CH_3COO^-

Single Bonds and Multiple Bonds: Sigma Bonds and Pi Bonds

There are two types of bonding which occur in the Lewis structures:

1. Sigma Bonding and
2. Pi Bonding.

The majority of covalent chemical bonds, (i.e. single bonds) are defined as sigma bonds. An example of such a bond occurs between the two hydrogen atoms in molecular hydrogen, H_2:

Figure 4.3

As previously discussed in Lewis Structures (p. 335), two atoms can achieve Noble Gas configuration by sharing electrons through the overlapping of valence orbitals.

In this case, the 1s orbitals of the hydrogens (each containing one unpaired electron) overlap to form a single covalent bond.

When the bond is located between the two nuclei, it is called a **sigma bond**.

In this regard, all single bonds are also sigma bonds.

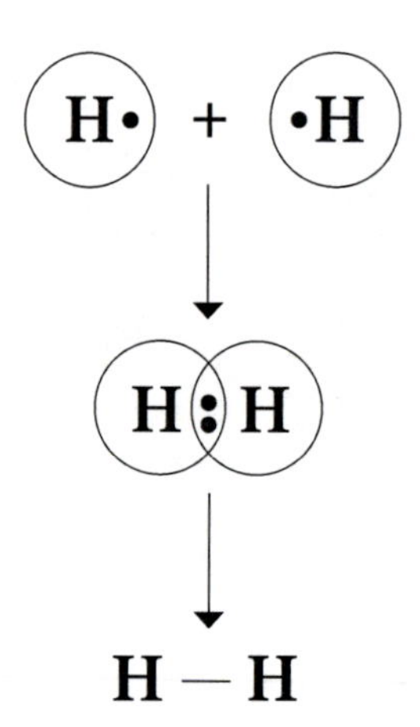

When a covalent compound contains only single bonds, it is a simple procedure to identify the number of sigma bonds within the structure:

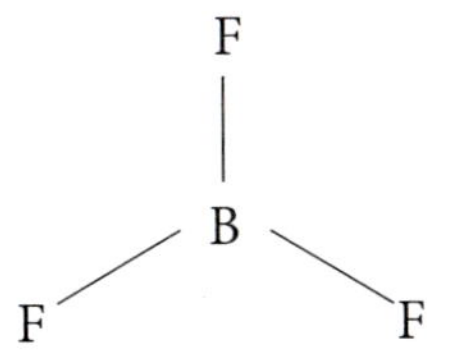

BF_3: the structure of boron trifluoride contains three single bonds:
i.e. three sigma bonds.

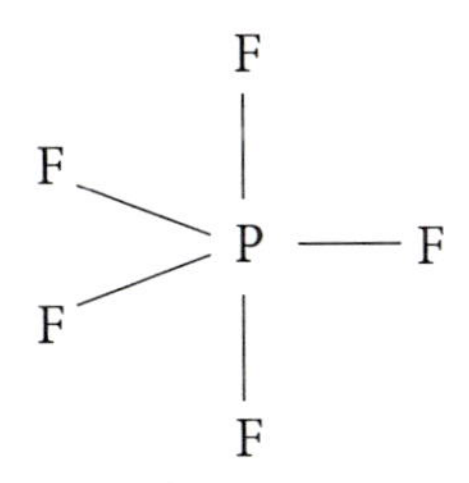

PF_5: the structure of phosphorous pentafluoride contains five single bonds:
i.e. five sigma bonds.

When a multiple bond between two atoms occurs, (i.e. a double bond or a triple bond), one of the bonds is a sigma bond, the remaining one or two bonds are defined as pi bonds.

Figure 4.4

Multiple Bonds Between Two Atoms

carbon to carbons double bond:
contains one sigma bond and one pi bonds

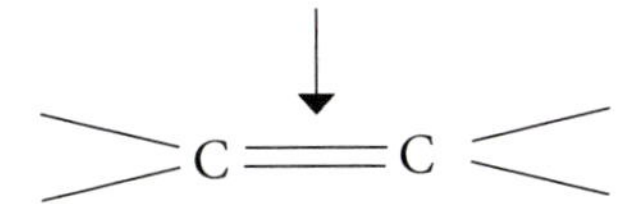

carbon to carbons triple bond:
contains one sigma bond and two pi bonds

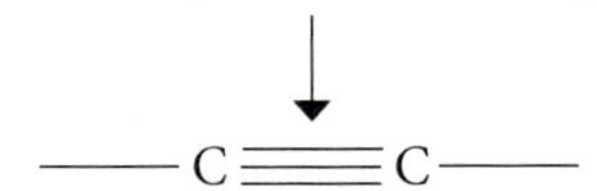

When indicating sigma and pi bonds, sometimes Greek letters are used:

sigma bond: σ bond

pi bond: π bond

Assignment 4.23 Determine the total number of sigma bonds and pi bonds that occur in the following compounds.

$$H-C\equiv N$$

sigma (σ) bonds _____

pi (π) bonds _____

$$H_2C=O$$

sigma (σ) bonds _____

pi (π) bonds _____

$$H-O-S(=O)_2-O-H$$

sigma (σ) bonds _____

pi (π) bonds _____

Sigma and Pi Framework

In an upcoming section, Hybridization of Atomic Orbitals, the bonding theory will be introduced which can better address the occurrence of sigma and pi bonding within a structure.

For now, the following physical description will need to suffice.

As previously described, a sigma bond occurs when a covalent bond is located between the two bonded atoms, in other words, the bond occurs in the same plane as the bonding atoms:

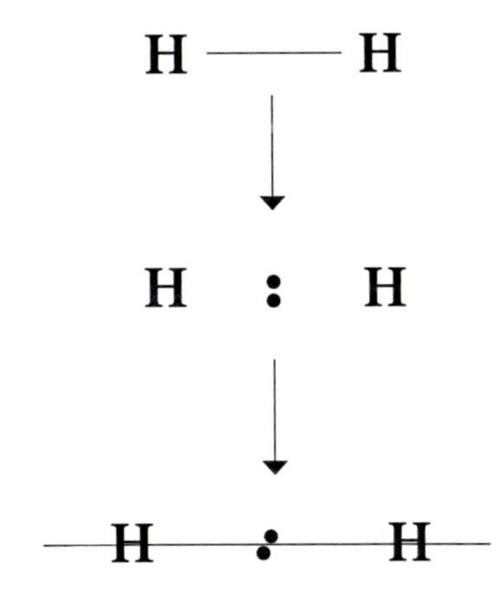

If we consider the compound, ethane, C_2H_6, the sigma bonding framework could be represented as:

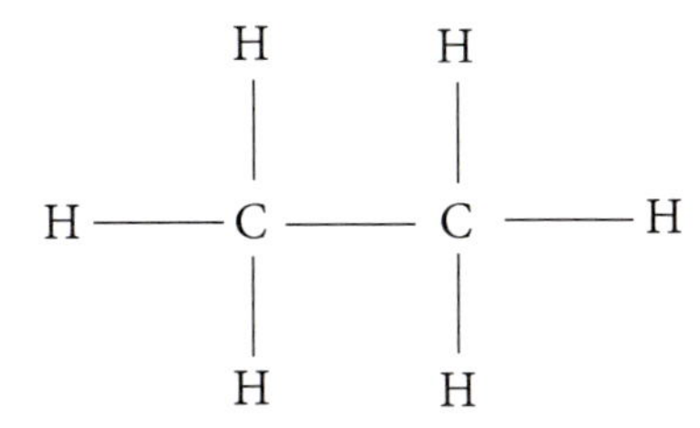

The seven sigma bonds all occur in the same plane as their two bonding atoms:

6 C-H bonds and 1 C-C bond

In contrast to sigma bonding, pi bonds occur outside the sigma framework.

The compound ethene, C_2H_4, contains one double bond, which consists of one sigma bond and one pi bond.

C=C
1 2

The five sigma bonds comprise a framework in which all the atoms exist in a single plane:

C=C
1 2

The pi bond results from the overlap of two valence p-orbitals, one from each carbon. Each p-orbital contains a single electron.

Recall that p-orbitals consist of two lobes, with opposite phases.

In order for the two p-orbitals to overlap, they must be in alignment, i.e. coplanar.

Pi bonds form when two coplanar p-orbitals on adjacent atoms overlap.

Note that both lobes of the p-orbitals overlap resulting in two distinct regions in which the pi bonding occurs.

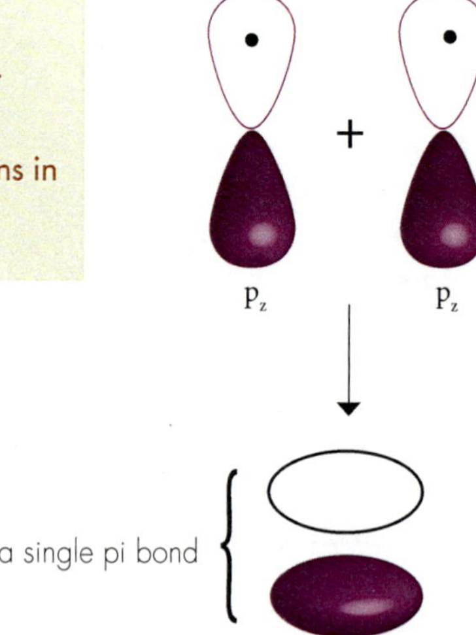

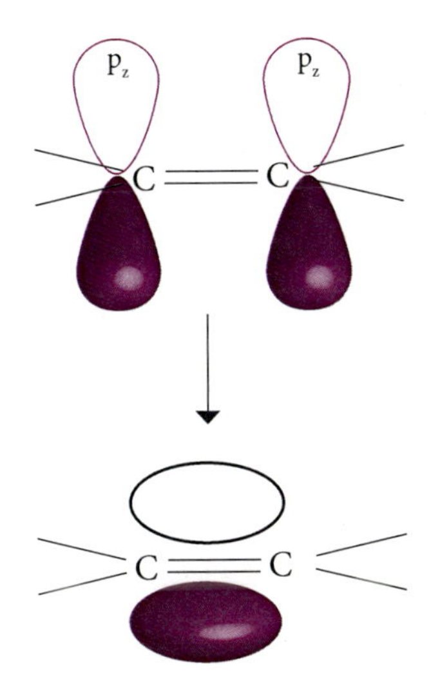

Since the double bond occurs between the two carbons, it is important to understand the orientation of this pi bonding in relation to the carbon – carbon sigma bond:

Superimposing the overlapping adjacent p-orbitals onto the sigma framework of the ethane, it is apparent that the pi overlap from both above and below the plane of the carbon-carbon sigma bond.

It's a common misconception to think of a single pi bond as two separate bonds when an illustration indicates the above and below p-overlap.

A triple bond forms in the same fashion as the double bond, with the exception that each of the adjacent atoms contributes two p-orbitals instead of just one.

Since the two p-orbitals are orthogonal to each other (i.e. at 90° orientations to each other), for triple bond as in acetylene, C_2H_2,

$$H—C\equiv C—H$$

the bonding orientation of the two p-orbitals on each carbon atom would look something like this:

In a two dimensional plane, the two p-orbitals of each carbon would be oriented at 90° to each other.

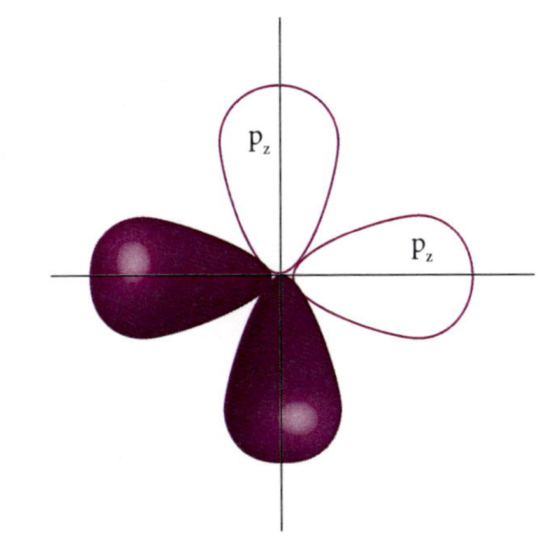

By looking through the sigma bonding plane of acetylene, the orientations of both pi bonds are apparent.

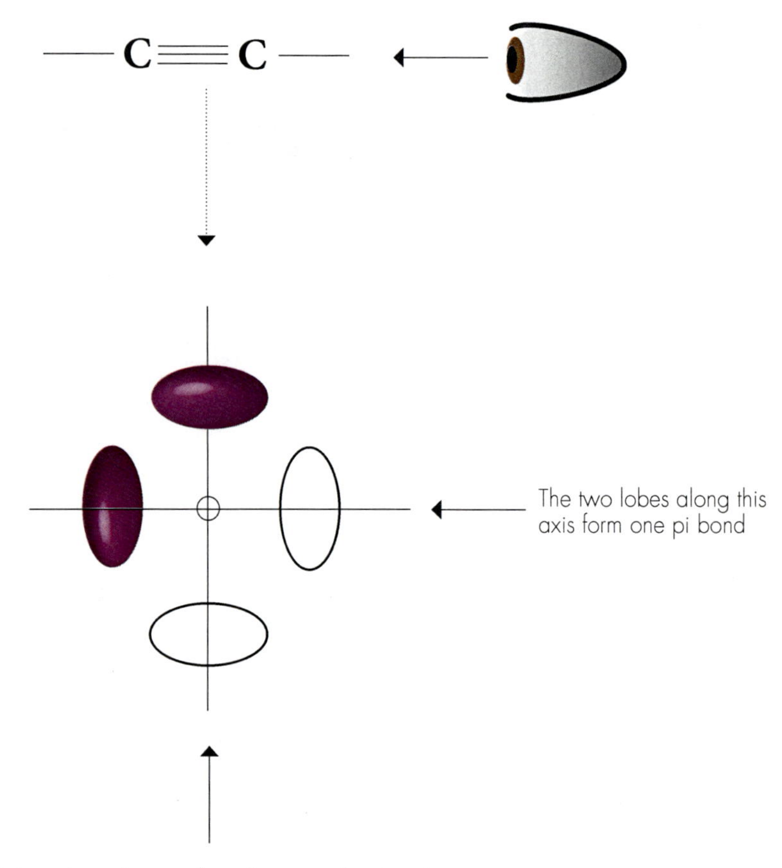
C≡C
The two lobes along this axis form one pi bond
The lobes along this axis form the second pi bond

Molecular Shapes

Given the chemical formula BF_3, we can put together the Lewis structure:

This structure indicates that

1. the three fluorine atoms are connected to the boron atom, and
2. the valence electrons are distributed into three bonding pairs of single bonds, and nine non-bonding lone pairs.

But what information do we need to determine the actual shape of the BF_3 molecule?

Three possible shapes are

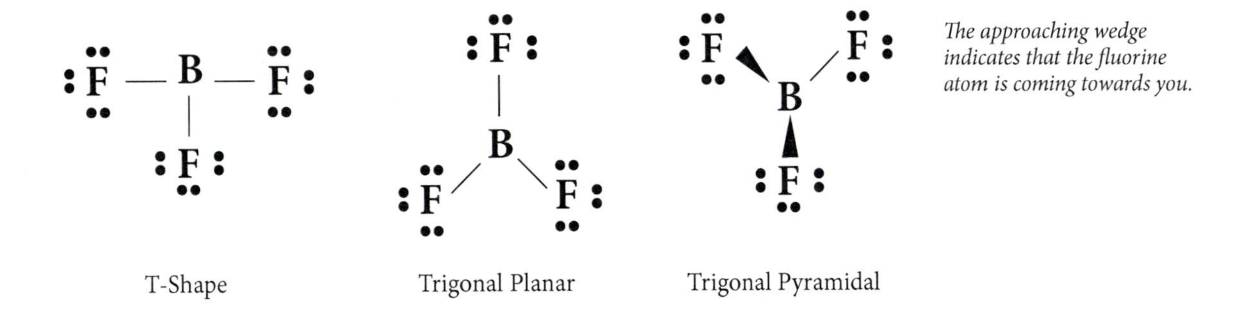

The approaching wedge indicates that the fluorine atom is coming towards you.

In order to discuss the molecular geometry of this structure, we must determine the proper placement of fluorine atoms around the boron atom.

Does the "T-shape" arrangement shown above represent the actual geometry for boron trifluoride, or are either the trigonal planar or trigonal pyramidal preferred?

To arrive at the best placement of the fluorine atoms around the boron atom, we need to discuss the nature and spatial arrangement of the valence electrons on the boron atom.

Figure 5.1

The Lewis structure gives only information regarding how the atoms are connected and the electron pair distribution.

BF_3 contains a total of six valence electrons existing as three B-F single bonds.

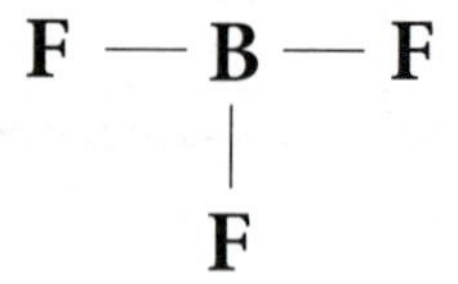

Figure 5.2

Recall that each of the B-H bonds represents two valence electrons being shared.

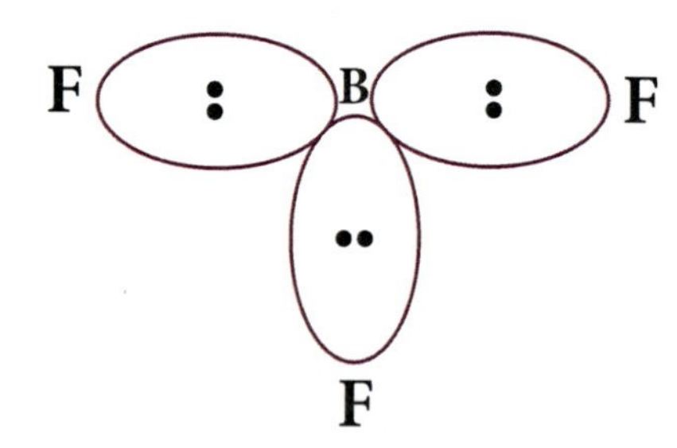

Figure 5.3

Since each covalent bonded pair of electrons is moving in these general areas at great speed, we may think of them as negative charge clouds which will repel each other.

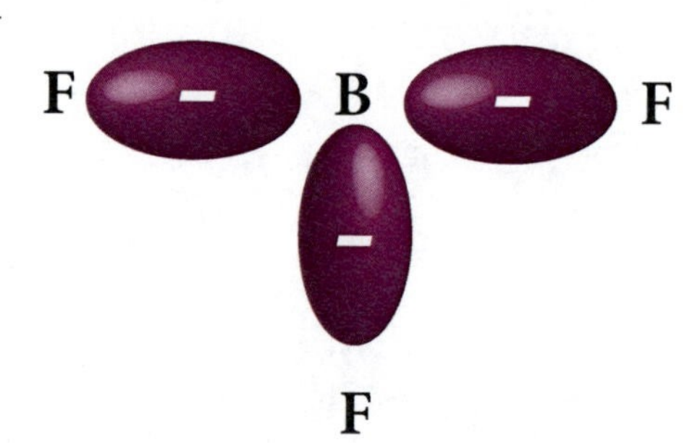

In determining the electron pair geometry, only consider the charge clouds surrounding the central atom.

More About Charge Clouds

N ——— N		N ⬬ N
single bond	=	one charge cloud

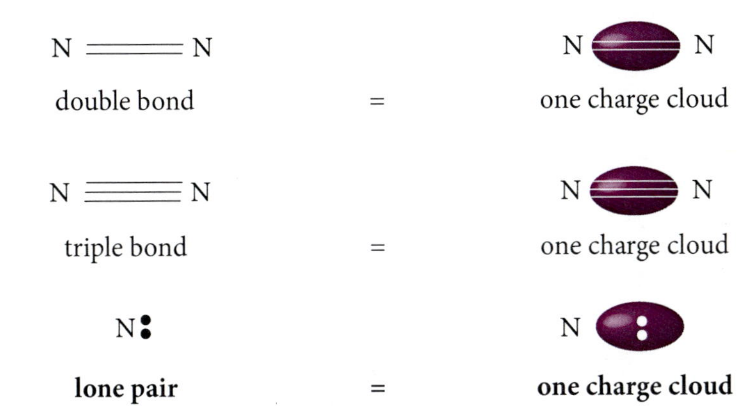

A single charge cloud may be the result of

1 a single bond
2 a double bond
3 a triple bond
4 a lone pair

How many charge clouds are surrounding the central atom in the following molecules?

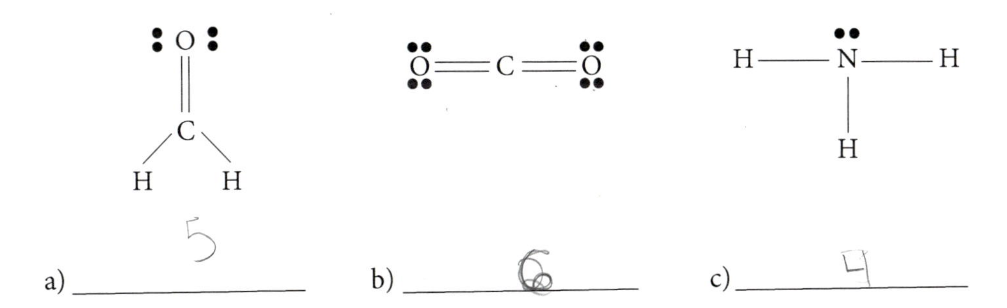

Now consider the **electron pair geometry** of BF_3.

Figure 5.4

According to electrostatic interactions, these charge clouds will want to be as far apart as possible.

Is it possible to relieve the two strong 90° interactions which occur in the T-shape geometry?

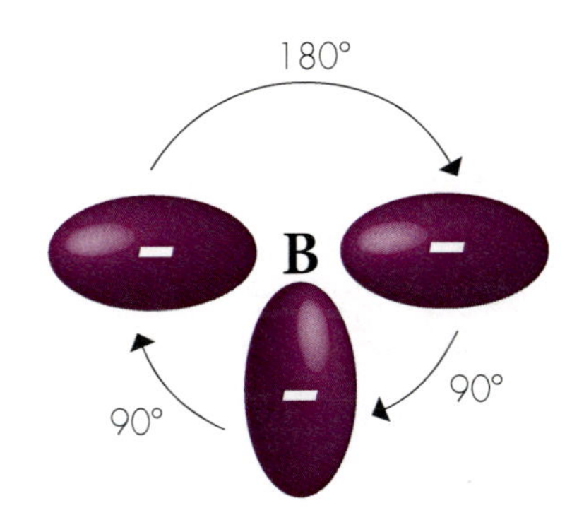

Figure 5.5.

By orienting the charge clouds into a pyramidal structure all angles open up.
But is this the best we can do?

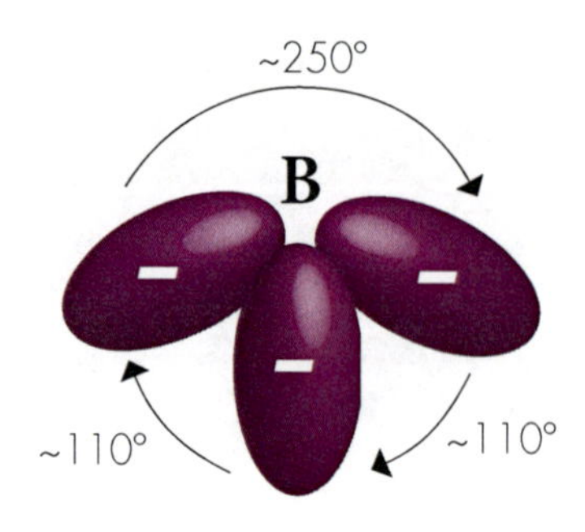

Figure 5.6

By pushing the right and left hand charge clouds up, three 120° separations can be obtained,

Given three charge clouds, this would be the most favorable arrangement.

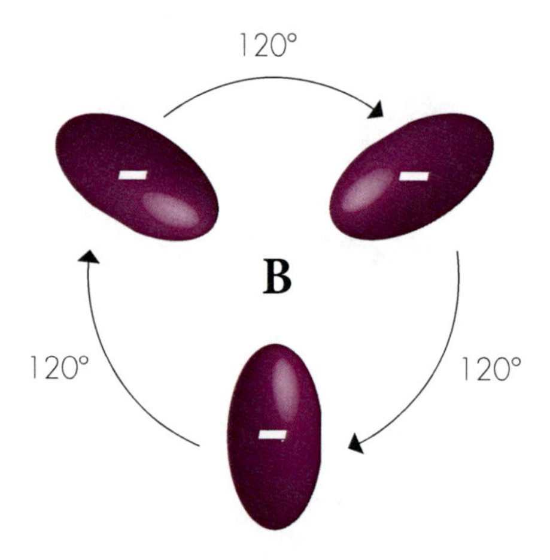

Filling in the three single bonds, we can see that both the electron pair geometry and the molecular geometry are the same, i.e. trigonal planar.

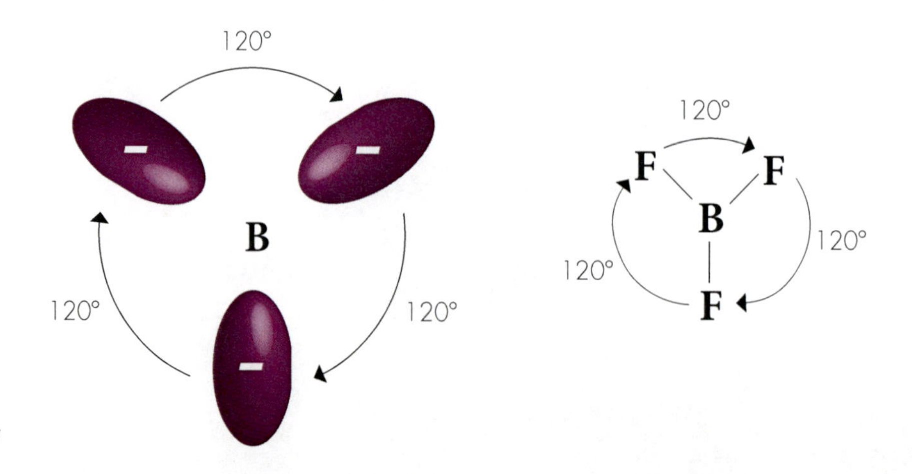

When referring to the arrangement of only the **charged clouds**, it is called

the electron pair geometry,

which shows us all the possible positions for both bonding and non bonding electrons.

When referring to the arrangement of **only the atoms around the central atom**, it is called

the molecular geometry

which refers to only the positions of atoms and not the lone pairs of electrons.

Sometimes the electron pair geometry is the same as the molecular geometry, andsometimes the two geometries seem very different.

Consider the Lewis structures of BF_3 and SO_2:

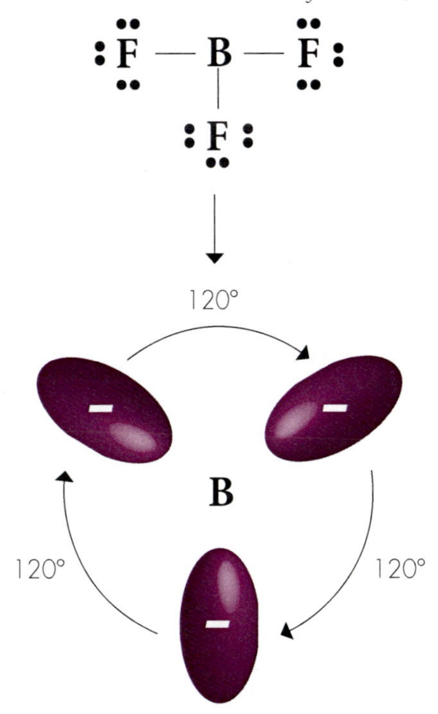

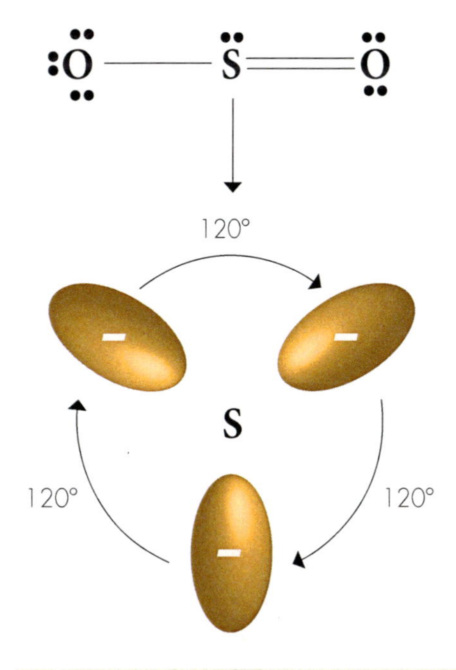

Now in deriving the molecular geometry of each molecule, place into each charge cloud of the molecules a pair of bonding or nonbonding electrons:

In BF_3, each charge cloud represents a boron-fluorine single bond.

In the SO_2, one charge cloud represents a sulfur - oxygen single bond, another charge cloud represents a sulfur - oxygen double bond, and the third charge cloud represents a lone pair of electrons.

Both the sulfur and the boron have three charge clouds surrounding them:

Boron:	Sulfur:
3- single bonds	1- single bond 1- double bond 1- lone pair

Each has an Electron Pair Geometry of trigonal planar.

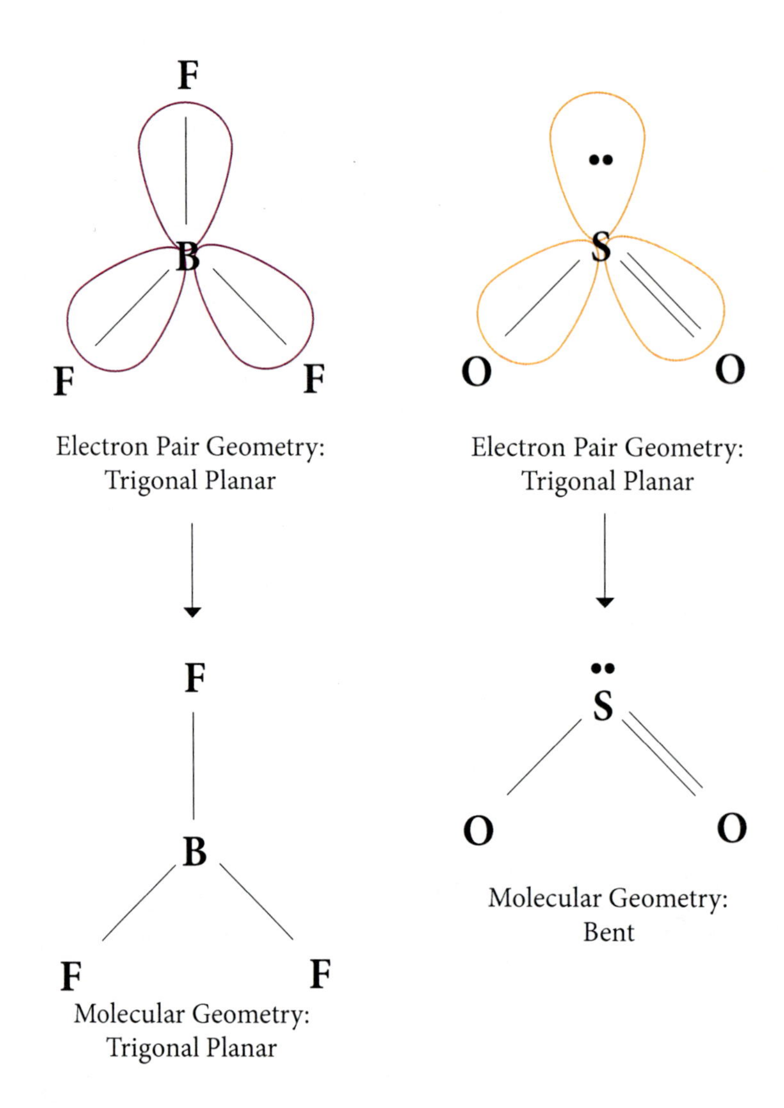

Assignment 5.1

Notice how the electron pair geometry provides the framework for placing all the bonding and non bonding electron pairs around the central atom.

The following pages provide exercises to help familiarize you with the various electron pair geometries along with their corresponding molecular geometries.

1. Using the following Lewis Structures, 1) count the number of bonding and non-bonding regions and determine the number of charge clouds surrounding the central atoms of each molecule.
2. With the number of charge clouds refer to Table A and select the correct Electron **Pair Geometry** from Table A.
3. With the proper electron pair geometry determine the Molecular Geometry after inserting all the bonding and non bonding electrons into the EPG structure.

SB = single bond **DB** = double bond **TB** = triple bond **LP** = lone pair
CC = charge clouds **EPG** = electron pair geometry **MG** = molecular geometry

H

H— C — H

H

SB ____________

DB ____________

TB ____________

LP ____________

CC ____________

EPG ____________

MH ____________

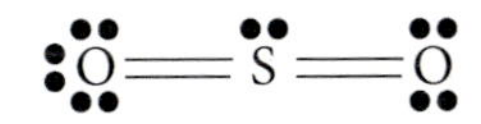

SB 2

DB 1

TB 0

LP 1

CC (on central atom)

EPG ____________

MH ____________

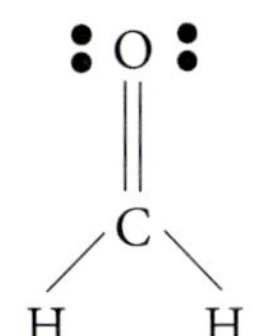

SB ____________

DB ____________

TB ____________

LP ____________

CC ____________

EPG ____________

MH ____________

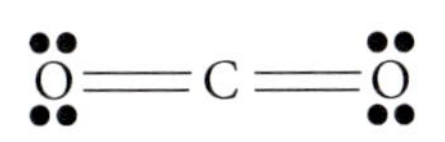

SB ____________

DB ____________

TB ____________

LP ____________

CC ____________

EPG ____________

MH ____________

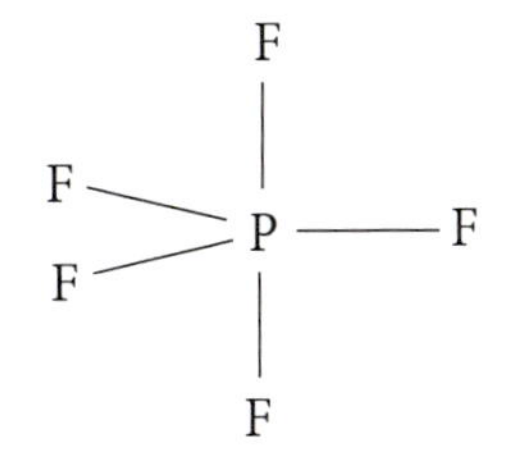

SB ____________

DB ____________

TB ____________

LP ____________

CC ____________

EPG ____________

MH ____________

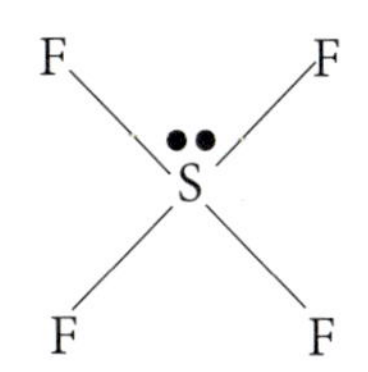

SB ____________

DB ____________

TB ____________

LP ____________

CC ____________

EPG ____________

MH ____________

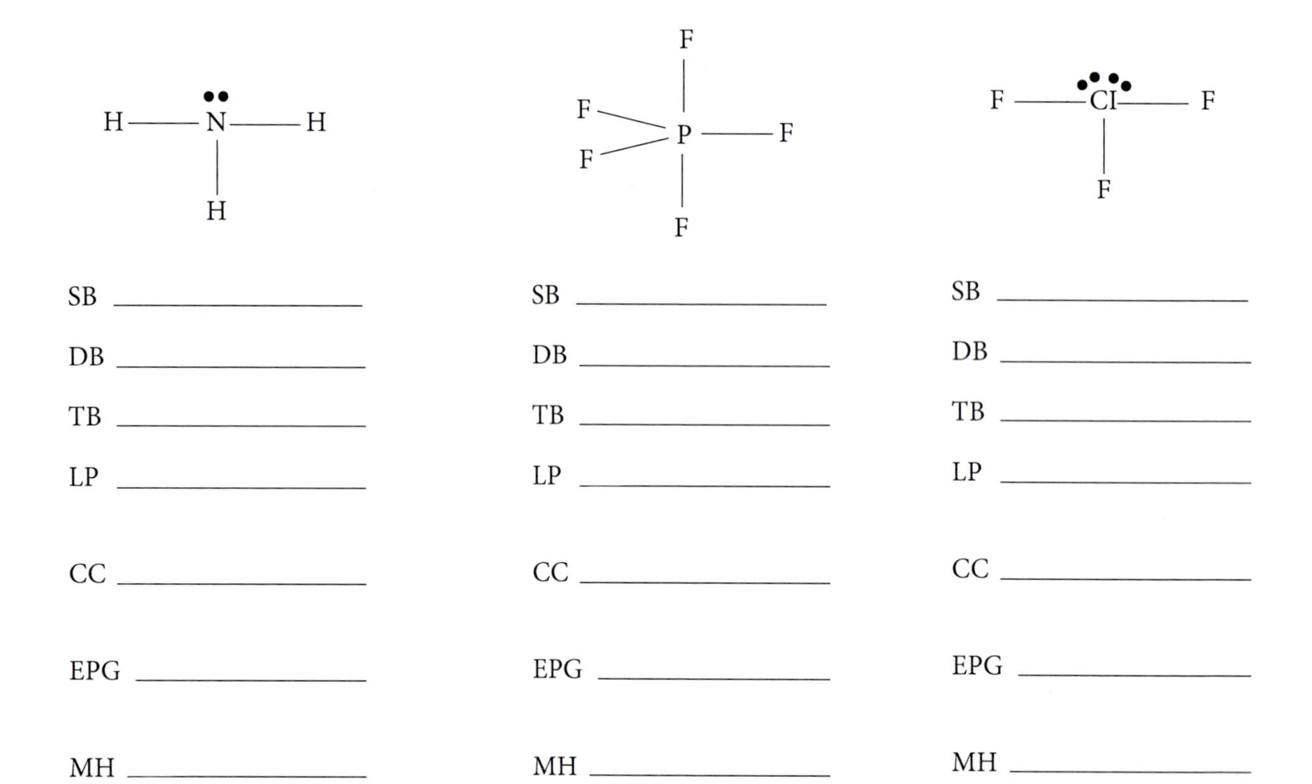

TABLE 5.1 — Summation for Determining Electron Pair Geometry and Molecular Geometry Around the Central Atom

Charge Clouds

2 Charge Clouds

Electron Pair Geometry: **Linear**

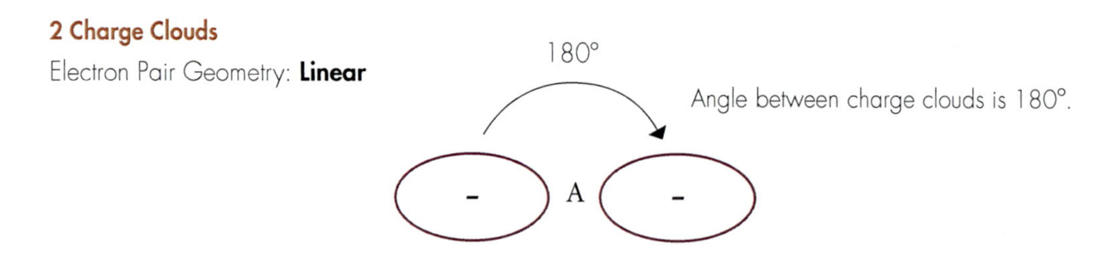

With one **lone pair** and **one bonding pair** of electrons:

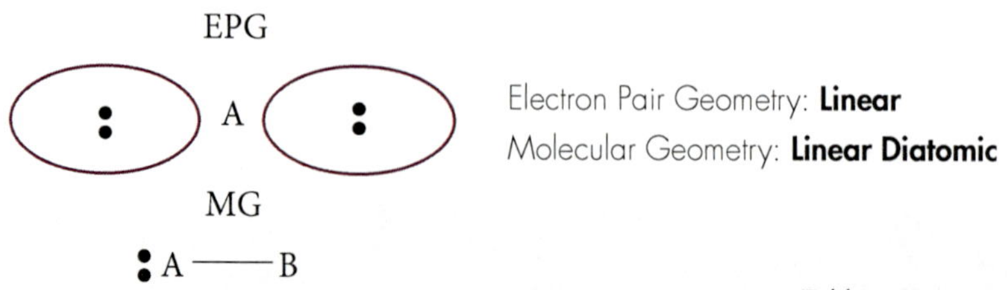

Electron Pair Geometry: **Linear**

Molecular Geometry: **Linear Diatomic**

(Table 5.1 continued)

With **two bonding** pairs of electrons:

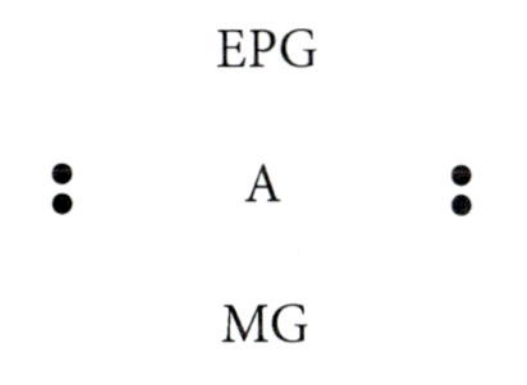

Electron Pair Geometry: **Linear**
Molecular Geometry: **Linear**

Again
Notice how the electron pair geometry provides the framework for placing all the bonding and non bonding electron pairs around the central atom.

3 Charge Clouds

Electron Pair Geometry: **Trigonal Planar**

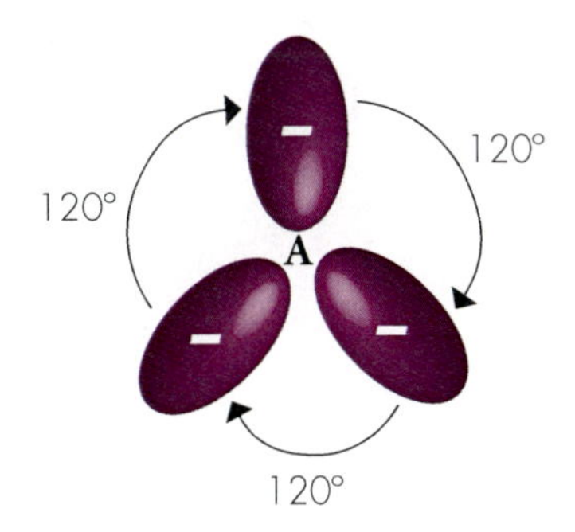

Angle between charge clouds is 120°.

With **three bonding pairs** of electrons:

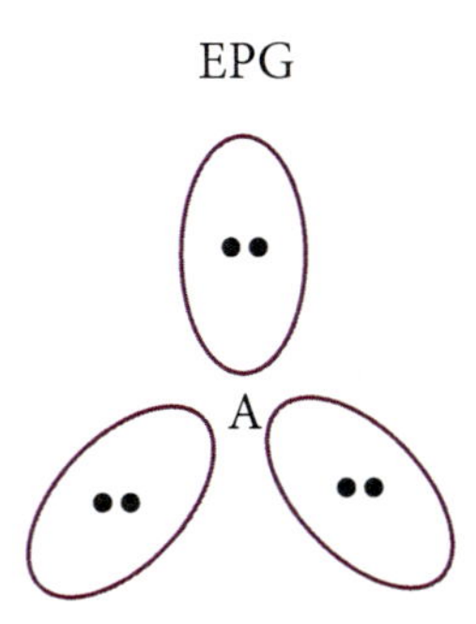

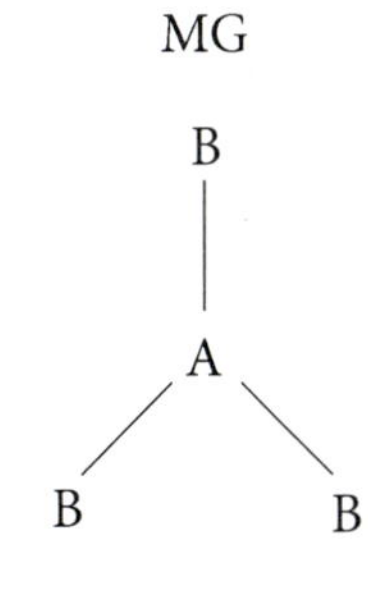

Electron Pair Geometry: **Trigonal Planar**
Molecular Geometry: **Trigonal Planar**

(Table 5.1 continued)

With **two bonding pairs** of electrons and **one lone pair** of electrons:

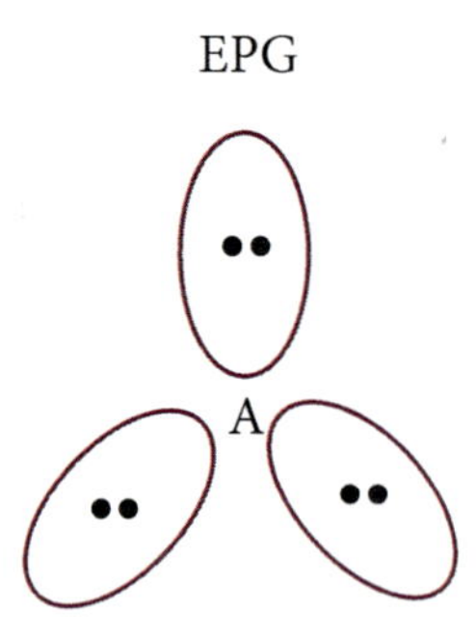

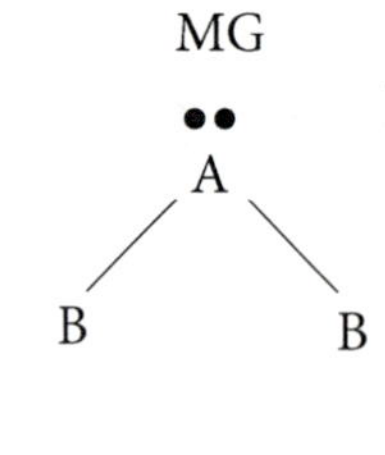

Electron Pair Geometry: **Trigonal Planar**
Molecular Geometry: **Bent**

Summary:
Trigonal Planar Electron Pair Geometry can have two molecular geometries: **Trigonal Planar** and **Bent**

4 Charge Clouds

Electron Pair Geometry: **Tetrahedral**

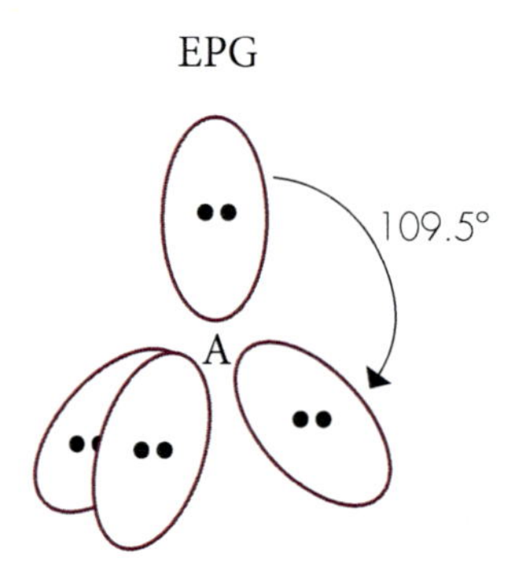

Angle between charge clouds 109.5°.

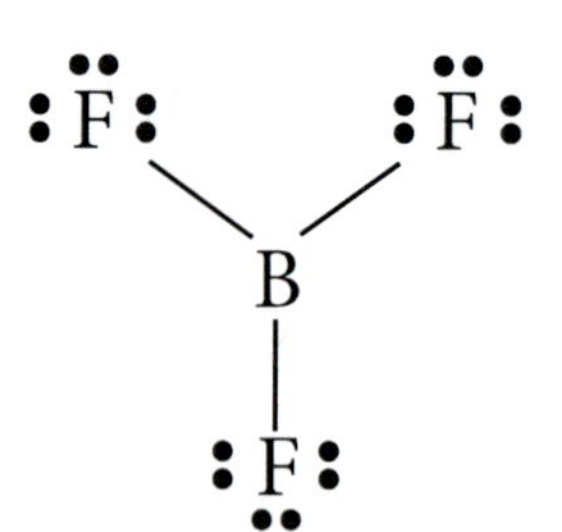

With **four bonding pairs** of electrons:

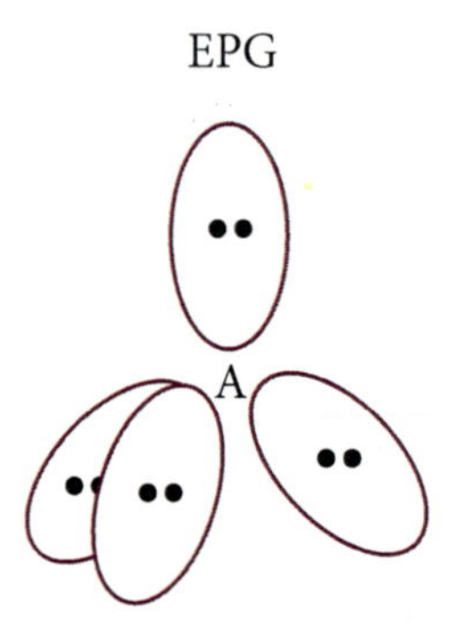

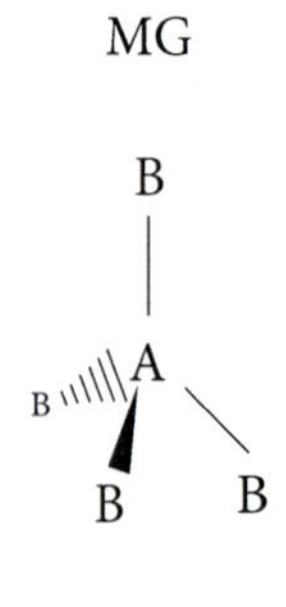

Electron Pair Geometry: **Tetrahedral**
Molecular Geometry: **Tetrahedral**

(Table 5.1 continued)

With **three bonding pairs** of electrons and **one lone pair** of electrons:

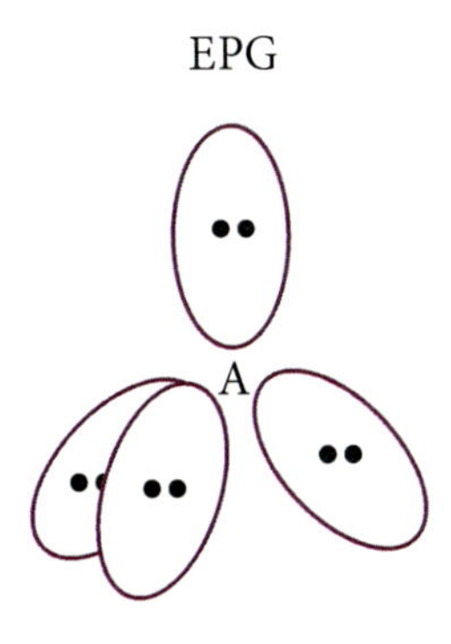

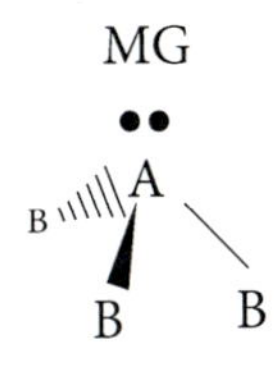

Electron Pair Geometry: **Tetrahedral**
Molecular Geometry: **Pyramidal**

With **two bonding pairs** of electrons and **two lone pairs** of electrons:

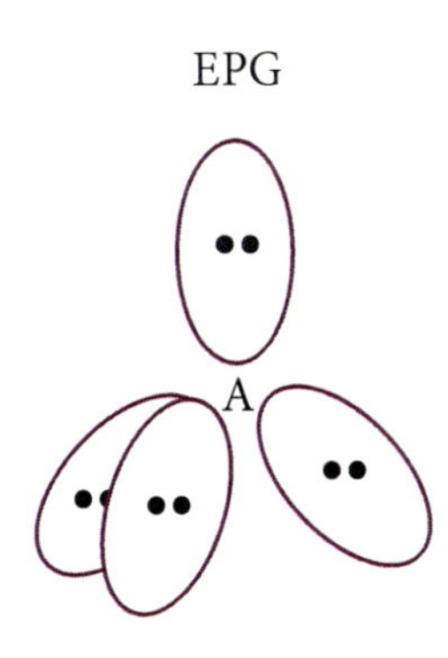

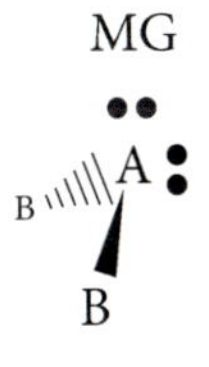

Electron Pair Geometry: **Tetrahedral**
Molecular Geometry: **Bent**

Summary:
Trigonal Planar Tetrahedral Electron Pair Geometry can have three molecular geometries:

Tetrahedral
Pyramidal
Bent

5 Charge Clouds
Electron Pair Geometry: **Trigonal Bipyramidal**

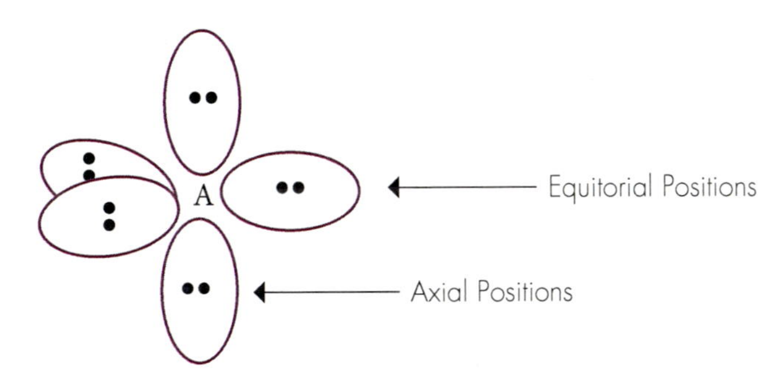

The three pairs of electrons in the equitorial positions form a trigonal plane.

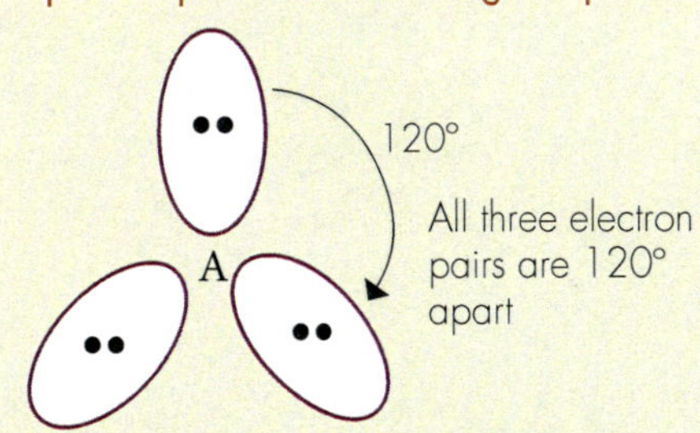

Whenever lone pairs are located on the central atom with a trigonal bipyramidal electron pair geometry, the lone pairs are placed at equitorial positions.

(Table 5.1 continued)

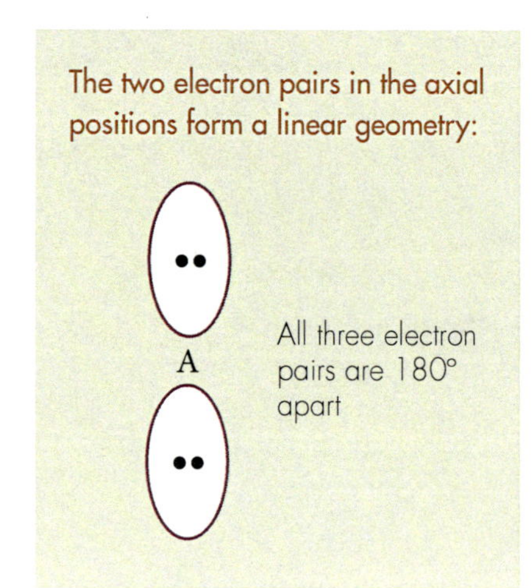

Summary:

In the trigonal bipyramidal electron pair geometry there are two different regions for the electron pair charge clouds:

the axial region containing two charge clouds 180° apart, and the equitorial region containing three charge clouds 120° apart.

With **five bonding pairs** of electrons:

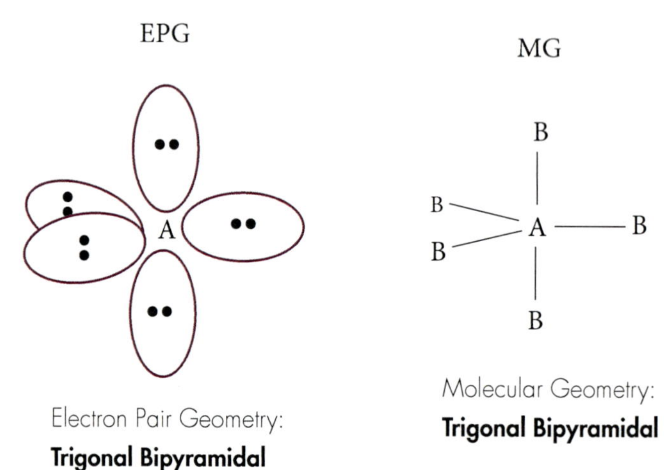

Electron Pair Geometry:
Trigonal Bipyramidal

Molecular Geometry:
Trigonal Bipyramidal

With **four bonding pair** of electrons and **one lone pair** of electrons:

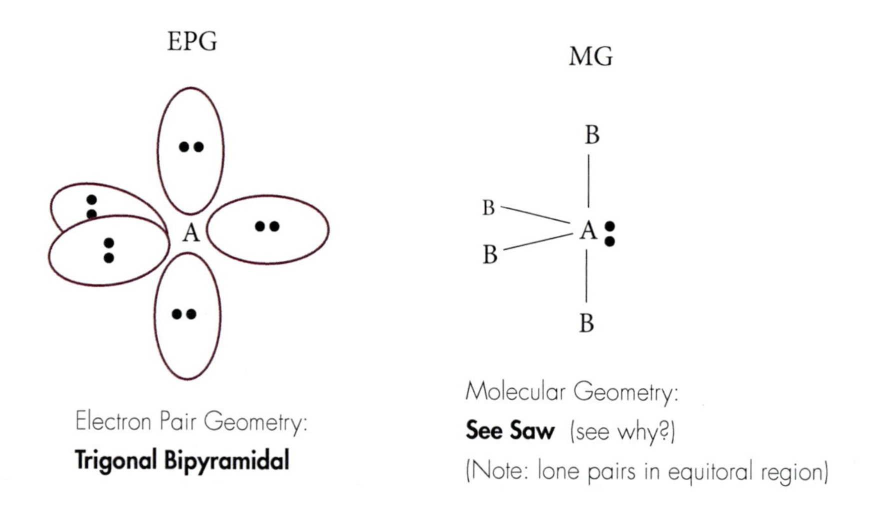

Electron Pair Geometry:
Trigonal Bipyramidal

Molecular Geometry:
See Saw (see why?)
(Note: lone pairs in equitoral region)

Note: *For the trigonal bipyramidal molecular geometry, the lone pairs are always placed in the equatorial region.*

(Table 5.1 continued)

With **three bonding pairs** of electrons and **two lone pairs** of electrons:

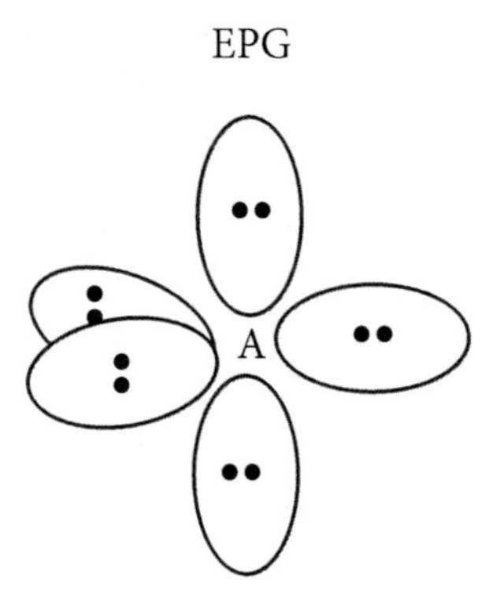

Electron Pair Geometry:
Trigonal Bipyramidal

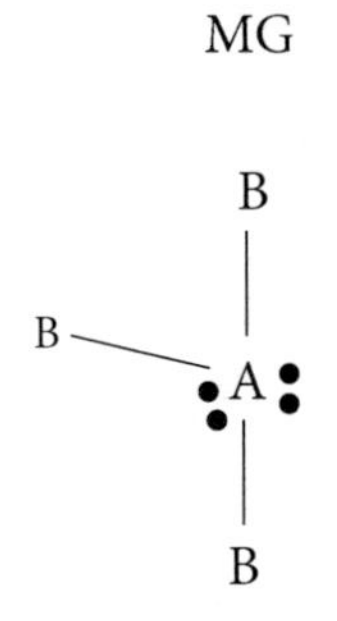

Molecular Geometry:
T-shape

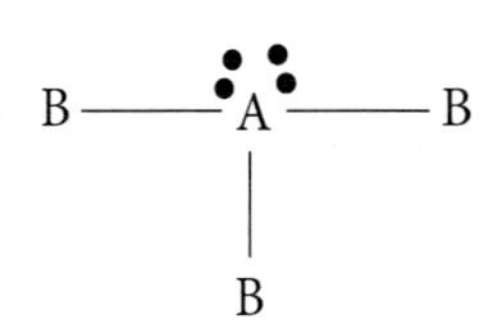

Rotate structure for better perspective

With **two bonding pairs** of electrons and **three lone pairs** of electrons:

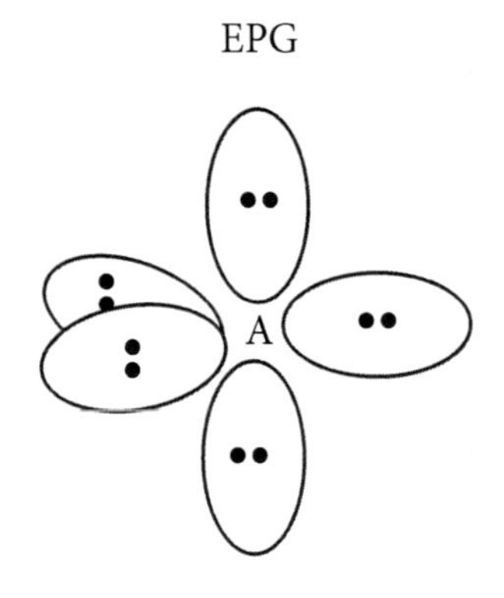

Electron Pair Geometry:
Trigonal Bipyramidal

Molecular Geometry:
Linear

Summary:
The Trigonal Bipyramidal Electron Pair Geometry can have four different Molecular Geometries:

Trigonal Bipyramidal

See Saw

T-Shape

Linear

(Table 5.1 continued)

6 Charge Clouds

Electron Pair Geometry: **Octahedral**

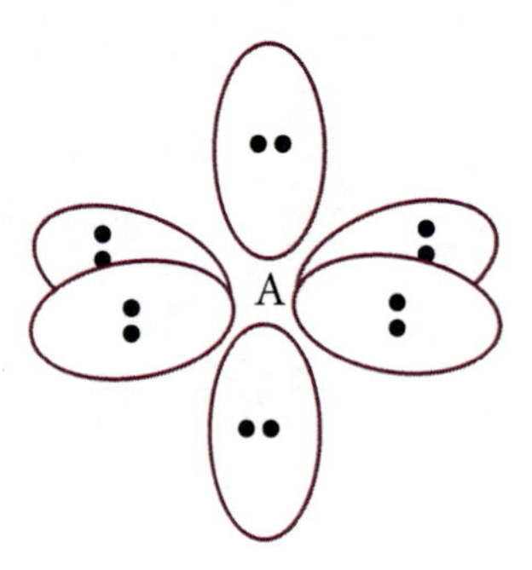

Angle between charge clouds 90°.

Note:

In the octahedral geometry all the regions are the same.

i.e. There are no axial and equitorial regions as in trigonal bipyramidal geometry.

With **six bonding pairs** of electrons:

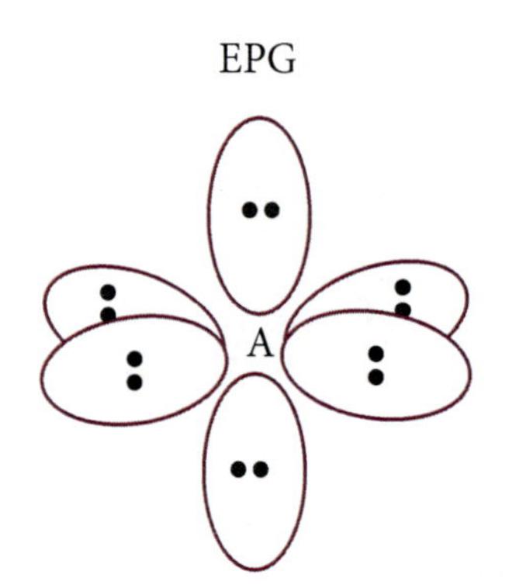

Electron Pair Geometry: **Octahedral**

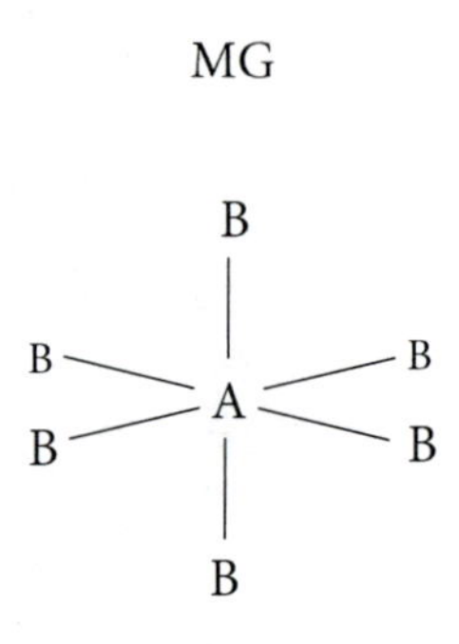

Molecular Geometry: **Octahedral**

With **five bonding pairs** of electrons and **one lone pair** of electrons:

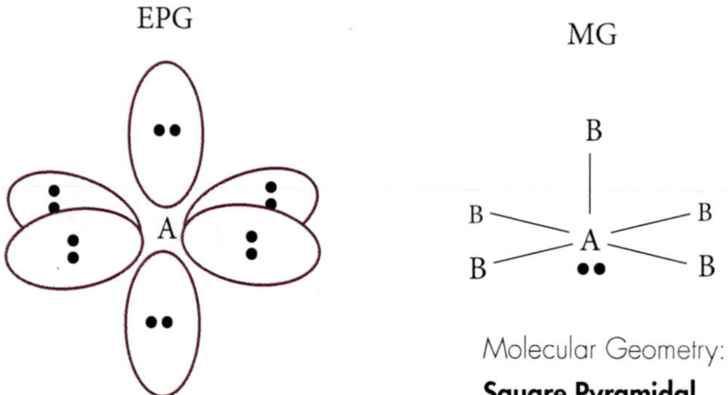

Molecular Geometry: **Square Pyramidal**

Electron Pair Geometry: **Octahedral**

(Table 5.1 continued)

With **four bonding pairs** of electrons and **two lone pair** of electrons:

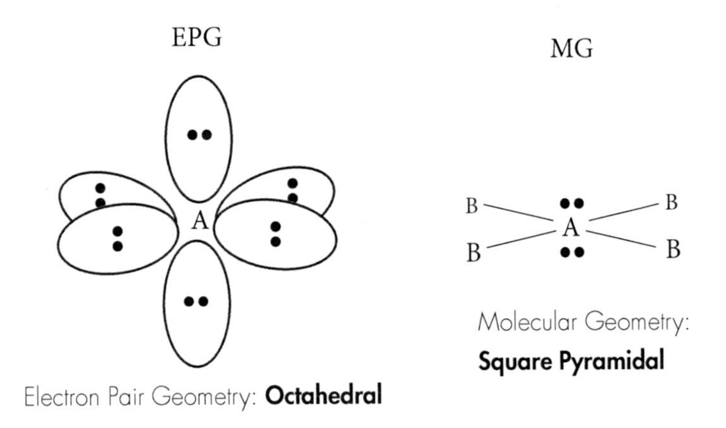

Molecular Geometry: **Square Pyramidal**

Electron Pair Geometry: **Octahedral**

Note:
The two lone pairs will orient themselves on opposite sides to achieve the least interaction.

With **three bonding pairs** of electrons and **three lone pair** of electrons:

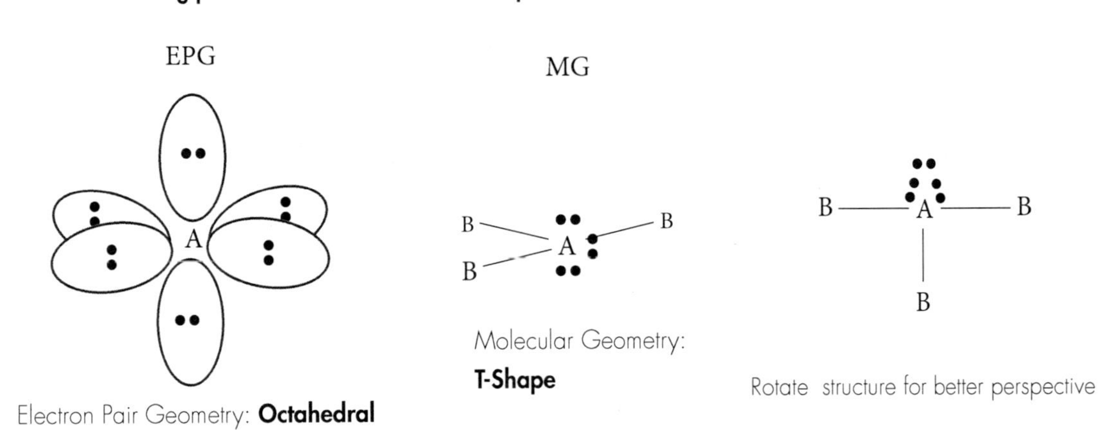

Molecular Geometry: **T-Shape**

Rotate structure for better perspective

Electron Pair Geometry: **Octahedral**

With **two bonding pairs** of electrons and **four lone pair** of electrons:

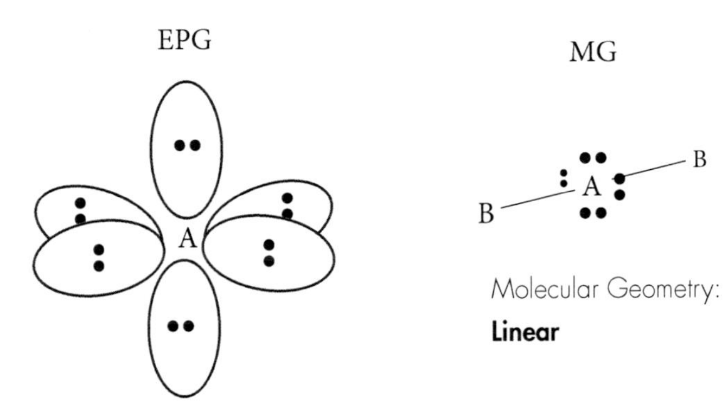

Molecular Geometry: **Linear**

Electron Pair Geometry: **Octahedral**

(Table 5.1 continued)

Summary:

Octahedrall Electron Pair Geometry can have five different Molecular Geometries:

Octahedral

Square Pyramidal

Square Planar

T-Shape

Linear

Assignment 5.2

1a. From the Lewis structure of iodide tetrachloride, ICl_4^-, determine the number of charge clouds surrounding the central atom, iodine.

1b. From the number of charge clouds in, ICl_4^-, determine the electron pair geometry of the molecule.

1c. From the electron pair geometry and the information we have from the Lewis structure (the lone pairs, the single, double, and triple bonds surrounding the iodine atom) determine the molecular structure of the iodine tetrachloride molecule.

2. Repeat the above exercise with SOF_2, SF_2, H_2O and H_3O^+, BrF_5, O_3, $SbCl_6^-$, and PCl_5.

3. What different electron pair geometries can support

   a) linear molecular geometry?

   b) T-shape molecular geometry?

   c) see saw molecular geometry?

4. In the trigonal bipyramidal molecular geometry how many 90° interactions occur?

Workspace

Assignment 5.3

1. For the following covalent compounds and polyatomic ions draw the Lewis Structures and determine the electron – pair geometries and molecular geometries. Use the most stable resonance structure when necessary.

BrO_3^-

Lewis Structure

Electron Pair Geometry:

Molecular Geometry:

PO_3^{3-}

Lewis Structure

Electron Pair Geometry:

Molecular Geometry:

Workspace

TeF_5^-

Lewis Structure

Electron Pair Geometry:

Molecular Geometry:

SeO_3

Lewis Structure

Electron Pair Geometry:

Molecular Geometry:

Workspace

IF_2^-

Lewis Structure

Electron Pair Geometry:

Molecular Geometry:

XeO_2F_4

Lewis Structure

Electron Pair Geometry:

Molecular Geometry:

Workspace

Polarity in Covalent Structures

Carbon monoxide possesses polar covalent bonding due to the difference in the **electronegativities** of the carbon atom (EN = 2.5) and the oxygen atom (EN = 3.5).

$$\overset{\delta^+}{:C} \equiv \overset{\delta^-}{O:}$$

The partial positive charge (δ+) and the partial negative charge (δ–) are the result of the electron density of the bonding valence electrons being drawn away from the atom of lower electronegativity (carbon) and toward the one with the higher electronegativity (oxygen).

By identifying a difference in electronegativity between two atoms covalently bonded together, it's easy to deduce that there is a certain amount of polarity within the bond.

But how does the existence of polar bonds within a molecule effect the overall polarity of a molecule?

In a diatomic molecule, such as carbon monoxide, there exists a single polar bonding region that is, in fact, the only bonding region of the molecule. Therefore, **carbon monoxide is a polar molecule**.

Let's extend the discussion, by adding one more oxygen atom to this molecule, which results in the covalent compound carbon dioxide, CO_2.

Once again, we consider the polar carbon – oxygen bond, but this time there will be two which are structurally opposed to each other.

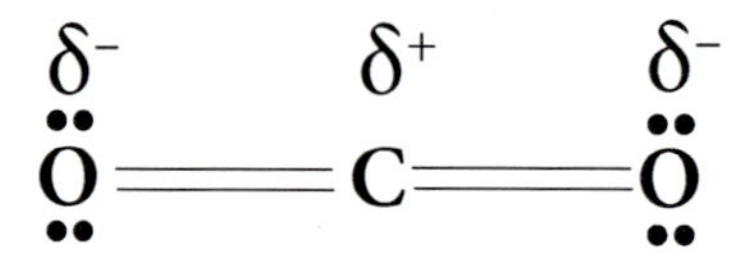

If we consider the polarities as two opposable vectors of equal magnitude in opposite directions, it becomes obvious that the two polarities cancel out and the carbon dioxide molecule is a non polar species.

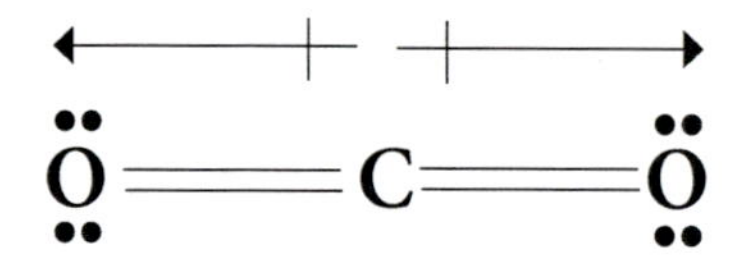

In this case, we can see that the existence of polar bonds does not result in a polar molecule.

By the same argument, we can conclude that carbon disulfide, CS_2, is also a non polar compound containing polar bonds.

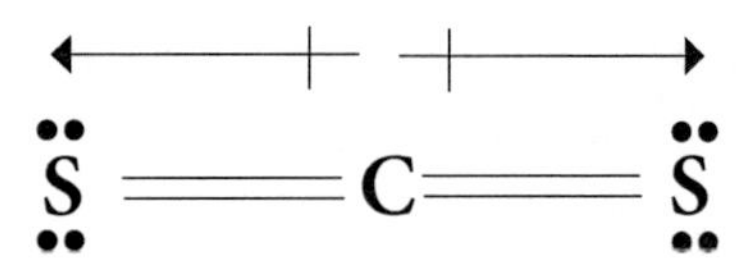

But, suppose we keep the carbon at the center, and change one of the outside atoms?

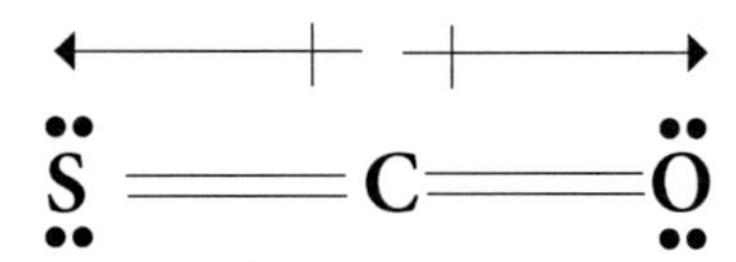

The compound still contains two polar that are in opposite directions, but the magnitudes of the vectors are now different since the sulfur and the oxygen have different electronegativities.

Therefore, the two polar bonds will not cancel each other out.

This **compound is polar**.

From the two non polar compounds, carbon dioxide and carbon disulfide, we can see that both compounds contain two equivalent bonds: carbon dioxide contains two carbon – oxygen bonds and carbon disulfide contains two carbon – sulfur bonds.

In order for the polarities of these two bonds to cancel, the two bonds need to be opposite each other.

Consider another triatomic molecule, sulfur dioxide, SO_2. Again, we have a molecule that contains two equivalent sulfur – oxygen bonds;

However, unlike the bonds of the carbon dioxide and the carbon disulfide, the bonds of the sulfur dioxide are not linear. It has a bent molecular geometry.

They are bent.

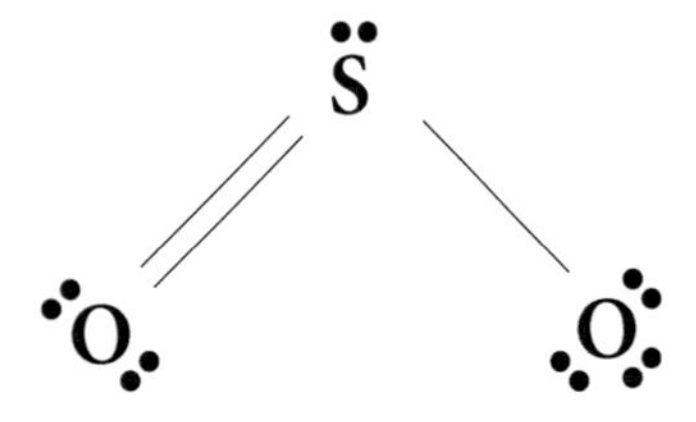

Since these polar bonds are not structurally opposite each other, they are not cancelled out as with CO_2 and CS_2.

Sulfur dioxide is polar.

Summary:

Two equivalent polar bonds in a linear molecular geometry completely cancel.

This results in a non polar molecule.

Two non equivalent polar bonds in a linear geometry do not completely cancel.

This results in a polar molecule.

Two equivalent polar bonds in a bent molecular geometry do not cancel.

This results in a polar molecule.

Assignment 5.4

Draw the proper molecular geometry and determine if the following molecules are polar or nonpolar.

$BeCl_2$

HCN

H_2O

XeF_2

Workspace

Considerations of Polarity Beyond Triatomic Species

Consider the non polar molecule boron trifluoride, BF_3. The molecular geometry of BF_3 is planar triangular.

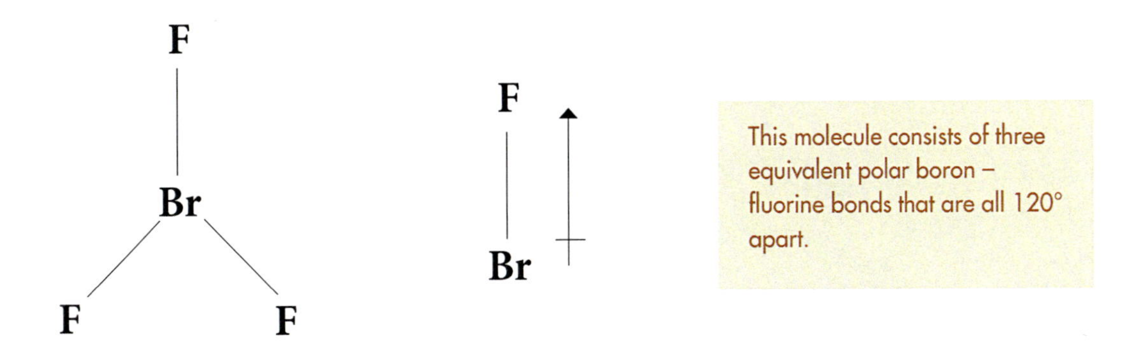

This molecule consists of three equivalent polar boron – fluorine bonds that are all 120° apart.

The cancellation of the polar B – F bonds may not be as apparent as with the CO_2 and CS_2 molecules. To better understand the net non polar character of this molecule, apply some vector arithmetic to the three polar bonds:

In vector addition, the three vectors are sequentially redrawn in a head to tail fashion maintaining their original angles.

Notice that the head of the third vector returns to the base of the first vector. This indicates a zero net polarity.

This same method of vector addition can be applied to CO_2 and CS_2 structures to explain their nonpolar natures.

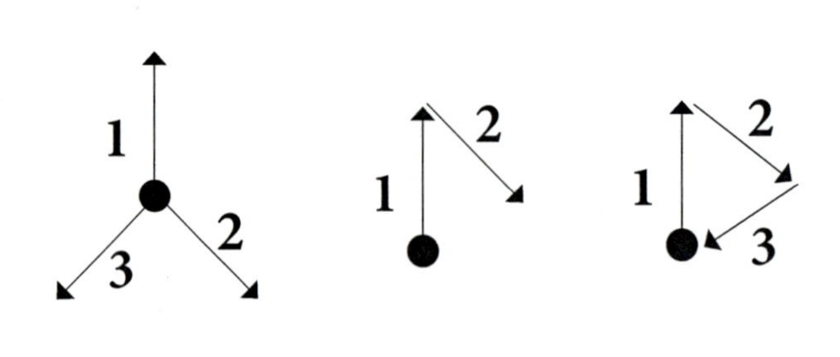

From the discussion of boron trifluoride we can infer that **molecules with planar triangular molecular geometries that contain three equivalent polar bonds are nonpolar.**

Assignment 5.5

Combining what we know about polarity of linear and planar triangular molecules explain the nonpolar natures of the following molecules:

PF_5

SF_6

XeF_4

Workspace

The one remaining molecular geometry to consider with respect to complete cancellation of polar bonding is the **tetrahedral** geometry.

For example, consider CCl_4

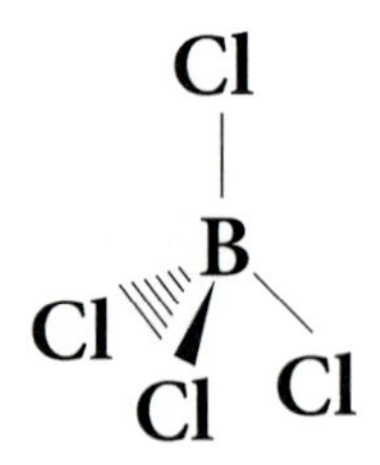

Unlike the previous examples, the complete cancellation of the four equivalent polar C – Cl bonds needs to be done by a more involved process of vector addition, but the result is the same as the **linear**, the **planar triangular**, the **trigonal bipyramidal** (combination of linear and planar triangular), the **octahedral** (combination of four linear), and the **square planar** (combination of two linear).

Summation

The two conditions to determine if a molecule containing polar bonds is, overall, a non polar structure are:

1. all the polar bonds within the molecule are equivalent (i.e. the same type of bonds) and
2. the molecular geometry which contain the polar bonds is one of the following:

   linear, planar triangular, tetrahedral, trigonal bipyramidal octahedral, or square planar

Assignment 5.6

Determine if the following molecules are polar or non polar.

$CFCl_3$

polar non polar

why

BrF_5

polar non polar

why

SF_4

polar non polar

why

PF_4Cl

polar non polar

why

Workspace

Hybridization of Atomic Orbitals

Some of the bonding angles that are observed in covalent compounds cannot be explained by the electron configuration and orbital orientations of the central atom's valence electrons.

In the preceding sections, various molecular geometries, (e.g linear, triangular planar, and tetrahedral) incorporated bonding angles of 180°, 120°, and 109.5° around a central atom.

Given what we learned about the shapes and orientations of the atomic orbitals, and the ground state electron configuration of the central atom, how are these bonding angles possible?

In order to address these issues, chemists describe a process called hybridization to explain these departures from ground state atomic configurations and orientations.

In the following section, the structure of methane, (tetrahedral molecular geometry), will be used to:

1. explain why the ground state configuration of carbon and its orbital descriptions are insufficient to explain the angles and energies of the C-H bonding orbitals in compounds such as methane, and
2. introduce the theory of the hybridization of carbon's atomic orbitals which allow the observed angles and energies. The section will also address how to recognize the different degrees of hybridization that occur.

Chemists have no hand in hybridization of atomic orbitals. It's just something that naturally happens to a central atom when needed, that is,

it allows the atom to achieve more stable bonding arrangements.

It's up to chemists to know what's going on.

Hybridization of atomic orbitals is a very important concept to understand, especially for those students, who intend to study Organic Chemistry. This particular topic seems to cause confusion and frustration among introductory general chemistry students, but the basic concept of how to "use" it is relatively easy to master. The more difficult aspect is grasping why it occurs.

"I hope this helps"

Consider the bonding angles of the tetrahedral compound methane, CH_4,

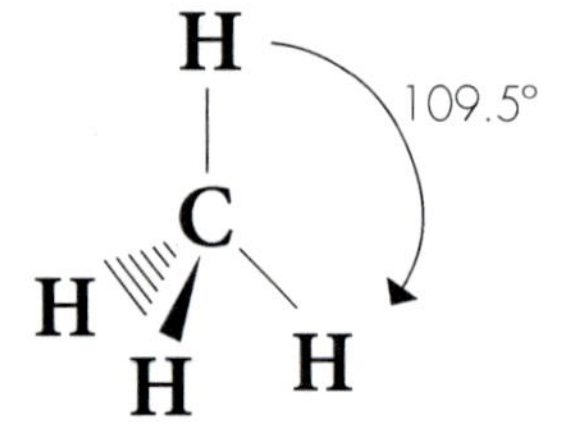

Considering the shapes and orientations of the atomic orbitals of carbon, are the 109.5° bonding angles in methane possible?

Is it possible to construct the methane geometry using the ground state valence electrons of carbon?

(In attempting this, we will find **two discrepancies.**)

Recall that the electron configuration for ground state carbon is $\mathbf{1s_2\ 2s_2\ 2p_2}$:

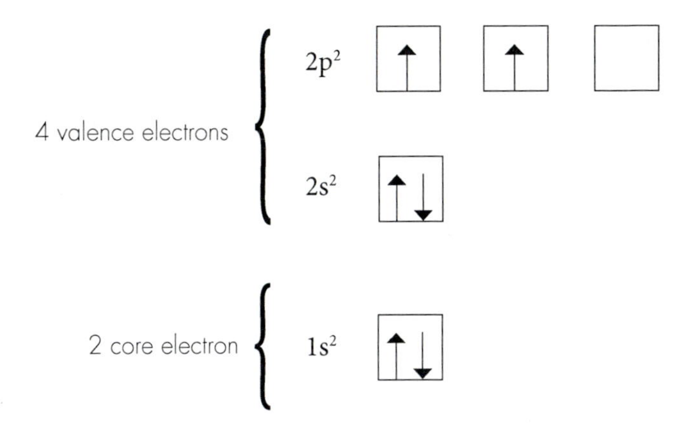

According to this configuration, the first two of the valence electrons, $2s^2$, are paired in the spherical s-type subshell orbital:

$2s^2$ =

Each of the remaining two valence electrons, $2p^2$, are in separate p-type subshell orbitals (dumb bell shaped):

$2p^2$ = +

carbon $2p_x$ carbon 2p orbitals 90°

In order to discuss what all this means, two mechanisms of covalent bond formation need to be addressed. Both mechanisms will be applied to the electron configuration of carbon's valence electrons.

Covalent Bond Formation I:

Electron-Electron Pairing (Review Lewis Structures)

Covalent bonds can be formed by each of the participating atoms donating one electron. In order for this to happen, both electrons must be unpaired. In regards to the carbon atom, each of the unpaired electrons in 2p orbitals can form a bonding pair with the single electrons of two separate hydrogen atoms:

For example,

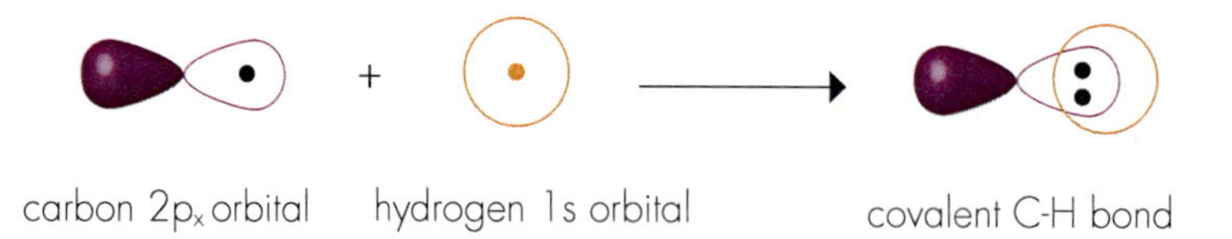

Therefore, if carbon has two available unpaired valence electrons in separate p-subshell orbitals, two covalent C-H bonds should be possible:

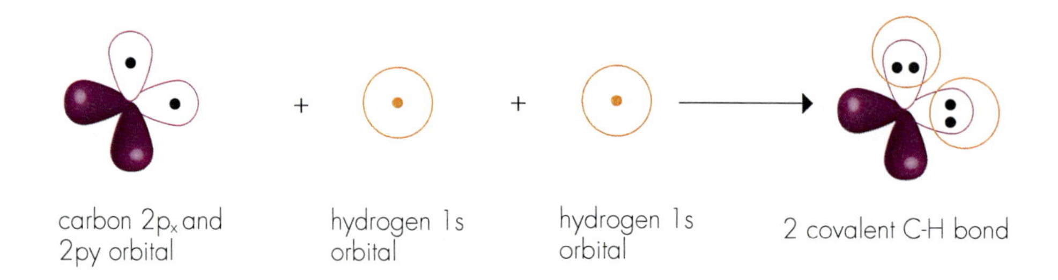

That is, two of the unpaired valence electrons in the ground state electron configuration of carbon should readily react with two hydrogen atoms to form two covalent bonds which are 90° apart:

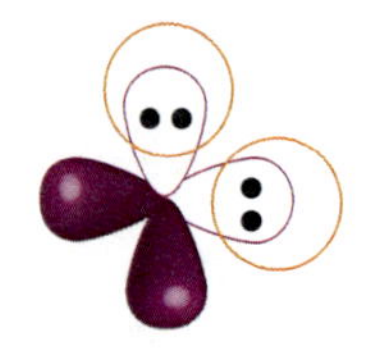

Discrepancy 1:

If this description of bonding did occur, two of the methane C-H bonds would have 90° angles instead of 109.5°.

Having dealt with the two valence electrons of carbon 2p orbitals, let's discuss the pair of valence electrons that reside in the carbon 2s orbital. Being paired electrons, can the carbon 2s2 electrons, in the valence shell, participate in bond formation?

Covalent Bond Formation II:
Coordinate Covalent Bond Formation

Besides electron-electron pairing, covalent bonding can also result when one of the participating atoms donates both of the bonding electrons:

$$\mathbf{A\!:} + \mathbf{B} \longrightarrow \mathbf{A^{+}}\!-\!\mathbf{B}$$

In order for this situation to arise,

1. one of the two atoms must have an available lone pair of electrons (**electron pair donor**), and
2. The other atom must be electron deficient (**electron pair acceptor**).

An example of such a combination would be the reaction between boron trifluoride, BF_3, and the fluoride ion, F-, to form the tetrafluoroborate ion.

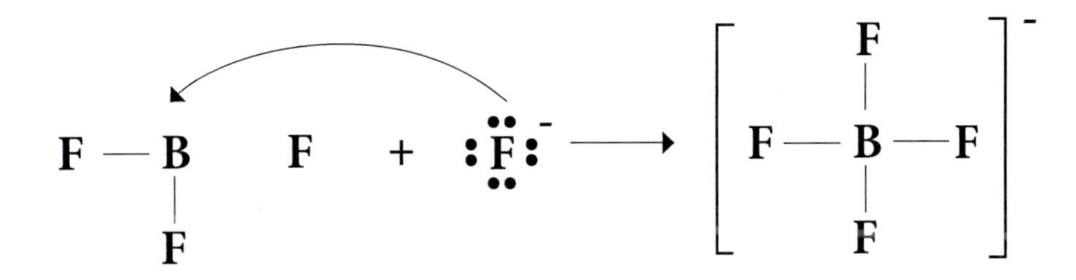

tetrafluoroborate ion

This type of covalent bond formation is called **coordinate covalent** bond formation. The formation of this type of covalent bond results in:

1. the electron pair donor becoming *more positive*, and
2. the electron pair acceptor becoming *more negative.*

It should be noted that once a coordinate covalent bond forms, it is no different from other covalent bonds.

In the case of the tetrafluoroborate ion, a negative charge develops on the boron atom.

Another example of such bonding occurs between ammonia, NH_3, and the hydrogen ion, H^+, in the formation of ammonium ion, NH_4^+.

Hypothetically, the pair of electrons in the carbon 2s orbital could provide a coordinate covalent bond with a hydrogen ion:

2s electron pair hydrogen ion C-H covalent bond

Discrepancy 2:

The two valence electrons paired in the carbon's 2s orbital could form a coordinate covalent bond to one hydrogen ion, which would result in a single C-H covalent bond and a positive charge on the carbon atom.

In comparison to methane, the hypothetical structure would look something like this:

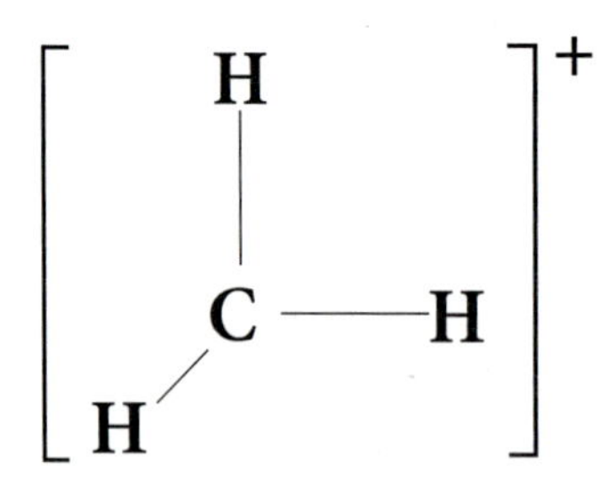

Hypothetical methane structure based on ground state electron configuration of carbon:

two C-H bonds are equal and 90° apart, one C-H bond at a lower energy, carbon atom has a positive charge

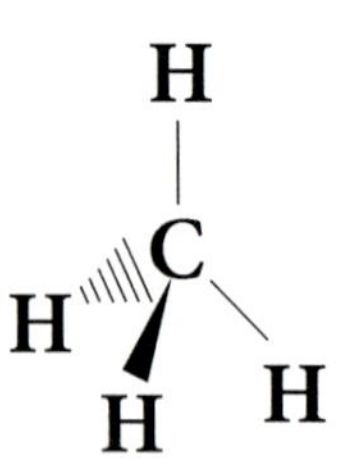

Actual structure of methane:

all four C-H bonds are equal in energy, all bonding angles are equal (109.5°), all atoms have neutral charge

Summary

Given the ground state electron configuration of carbon, two types of bonding would be possible in the formation of methane:

1. **electron-electron pairing** between the two valence electrons in the carbon's 2p orbitals and two hydrogen atoms, that would result in *two covalent C-H bonds, at 90° angles from each other,* and
2. one **coordinate covalent bond** formation between the remaining two valence electrons in the carbon's 2s orbital and one hydrogen ion, that would result in *a single C-H covalent bond and a positive charge on the carbon atom.*

Note:

Since the bond arising from the carbon 2s orbital is at a different energy than the bonds formed from the carbon 2p orbitals, the energies and bond lengths of both types of bond will be different.

Before proceeding to the description of hybridization, work this exercise:

Assignment 5.7

Boron trifluoride has a trigonal planar molecular geometry, in which all the B-F bonds are 120° apart. Using the two methods of forming covalent bonds discussed above, draw the structure you may expect from the ground state electron configuration of boron.

The Process of Hybridization of Atomic Orbitals

To explain the actual structure of methane, two separate steps must occur:

Step 1

Since four covalent bonds are formed, four unpaired electrons need to be available in the valence shell of the carbon atom.

(Note: Only electron-electron pairing will be considered.)

This will be accomplished by the **promotion** step in which one of the 2s paired electrons is promoted to the one empty 2p orbital.

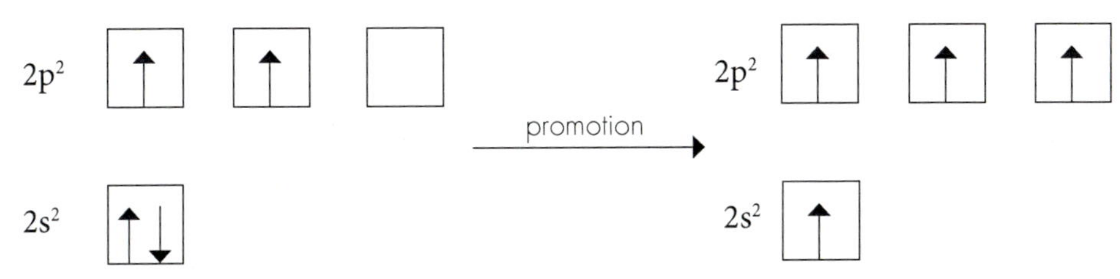

By promotion, four unpaired electrons are now available on the carbon atom to form four covalent bonds by electron-electron pairing. But, one more step is still needed.

Problem 5.1

Using the previous arguments of orbital energies and orientations, why won't promotion by itself, answer the bonding questions of methane?

Step 2

To create four equal bonding orbitals with the needed orientations, the s and p atomic orbitals are mixed together, i.e. **hybridization** occurs.

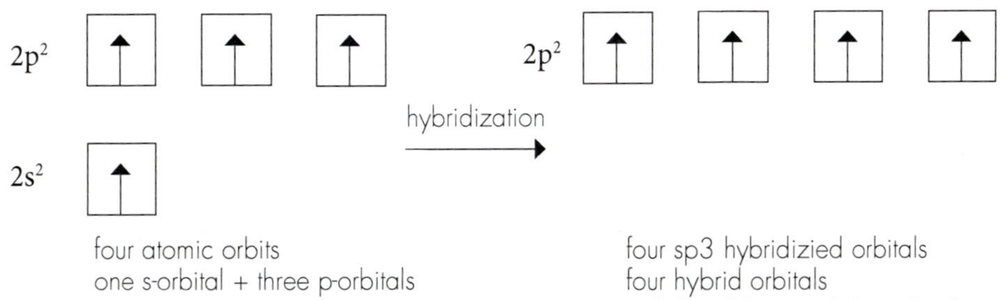

The four equivalent hybridized orbitals on carbon will be able to achieve the four 109.5° bonding angles observed in methane.

How is this so?

The Nature of Hybrid Orbitals

The first aspect of a hybrid orbital to consider is its shape. Unlike the s, p, and d atomic orbital shapes,

the hybrid's basic shape is unique.

The large top lobe represents the bonding region of the orbital. This region is more extensive than the bonding regions of the atomic orbitals.

Hybrid orbitals occur by the mixing of various combinations of s, p and also d atomic orbitals. Regardless of the combination of atomic orbitals, the basic shape of the hybrid orbital stays the same. What changes are the angles of the hybrid orbitals. For example, the mixing of one s-orbital and one p-orbital results in two sp hybrid orbitals that provide a 180° angle.

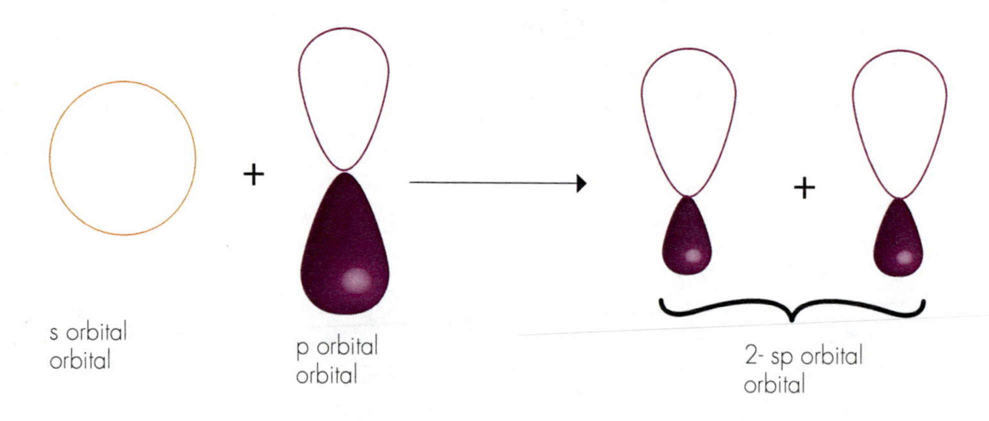

Hybridization of Atomic Orbitals:
Linear Geometry

E.g. beryllium chloride, $BeCl_2$

For linear geometry, two unpaired electrons in equal orbitals are needed.

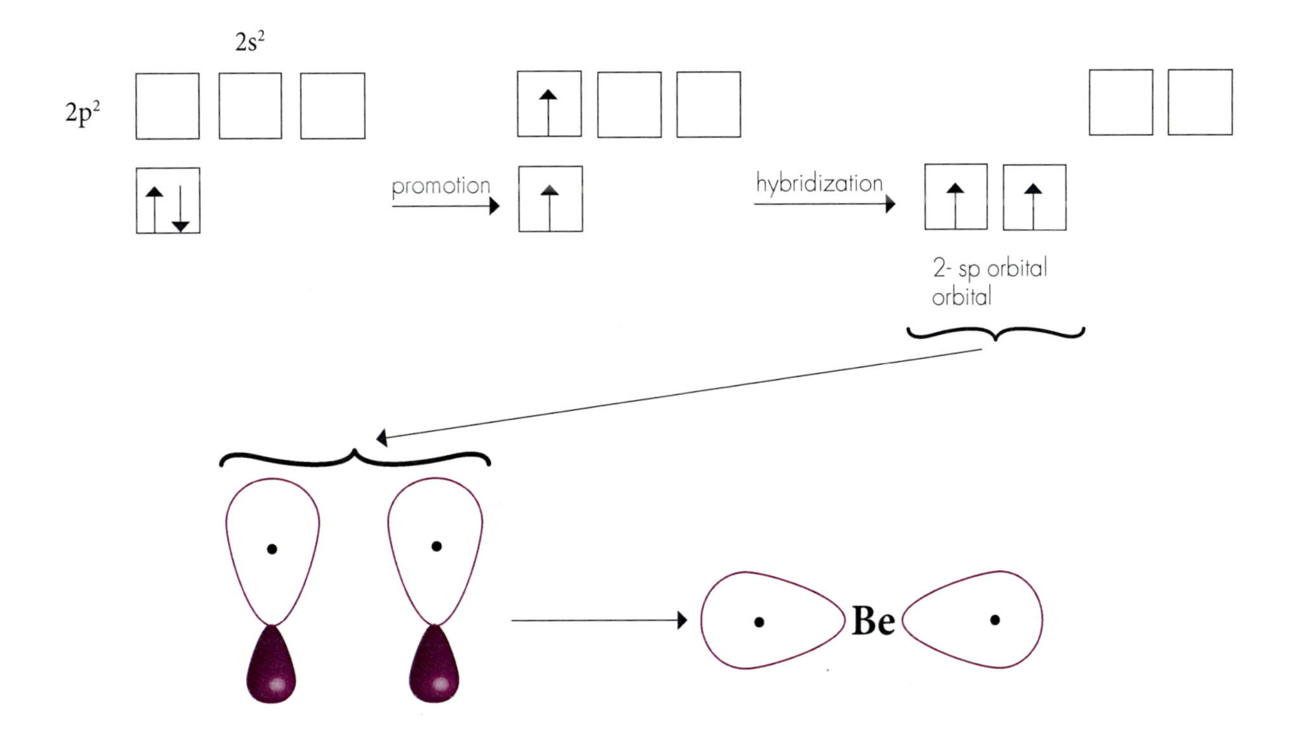

The two sp hybrid orbitals orient themselves in a linear arrangement with the bonding regions pointing in opposite directions:

This will allow the linear geometry to occur.

Note that prior to bonding, hybrid orbitals will always contain one electron.

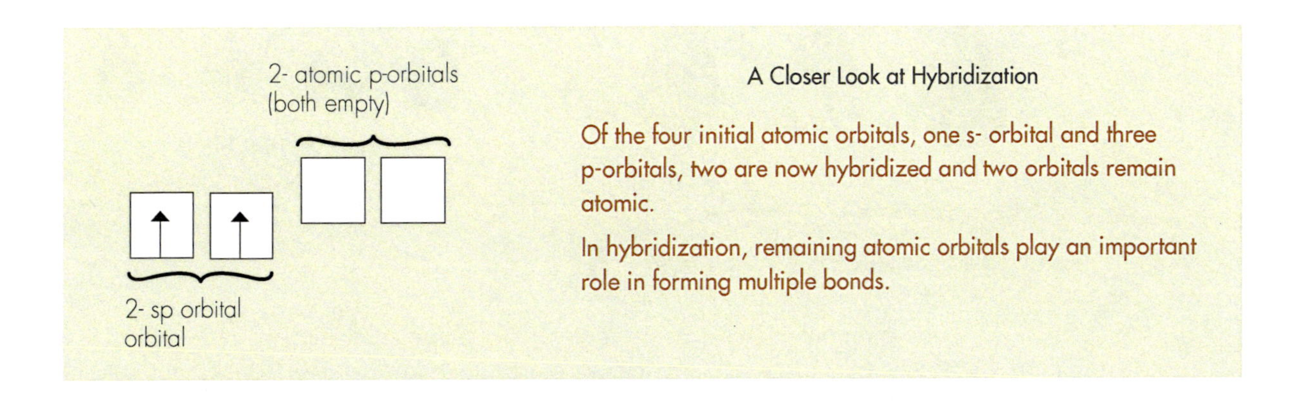

A Closer Look at Hybridization

Of the four initial atomic orbitals, one s- orbital and three p-orbitals, two are now hybridized and two orbitals remain atomic.

In hybridization, remaining atomic orbitals play an important role in forming multiple bonds.

The two sp-hybrid orbitals of beryllium overlap with an unpaired 2p-atomic orbital of each chlorine atom.

The available unpaired electron in each chlorine atom arises from the ground state valence configuration:

$$2s^2 \quad 2p^5$$

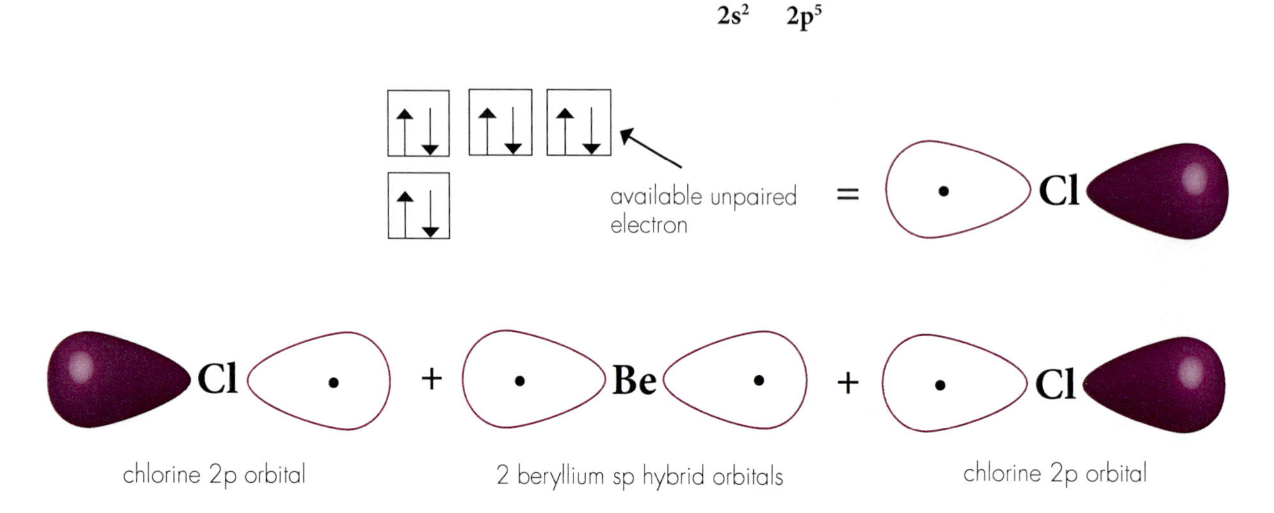

Note that only the bonding orbitals of beryllium are hybridized.

The bonding orbitals of the chlorines remain atomic.

Summary

To accommodate the tetrahedral geometry of methane, four sp3 hybridized orbitals were formed by mixing **one s-orbital and three p-orbitals** in carbon's valence shell:

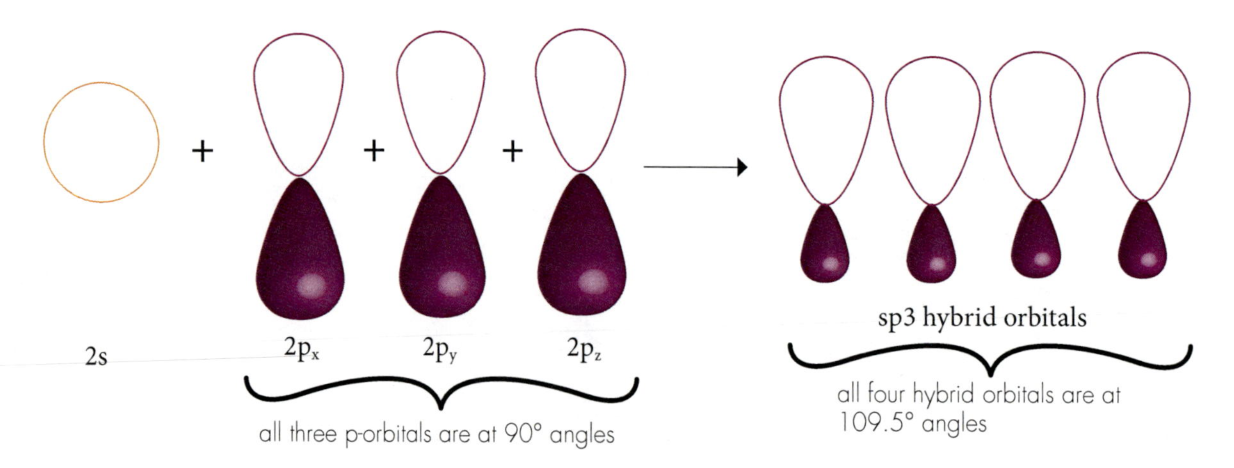

To accommodate the linear geometry of beryllium chloride, Two sp hybridized orbitals were formed by mixing **one s-orbital and one p-orbital** In the beryllium's valence shell:

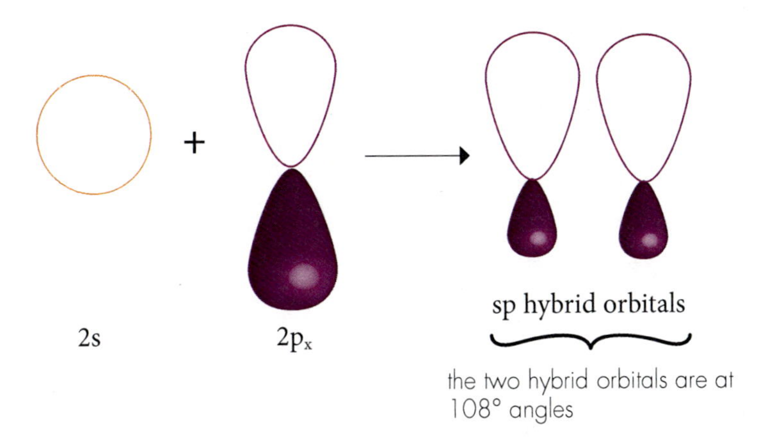

Notice that in forming four sp3 hybrid orbitals four atomic orbitals were mixed.

And in forming two sp hybrid orbitals, two atomic orbitals were mixed.

Rule of Thumb:

The number of atomic orbitals mixed equals the number of resultant hybridized orbitals.

Assignment 5.8

Using the method for determining the degree of hybridization for the carbon orbitals in methane and the beryllium orbitals in beryllium chloride, work the following exercise.

Boron trifluoride has a planar triangular molecular geometry:

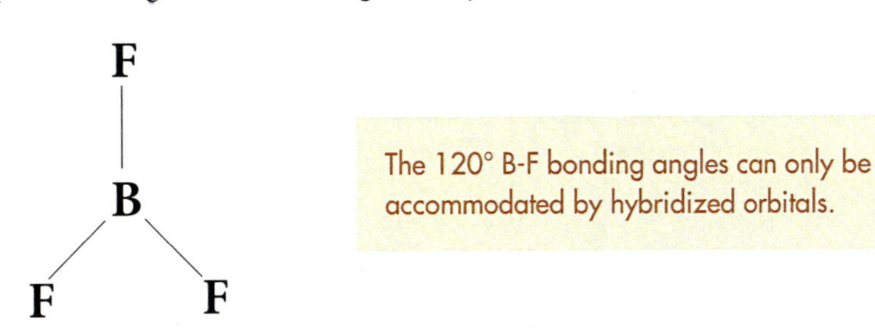

The 120° B-F bonding angles can only be accommodated by hybridized orbitals.

1. How many hybrid orbitals are needed?

   Answer ____________

2. Fill in the orbital-box diagrams below.

promotion →

hybridization →

What would this degree of hybridization be called? (e.g. sp, sp^3...)

Answer ____________

Hybrid Poem

Now don't despair,
but lionize
that electrons work
to emit some light or oxidize.
And notice the bends
hybrid orbitals take
in apology for angles
atomic ones can't make.
If you think this stuff
is beyond your realm,
don't jump overboard,
man the helm.
Orbitals mix!
Now realize
it's more than corn
that can hybridize!

Jerry Mundell

Geometries that Require Five and Six Hybrid Orbitals: Hybridization of s, p, and d Orbitals

Mixing of the s-orbitals and p-orbitals has yielded the proper number of hybrid orbitals to accommodate linear (2 - sp hybrid orbitals), planar triangular (3 - sp^2 hybrid orbitals), and tetrahedral (4 - sp^3 hybrid orbitals electron pair geometries.

In dealing with central atoms which occur after the second period on the Periodic Table, trigonal bipyramidal (e.g. PF_5) and octahedral geometries (e.g. SF_6) can occur.

Requiring five and six hybrid orbitals respectively, the ground states of these central atoms have empty valence d-orbitals which are available to mix with the valence s-orbitals and p-orbitals.

The Lewis structure of PF_5 requires five hybrid orbitals to yield the one axial F-P-F angle of 180°, and the three equatorial F-P-F angles of 120°.

As with the lower degrees of hybridization, (i.e. sp, sp^2, and sp^3), the initial step of promotion is the same except that now it employs an available d-orbital.

The ground state electron configuration of phosphorous is:

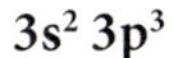

$$3s^2\ 3p^3$$

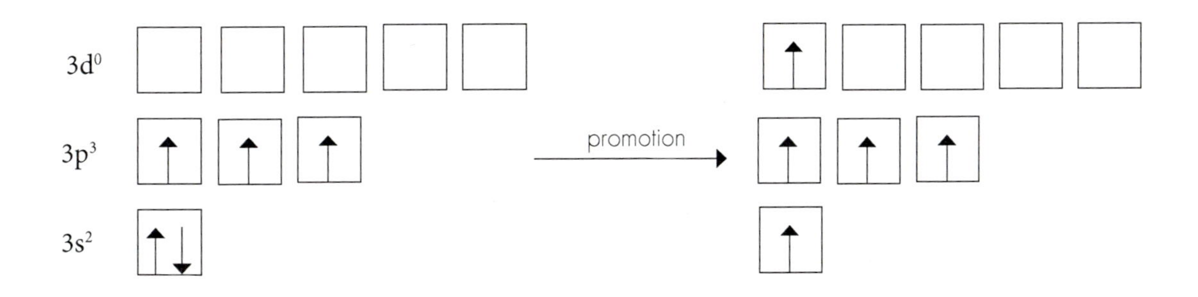

The mixing of the five atomic orbitals (one s-orbital, the three p-orbitals, and the one d-orbital) produces five sp^3d hybrid orbitals, each containing one unpaired electron. Note that after hybridization four empty atomic d-orbitals remain that will not participate in the trigonal bipyramidal bonding.

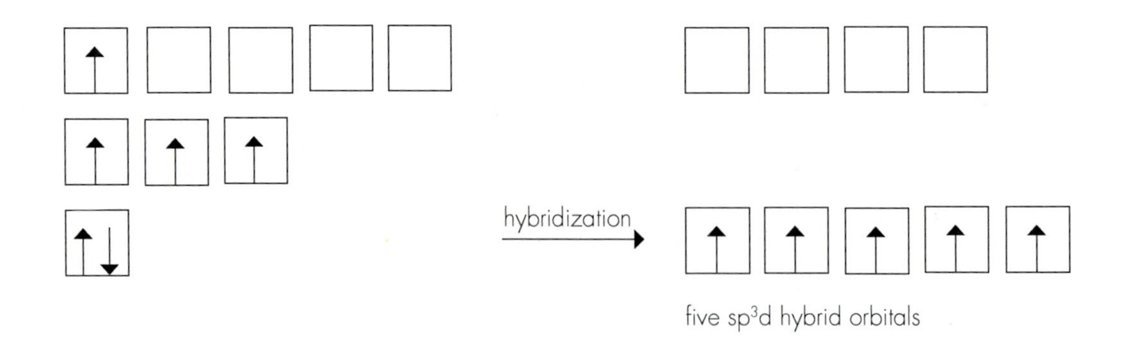

Problem 5.2

Draw the Lewis structure of sulfur hexafluoride, SF_6. Determine the electron pair geometry and molecular geometries of SF_6. Starting with the ground state electron configuration of sulfur, use orbital-box diagrams to determine the degree of hybridization and the number of hybridizied orbitals needed for this structure.

Hybridization and Multiple Bonds

The section, Single Bonds and Multiple Bonds: Sigma and Pi Bonds introduced the structural significance of sigma and pi bonding. What is normally referred to as a single bond is technically a sigma bond in which the bond is located between the covalently bonded atoms and also within the same plane. As previously discussed, the double bond of ethene, C_2H_4, exists between the two carbons. The sigma framework of ethene, below, indicates all sigma bonding in this compound,

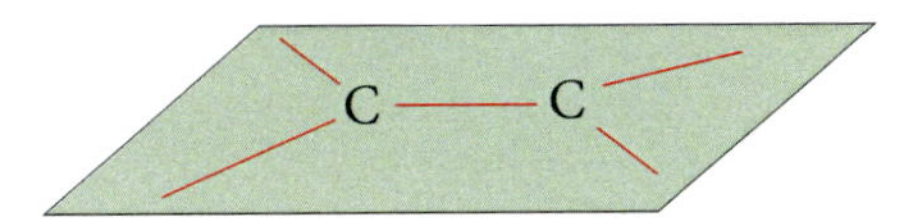

and the single pi bond, which results from the side to side overlap of two coplanar p-orbitals, exists as two distinct regions of bonding both above and below the plane of the carbon – carbon sigma bond.

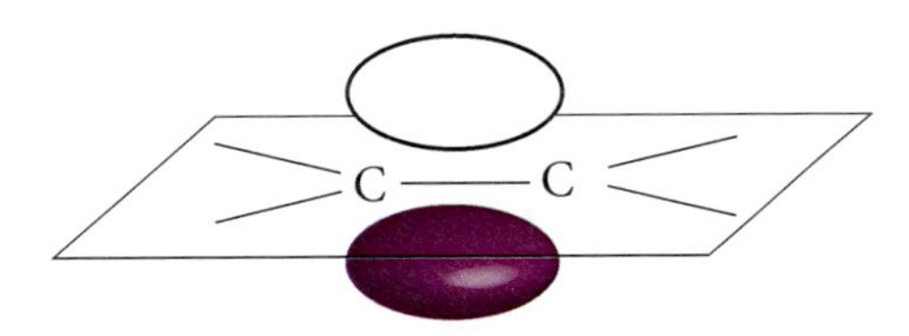

For a more thorough understanding of ethene's sigma and pi bonding, consider the degree of hybridization needed for each carbon.

When dealing with structures containing multiple bonds, the following rule is helpful:

> **When determining the number of hybridized orbitals needed, count the number of sigma bonds in the structure on the atom of interest.**

In the case of ethene, there are two atoms of interest, carbon 1 and carbon 2:

C═C — Lewis structure of ethene

C—C — sigma framework of ethene

It is apparent that each carbon requires three sigma bonds and bonding angles of 120°.

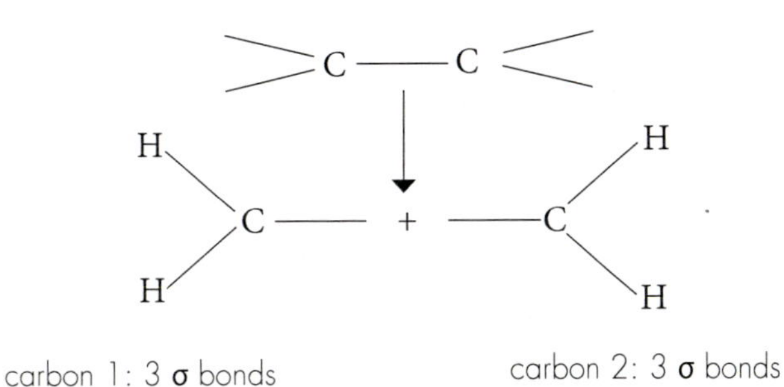

carbon 1: 3 σ bonds

carbon 2: 3 σ bonds

Since the bonding descriptions of both carbons are identical, we will only need to determine the hybridization once. As in the hybridization methane, CH_4, the ground state configuration of carbon goes through the promotion of one of the paired 2s electrons to the empty 2p orbitals.

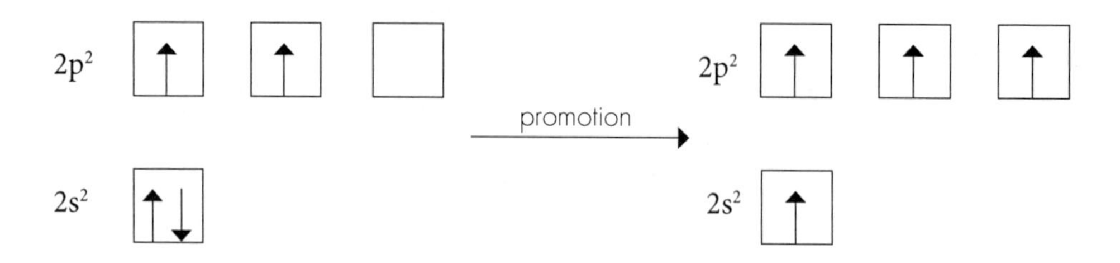

However, unlike methane, which requires four hybridized orbitals, each carbon on ethene requires only three hybrid orbitals. In this regard, the following hybridization would occur:

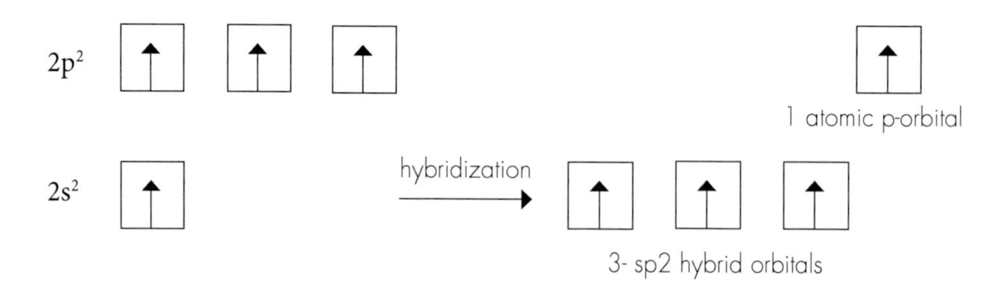

The resultant three sp2 hybrid orbitals allow the formation of the three sigma bonds on both carbons, and the one atomic p-orbital on each carbon allows the formation of the single pi bond.

In the case of the triple bond in acetylene, C_2H_2, each carbon requires only two hybrid orbitals for the two sigma bonds:

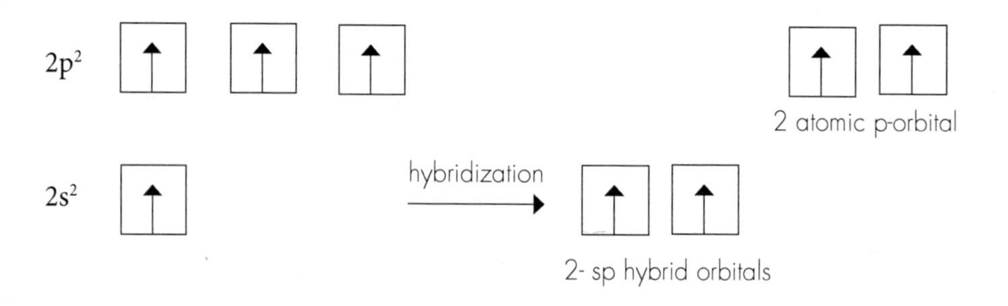

With this configuration, the two remaining atomic p-orbitals allow for the two pi bonds of the triple bond.

Types of Chemical Reactions

Aqueous Reactions and Solution Stoichiometry

One of the most important properties of water is its **ability to dissolve a wide variety** of substances.

Aqueous Solutions:

Solutions in which water is the dissolving medium

$$CaCO_{3\,(s)} + H_2O_{(l)} + CO_{2\,(aq)} \longrightarrow Ca(HCO_3)_{2\,(aq)}$$

In the following pages, we will look at the properties of solutes in aqueous solution. The solutes of interest will be:

Ionic Compounds in Water

Molecular Compounds in Water

Strong and Weak Electrolytes

Solution Composition

A solution is a homogeneous mixture of two or more substances.

The solvent is the substance usually present in greater quantity.

The solutes are the other substances which are dissolved in the solvent.

Molecular Compounds in Water

When dissolved in water, molecular compounds maintain their structural integrity.

$$\underset{\text{pure ethyl alcohol}}{CH_3CH_2OH_{(l)}} \xrightarrow{H_2O} \underset{\text{aqueous ethyl alcohol}}{CH_3CH_2OH_{(aq)}}$$

i.e. Solutions contain molecules dispersed throughout. Some molecules interact so strongly with water that they are pulled apart and form ions.

Hydrogen chloride, HCl, ionizes in water to form hydrochloric acid.

$$HCl_{(g)} \longrightarrow H^+_{(aq)} + Cl^-_{(aq)}$$

Ionic Compounds in Water

ter is a very effective solvent for ionic compounds Ion-dipole interactions to occur. Ions in aqueous solutions are surrounded by water molecules.

Strong and Weak Electrolytes:

The term electrolyte refers to the ions in solutions. The origin of the term comes from the ability of these ionic solutions to conduct electricity.

Strong Electrolytes:

Ionic and some molecular compounds that completely dissociate in water.

Weak Electrolytes:

Molecular compounds that only partially ionize in water, and because of this, produce a small concentration of ions in water.

Double Replacement Reactions
(Metathesis Reactions)

In **double replacement reactions** positive ions and negative ions exchange partners:

$$AX + BY \longrightarrow AY + BX$$

Precipitation Reactions:

A double replacement reaction that results in the formation of an insoluble product.

For a precipitation reaction to occur, a product must be formed which will cause a net change in the concentration of ions in the solution. That is, ions are removed from the solution.

The formation of an insoluble product.

The formation of either a weak electrolyte or non electrolyte.

The formation of a gas that escapes from the solution.

$$2KI_{(aq)} + Pb(NO_3)_{2\,(aq)} \longrightarrow PbI_{2\,(s)} + 2\,KNO_{3\,(aq)}$$

Note: *The solubility of a substance is the amount of substance that can be dissolved in a given quantity of water at 25 °C. Any substance with a solubility of less than 0.01 mol/L will be considered insoluble.*

Further Considerations of Ionic Compounds
Ionic Compounds: Solid and Aqueous States

Unlike covalent compounds, such as:

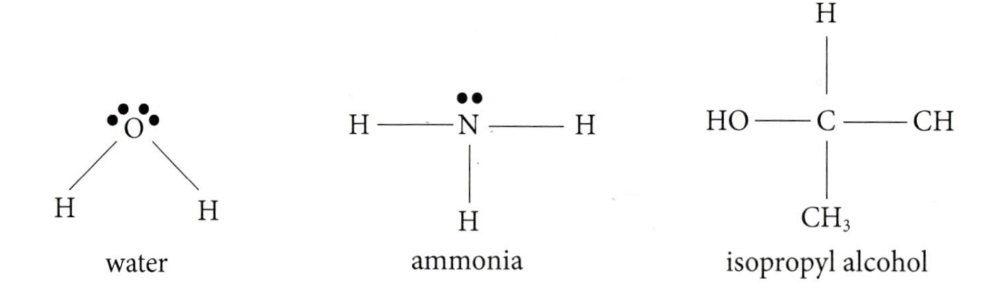

that exist as discreet molecules, ionic compounds, in the solid state, exist in extended **crystal lattices**. In these structures the positive charged cations and the negative charged anions occur in alternating patterns.

Using sodium chloride as an example, it should be apparent that the chemical formula, NaCl does not represent a discrete molecule, but instead the overall ratio of sodium ions to chloride ions in the structure.

The two dimensional array below represents the alternating sequence of sodium ions and chloride ions.

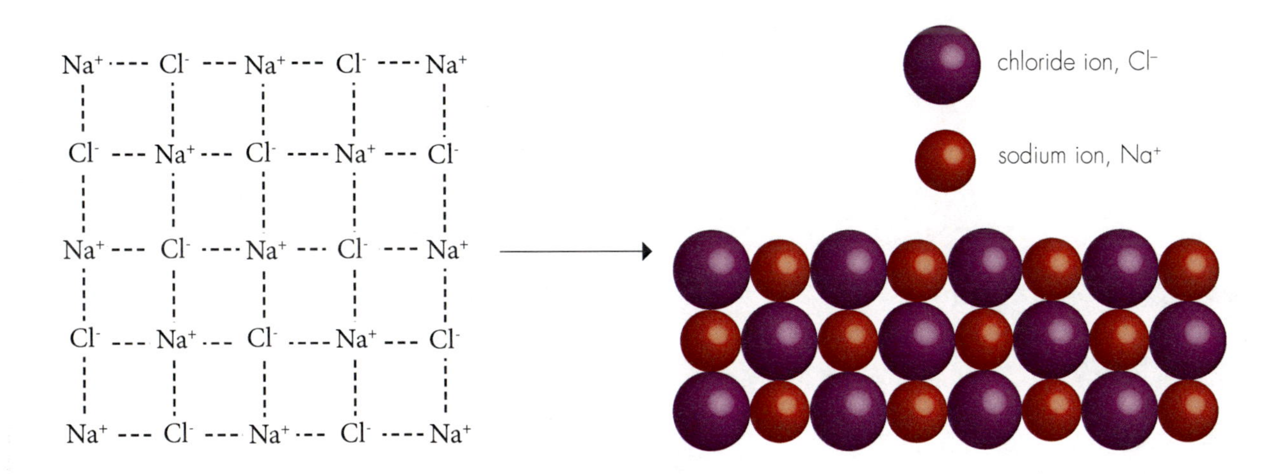

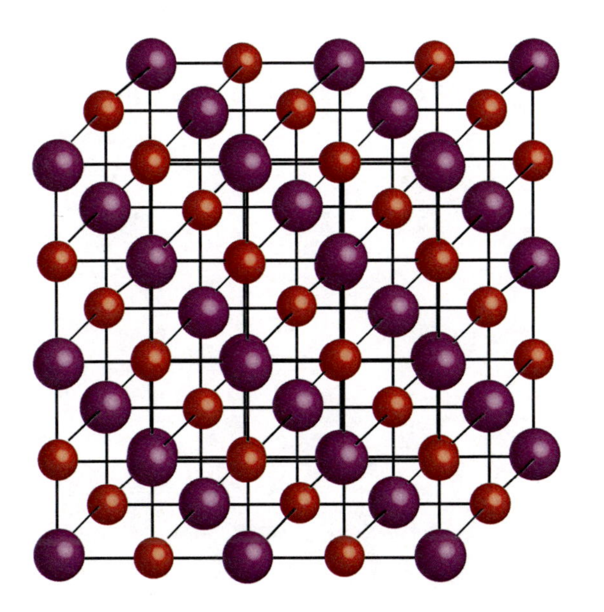

To the left, in a ball and stick representation, is the three dimensional array of sodium chloride. (In the ball and stick model, the relative distances between the ions are exaggerated.)

Unlike covalent bonds, the ions maintain their positions through electrostatic interactions known as ionic bonds.

This crystal lattice bonding arrangement makes for a very stable environment for the ions.

In the sodium chloride structure, each sodium ion is surrounded by six chloride ions, and each chloride ion is surrounded by six sodium ions.

This bonding arrangement is so strong that NaCl melts at 801° C.

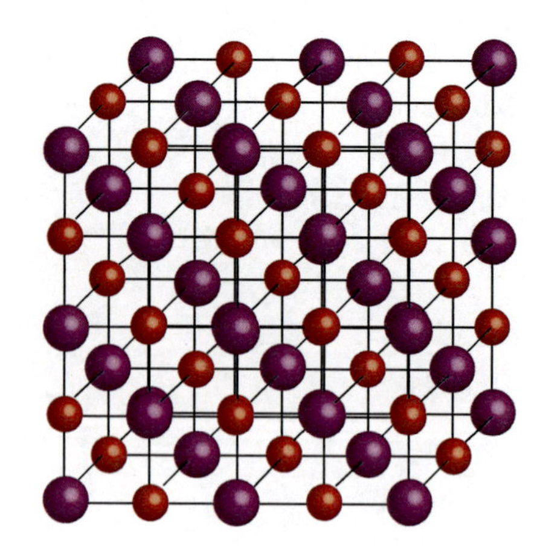

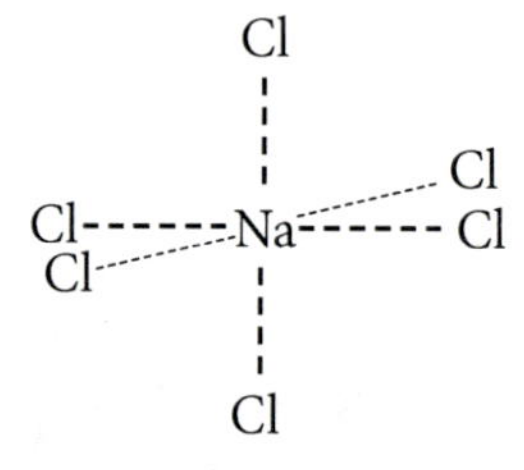

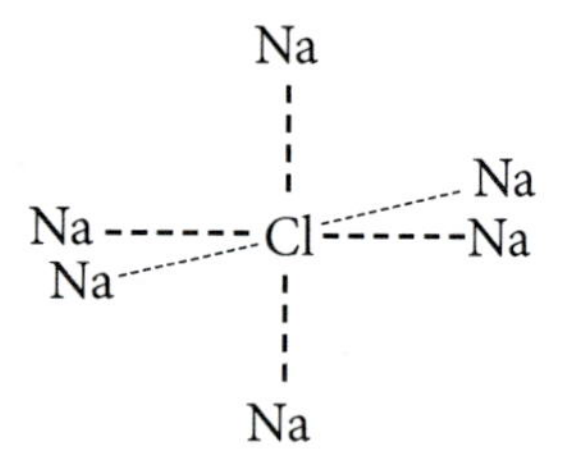

Solubility of Ionic Compounds

Within the sodium chloride crystal lattice, it takes a great deal of heat to overcome the electrostatic interactions which allow the phase change from solid to liquid, yet the same solid is readily soluble in water. How does this happen?

To answer this question, we need to take a look at the interactions, (the intermolecular forces), that exist between water molecules and the sodium ions and the chloride ions.

Since oxygen is much more electronegative than hydrogen, the oxygen region of the molecule is electron rich (δ-) and the hydrogen region is electron deficient (δ+).

Because of the two regions, a molecule of water can be thought of as a magnet.

Molecules that have this polar quality, possess permanent dipoles and are called dipolar.

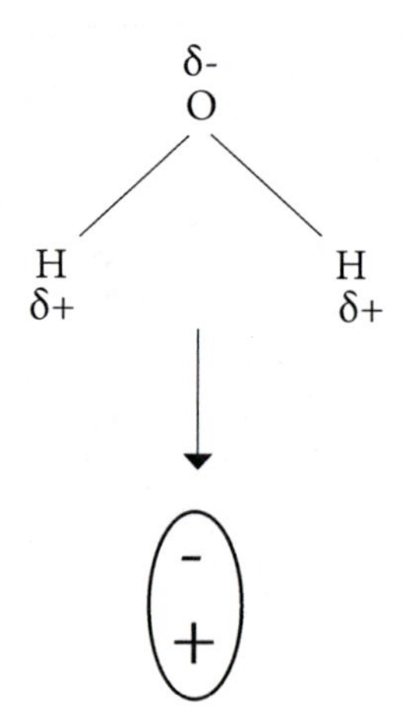

When water comes in contact with solid sodium chloride, the dipolar water molecules that are close to the surface of the ionic compound begin to orient themselves as such:

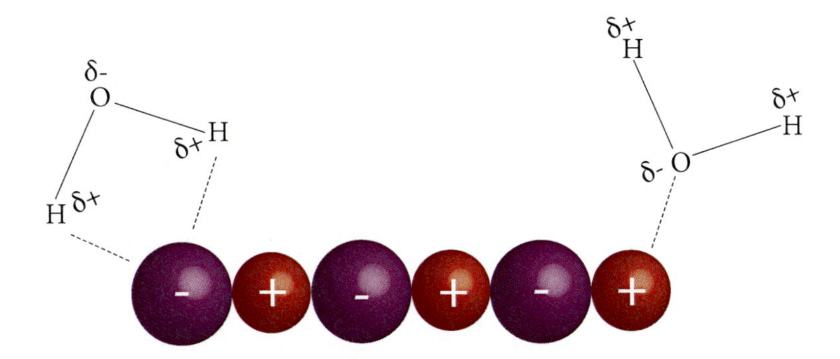

The negative region of the water molecules become oriented towards the positive sodium ions while the positive regions of the water molecules become oriented towards to negative chloride ions. As the attractions (the intermolecular forces) between the dipoles of the water molecules and the ions increase, the ions break away from the crystal lattice and are pulled into solution. Moving away from the lattice, more water molecules surround each ion. This process is called **hydration**.

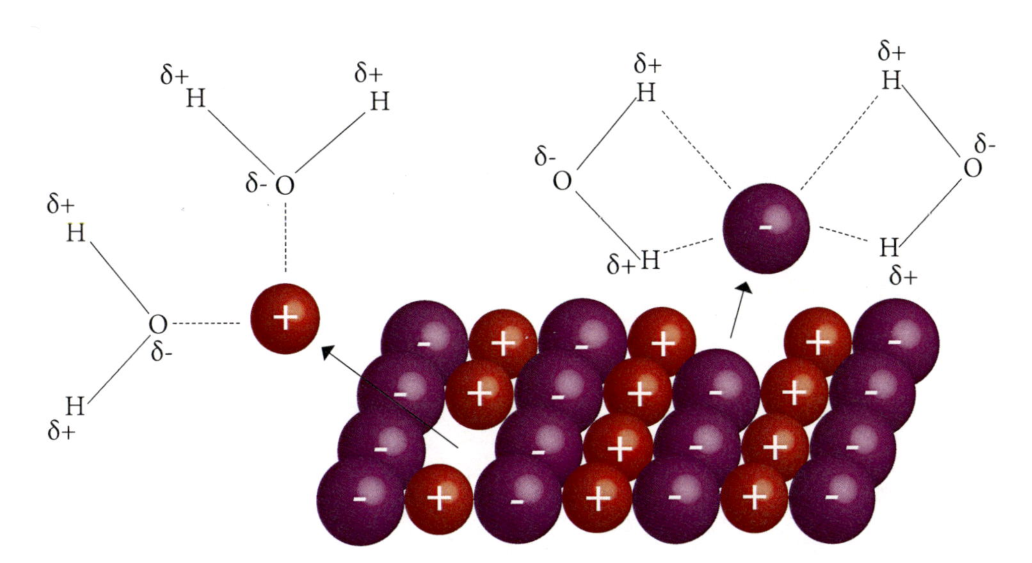

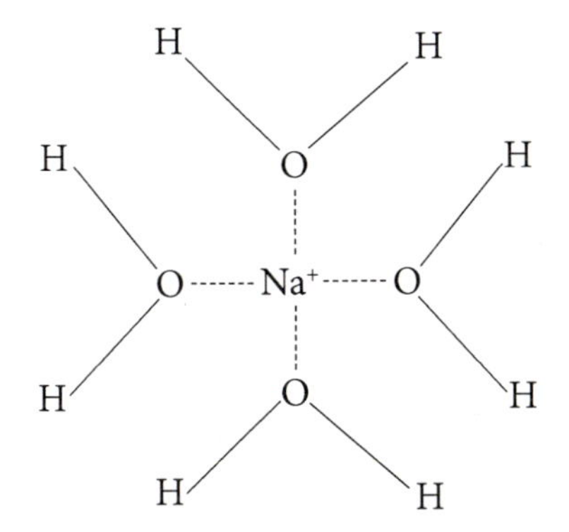

Fully hydrated, the sodium ion has six water molecules associated with it.

This figure to the left,shows how four of the waters would be oriented in the plane of the paper.

Another water directly above the sodium and another directly behind it, would complete the hydration.

With the fully hydrated sodium ions and chloride ions, we now have a picture of the dissolution of this ionic compound. But still, the question hasn't been answered; why would a structure as stable as solid sodium chloride, dissolve so easily in water?

It's a matter of relative stabilities. Fully hydrated, the sodium ions and the chloride ions are more stable than in the crystal lattice. Therefore, salt water solutions abound. By this same argument can we say that all ionic compounds are readily soluble in water? The answer is no.

Depending on the relative sizes and charges of the ions, some ionic compounds are more stable in their crystal lattices than hydrated in water. These ionic compound are insoluble in water. Examples of insoluble ionic compounds are:

$AgCl$	$BaSO_4$	PbI_2
silver chloride	barium sulfate	lead (II) iodide

The following tables indicate which combinations of ions are soluble in water and which are not soluble in water.

Soluble Ionic Compounds

All compounds containing
NO_3^-, Li^+, Na^+, K^+, and NH_4^+

Most compounds containing
Cl^-, Br^-, and I^-

major exceptions are combinations with
Ag^+, Pb 2+, and Hg^+

Most compounds containing
SO_4^{2-}

major exceptions are combinations with
Ca^{2+}, Sr^{2+}, Ba^{2+}, Ag^+, and Pb^{2+}

Insoluble Ionic Compounds

Most compounds containing
OH^-

major exceptions are combinations with
Ba^{2+}, Li^+, Na^+, K^+, and NH_4^+

Most compounds containing
CO_3^{2-}, PO_4^{3-}, SO_3^{2-}, and S^{2-}

major exceptions are combinations with
Li^+, Na^+, K^+, and NH_4^+

Assignment 6.1 Using the two preceding Solubility Tables, Which ionic compounds are water soluble and are not.

K_2CO_3	soluble	insoluble
$Ba_3(PO_4)_2$	soluble	insoluble
NH_4OH	soluble	insoluble
$CaSO_3$	soluble	insoluble
$SrSO_4$	soluble	insoluble
$AgBr$	soluble	insoluble
KBr	soluble	insoluble
NH_4NO_3	soluble	insoluble

Ionic Equations

Equations describing the solutions reactions of ionic compounds can be written one of three ways:

molecular equations:

Express reactions as if all the reactants and products are undissociated, i.e. they appear as molecules rather than dissolved ionic species.

complete ionic equations:

Indicate explicitly whether the dissolved substances are present as ions or molecules.

net ionic equations:

The net ionic equation only includes ions and molecules directly involved in formation of the product.

The ions that do not participate in product formation, and appear on both sides of the equation, are called spectator ions.

Refer to the sodium and chloride ions in the above Complete Ionic Equation. They are the spectator ions in that reaction.

Molecular equation

$$HCl_{(aq)} + NaOH_{(aq)} \longrightarrow H_2O_{(l)} + NaCl_{(aq)}$$

Complete ionic equation

$$H+_{(aq)} + Cl -_{(aq)} + Na^+_{(aq)} + OH^-_{(aq)} \longrightarrow H_2O_{(l)} + Na^+_{(aq)} + Cl^-_{(aq)}$$

Net ionic equation

$$H^+_{(aq)} + OH^-_{(aq)} \longrightarrow H_2O_{(l)}$$

Solubility and Net Ionic Equations — Assignment 6.2

Mixing a solution of iron (III) chloride with a solution of sodium hydroxide produces a mixture of iron (III) hydroxide and sodium chloride.

a) Write the balanced equation for this reaction.

b) Ionic compounds may or may not be soluble in water. Use a solubility table to determine which of the ionic compounds in this reaction are soluble and which are insoluble.

Soluble Ionic Compounds

Insoluble Ionic Compounds

c) Using subscripts (s) and (aq) indicate in the balanced equation whether the ionic compounds are soluble or insoluble.

d) Rewrite the above balanced equation in the form of a complete ionic equation.

Complete Ionic Equation

e) In the above complete ionic equation pick out the ions that aren't participating in product formation. That is, the ions which are unchanged on both sides of the equation. These ions are called spectator ions.

Spectator Ions

f) Rewrite the complete ionic equation without the spectator ions. This type of chemical equation is called the net ionic equation. It indicates only those species directly involved in product formation.

Net Ionic Equation

____

____

Part 2 of Solubility and Net Ionic Equations

Assignment 6.3

A solution of silver(I) nitrate is mixed with a solution of potassium iodide.

a. What are the products? ____________ ______________

b. Write the balanced equation for this reaction.

c. Identify the spectator ions in this reaction.

d. Write the net ionic equation for this reaction.

Acids, Bases and Salts

Acids:

Substances that ionize to form hydrogen ions in aqueous solution.

$$HCl \longrightarrow H^{+}_{(aq)} + Cl^{-}_{(aq)}$$

Bases:

Substances that increase OH^{-} ion concentration in aqueous solution.

Bases react with H^{+}

$$OH^{-}_{(aq)} + H^{+}_{(aq)} \longrightarrow H_2O_{(l)}$$

Strong and Weak Acids and Bases

Strong Acids and Strong Bases:

Strong electrolytes
Completely ionize in aqueous solution

Strong Acid:

$$HNO_{3\,(aq)} \longrightarrow H^{+}_{(aq)} + NO_3^{-}{}_{(aq)}$$

Strong Base:

$$KOH_{(aq)} \longrightarrow K^{+}_{(aq)} + OH^{-}_{(aq)}$$

Weak Acids and Bases:

Those that are weak electrolytes (i.e. partly ionized)

Weak Acid:

$$CH_3CO_2H_{(aq)} \longrightarrow H^{+}_{(aq)} + CH_3CO_2^{-}$$

Neutralization Reactions and Salts:

The reaction which occurs when equal amounts of acid and base are mixed together.

$$HCl_{(aq)} + NaOH_{(aq)} \longrightarrow H_2O_{(l)} + NaCl_{(aq)}$$

(acid) (base) (water) (salt)

Reactions in Which a Gas Forms:

Sometimes the product of a metathesis reaction is a gas that has a low solubility in water

Metal sulfide and strong acid:

$$2HCl_{(aq)} + Na_2S_{(aq)} \longrightarrow H_2S_{(g)} + 2NaCl\ (aq)$$

Carbonates/Bicarbonates and acids:

$$HCl_{(aq)} + NaHCO_{3\,(aq)} \longrightarrow NaCl_{(aq)} + H_2CO_{3\,(aq)}$$

$$H_2CO_{3\,(aq)} \longrightarrow H_2O_{(l)} + CO_{2\,(g)}$$

Introduction to Oxidation-Reduction Reactions

In some chemical reactions, electrons "shift" around, that is, they may actually transfer from one atom or ion to another. Other times, within covalent bonds, they may just be "pulled" towards another atom, increasing the positive charge on the more electronegative atom. These reactions are called oxidation-reduction reactions or redox reactions.

In order to recognize a redox reaction, a chemist has to be able to observe the transfer of electrons in the reaction. This can accomplished by assigning oxidation numbers to the atoms on both the reactant and product sides of the reaction. In reactions involving ions, the oxidation numbers are the ionic charges.

For example, when a zinc rod is placed into a solution of copper sulfate, electrons are transferred from the metallic zinc to the copper ions. As a result, soluble zinc ions are formed as metallic copper precipitates out of solution:

$$Zn_{(s)} + Cu^{2+}{}_{(aq)} \longrightarrow Zn^{2+}{}_{(aq)} + Cu_{(s)}$$

In this reaction, electron transfers are readily observed by dividing the reaction into two half-reactions:

1. metallic zinc, with a charge (i.e. oxidation number) of zero, loses two electrons to form the zinc (II) ion:

   $$Zn_{(s)} \longrightarrow Zn^{2+}{}_{(aq)}$$

   The process in which a chemical species loses electrons is called oxidation.

   Thus, in this reaction, zinc is **oxidized**.

2. copper (II) ion gains two electrons from metallic zinc to form metallic copper:

   $$Cu^{2+}{}_{(aq)} \longrightarrow Cu_{(s)}$$

   The process in which a chemical species gains electrons is called reduction.

   Thus, copper (II) ion is **reduced**.

In Summation

Oxidation:

When a substance loses electrons i.e. An atom, molecule or ion becomes more positively charged.

In a redox reaction, the species that undergoes oxidation is also called the reducing agent.

Reduction:

When a substance gains electrons i.e. When a atom, molecule, or ion becomes more negatively charged.

In a redox reaction, the species that undergoes reduction is also called the oxidizng agent.

Oxidation Numbers

Determining the oxidation numbers of both the reactants and products is most important in observing the redox behavior of the chemical species involved in the reaction. Two different approaches to oxidation numbers will be discussed which will depend on whether the species is an ionic compound or a covalent compound:

Oxidation Number Rules

1. The oxidation number of a species in its elemental form is zero.
2. The oxidation number of a monatomic ion is the same as its charge.
3. In binary compounds the element with the greater electronegativity is assigned a negative oxidation number equal to the charge it would have if it were a simple ion.

The sum of the oxidation numbers equals zero for an electrically neutral compound and equals the overall charge for an ionic species.

Binary Ionic Compounds:

Oxidation numbers will be the same as the charge on the ionic species. When the ionic charge becomes more positive, the species is undergoing oxidation. Development of a more negative charge indicates a reduction process.

Covalent Compounds:

Since molecules do not carry a net charge, assigning oxidation numbers becomes more difficult. The following rules show how this is done. Keep in mind that in covalent compounds, electrons are not actually transferred, but rather "shifted" either towards or away from the parent atom, resulting in an "assigned" charge or oxidation number.

Assign oxidation numbers in the following reactions:

$$2Fe_{(s)} + 3Cl_{2\,(g)} \longrightarrow FeCl_{3\,(s)}$$

$$CH_{4\,(g)} + O_{2\,(g)} \longrightarrow H_2O_{(g)} + CO_{2\,(g)}$$

Oxidation of Metals by Acids and Salts:

Many metals react with acids to form salts and hydrogen gas.

$$Mg_{(s)} + 2HCl_{(aq)} \longrightarrow MgCl_{2\,(aq)} + H_{2\,(g)}$$

(Table 5.1 continued)

Single Replacement Reactions:

when one element displaces another element from a compound.

$$A + BC \longrightarrow AC + B$$

$$CuSO_{4\,(aq)} + Zn_{\,(s)} \longrightarrow Cu_{\,(s)} + ZnSO_{4\,(aq)}$$

$$Sn_{\,(s)} + 2AgNO_{3\,(aq)} \longrightarrow Sn(NO_3)_{2\,(aq)} + 2Ag_{\,(s)}$$

Assignment 6.4

Part 1 Redox Involving Ionic Species

1. Consider the following reaction:

$$2\ Ag^{+}_{\,(aq)} + Cu_{\,(s)} \longrightarrow 2\ Ag_{\,(s)} + Cu^{2+}_{\,(aq)}$$

The first rule of redox chemistry is to refer to charges as "**oxidation numbers**"

a. How do the charges (i.e. oxidation numbers) of the species change?

1) Ag:

2) Cu:

b. Which species loses electrons? __________ (i.e. oxidized)

How many? ____________________

c. Which species gains electrons? _________ (i.e. reduced)

How many ? ______________________

2. Consider the following reaction:

$$2\ Mg_{\,(s)} + O_{2\,(g)} \longrightarrow 2\ MgO_{\,(s)}$$

a. Which species is oxidized and by how many electrons ?

____

b. Which species is reduced and by how many electrons?

____

Part 2 Redox Involving Covalent Species

Assignment 6.5

Redox reactions do not always involve the formation of ionic species. In dealing with covalent compounds, electron density shared between two atoms is merely shifted from one atom to another (the more electronegative one).

1. Assign oxidation numbers to the following atoms:

   (Note the more electronegative atom is assigned its ionic charge number.)

   a. CO C___ O___

   b. CO_2 C___ O___

   c. SO_2 S___ O___

   d. SO_3 S___ O___

   e. SO_4^{2-} S___ O___

2. Consider the following reaction:

$$CH_{4\,(g)} + 2\,O_{2\,(g)} \longrightarrow CO_{2\,(g)} + 2\,H_2O_{\,(l)}$$

   a. What are the oxidation numbers for each of the atoms?

   CH_4 C____ H____

   O_2 O____

   CO_2 C____ O____

   H_2O H____ O____

   b. In this reaction, which species are oxidized and which are reduced?

Additional Reactions

Reactions in Which a Weak Electrolyte or Nonelectrolyte Forms

Primary example is formation of water in neutralization reactions and reactions of metal oxides and acids.

Neutralization reaction:

$$Mg(OH)_{2\,(s)} + 2HCl_{\,(aq)} \longrightarrow MgCl_{2\,(aq)} + 2H_2O_{\,(l)}$$

Metal oxide and acid:

$NiO_{(s)} + 2HNO_{3\,(aq)} \longrightarrow Ni(NO_3)_{2\,(aq)} + H_2O_{(l)}$

Other Types of Chemical Reactions

Combination and Decomposition Reactions

a) Combination Reactions:

two or more substances react to form one product

$$2\ Mg_{(s)} + O_{2\,(g)} \longrightarrow MgO_{(s)}$$

b) Decomposition Reactions:

one substance undergoes a reaction to produce two or more substances

$$NaN_{3\,(s)} \longrightarrow 2\ Na_{(s)} + 3\ N_{2\,(g)}$$

Decomposition Reactions

$$2\ HgO_{(s)} \longrightarrow 2\ Hg_{(l)} + O_{2\,(g)}$$

$$2\ PbO_{2\,(s)} \longrightarrow 2\ PbO_{(s)} + O_{2\,(g)}$$

$$2\ KClO_{3\,(s)} \longrightarrow 2\ KCl_{(s)} + 3O2_{(g)}$$

Assignment 6.5

Identify the following types of reactions. Note that some reactions maybe more than one type.

1. $2\ KClO_{3\,(s)} \longrightarrow 2\ KCl_{(s)} + 3\ O_{2\,(g)}$

Answer

____

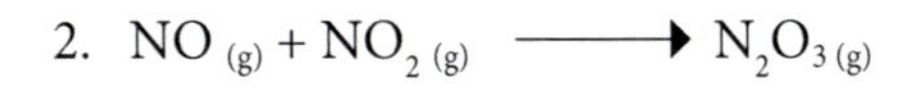

2. $NO_{(g)} + NO_{2\,(g)} \longrightarrow N_2O_{3\,(g)}$

Answer

____

3. $Ba(OH)_{2\,(aq)} + H_2SO_{4\,(aq)} \longrightarrow BaSO_{4\,(s)} + 2\ H_2O_{\,(l)}$

Answer

---

4. $2\ PbS_{\,(s)} + 3\ O_{2\,(g)} \longrightarrow 2\ PbO_{\,(s)} + 2\ SO_{2\,(g)}$

Answer

---

5. $2\ KF_{\,(aq)} + Sr(NO_3)_{2\,(aq)} \longrightarrow SrF_{2\,(s)} + 2\ KNO_{3\,(aq)}$

Answer

---

6. $CO_{\,(g)} + 2\ H_{2\,(g)} \longrightarrow CH_3OH_{\,(l)}$

Answer

---

7. $AgNO_{3\,(aq)} + KBr_{\,(aq)} \longrightarrow AgBr_{\,(s)} + KNO_{3\,(aq)}$

Answer

---

8. $Si_{\,(s)} + 2\ Cl_{2\,(g)} \longrightarrow SiCl_{4\,(g)}$

Answer

---

9. $2\ HCl_{(aq)} + K_2CO_{3\,(aq)} \longrightarrow 2\ KCl_{(aq)} + H_2O_{(l)} + CO_{2\,(g)}$

Answer

____

10. $2\ F_{2\,(g)} + O_{2\,(g)} \longrightarrow 2\ OF_{2\,(g)}$

Answer

____

11. $CaCl_{2\,(aq)} + 2\ NaF_{(aq)} \longrightarrow CaF_{2\,(s)} + 2\ NaCl_{(aq)}$

Answer

____

12. $Cu_{(s)} + HNO_{3\,(aq)} \longrightarrow Cu(NO_3)_{2\,(aq)} + NO_{2\,(g)} + H_2O_{(l)}$

Answer

____

p2

The Mathematics of Chemistry

Chapter 7

By now you should be able to read a stated chemical observation, translate it into a balanced chemical equation, and understand the nature of the components, both reactants and products involved in the chemical reaction. All of this has been addressed at the particulate level. That is, for example, how one molecule of methane combines with two molecules of oxygen to produce one molecule of carbon dioxide and two molecules of water.

In Part 2: Applications, we take the concepts from the first part of this book and learn how to use this information in a chemical laboratory. Remember that, in essence, chemistry is a practical science, and to better understand it and properly use it, we must be able to apply. Chapter 7 deals with our much needed mathematical tools. In Chapter 8 you will learn the use of these tools in solving chemical problems.

Math Remediation It is normal for students entering an introductory chemistry course to be concerned about the needed math skills required for the course's problem solving. Listen up, because for a majority of a first semester general chemistry course, most calculations will only require knowledge of four basic mathematical operations: addition, subtraction, multiplication, and division. The challenge of how to use these operations in problem solving is the goal of Chapters 7 and 8.

Scientific Calculators To begin this part of the course, a scientific calculator is needed. A good, reliable scientific calculator can be purchased for under $20. There are many different brand-named calculators available, each with slightly different keypads. If possible, students and instructor should use the same brand and model calculator to simplify classroom instruction. Rather than addressing the operation of calculators in a separate section, specific calculator use will be incorporated into discussions where needed.

7.1 Percentages, Parts per Million, and Parts per Billion

In our daily lives, we read and hear about 85 percent of voters supporting a certain bill, or drinking water containing a certain parts per million or parts per billion of toxic substances. What do these numbers actually mean and how are they calculated? In this section, we will not only become more familiar with their meaning, we will also see that the calculations involved are very similar. Let's start with percentages.

Percentages

In a newspaper we read that in a certain city, 65 percent, 65%, of the residents support the building of a factory alongside the city's river. Let's say that 1500 people were polled on this issue and 950 of the residents supported this issue. To perform the calculation we first take the number of people who support the issue and divide that number by the total number of people polled.

$$\frac{950 \text{ (people in favor)}}{1500 \text{ (total people polled)}} = 0.65$$

The 0.65 represents a fraction of all people polled who favor the factory construction. Normally, we think of fractions as rational numbers (numbers expressed as ratios) that might be expressed, for example, as ½ (a 1 to 2 ratio), ¼(a 1 to 4 ratio), or ⅛ (a 1 to 8 ratio). These same fractions can also be expressed as 0.5, 0.25, and 0.125, that is, as decimals. Fractions can easily be converted into decimals by dividing the bottom number (the denominator) into the upper number (the numerator).

Example 7.1 Express a Fraction in Decimal Form

Problem

Express the fraction ¾ as a decimal.

Solution

Divide the denominator into the numerator.

$$\frac{3}{4} = 0.75$$

Using a calculator

we obtain

$$0.75$$

Practice Problem

7.1 Express the following fractions as decimals:

⅞ = ________   ⅚ = ________   ⅔ = ________

$^{15}/_{16}$ = ________ $1\,^{1}/_{20}$ = ________ $^{1}/_{1000}$ = ________

Continuing our discussion, we all know the next step of multiplying this fraction by 100 in calculating the percentage.

$$0.65 \times 100 = 65\%$$

But why do we do this? Why not just use the fraction by itself? Why multiply by 100, let alone 10 or 1000? By multiplying by 100, we are saying that out of every 100 people polled, 65 people support the factory. We could easily have multiplied the fraction by 10 or 1000. Note that traditionally we use groupings of 100 in most of our everyday analysis. If we had multiplied by 10, the analysis would state that 6.5 people out of every 10 supported the issue. If we multiplied by 1000, the result would read that for every 1000 people questioned, 650 were in favor. Traditionally, we have settled on 100, which is why it is called a **percent,** that is *parts per 100.* Let's try a few simple examples in calculating percentages.

Percent:
A fraction of the whole expressed as parts per hundred.

Example 7.2 Calculate Fraction and Percent

Problem

A car dealer in Cleveland, Ohio had 350 used cars on his lot last April. That month, he sold 21 cars. a) What fraction of his inventory did he sell that month? b) For every 50 cars on his lot, how many did he sell that month? c) What percentage of his total inventory did he sell?

Solution

a) We are told the dealer sold 21 cars out a total of 350 cars. To determine the fraction of sold inventory, divide the number of cars sold (21) by the total number of used cars (350) in his inventory.

$$\frac{21}{350} = 0.06$$

b) We are asked how many cars were sold for every 50 cars on the lot. Multiply the fraction of sold cars (the answer from a) by 50.

$$0.06 \times 50 = 3$$

c) To determine the percentage of inventory sold, multiply the fraction of cars sold (from a) by 100. Remember, percent is $\frac{\text{part}}{\text{whole}} \times 100$. In part a) we determined the part/whole or the fraction as 0.06. Now we multiply by 100 to calculate the percent

$$0.06 \times 100 = 6\%$$

The dealer sold 6% of his inventory.

Practice Problems

7.2 Convert the following information into fractions and percentages: pencast

a) A penny collection contains 250 pennies. 50 of the pennies are dated 2005.

Fraction of pennies (2005) ________________

Percentage of pennies (2005) ________________

b) Of the 560 cars that traveled down main street one day, 175 were Fords.

Fraction of cars that were Fords ________________

Percentage of cars that were Fords ________________

c) An English class has 40 students in it. Yesterday, 2 students were absent and 10 students came late.

Fraction of students who were late ________________

Percentage of students who were late ________________

Fraction of student who were absent ________________

Percentage of students were absent ________________

7.3 Alex gets home from school at 4 PM, and spends 3 hours online talking to his friends before going to bed at midnight.

a) What fraction of his total time spent at home in the evening, does Alex spend online?

b) Of his total time spent at home, what part of each hour does Alex spend online?

c) What percentage of his home time is online?

Note that the calculated percentages add up to 100% and the fractions add to 1. This is true for all calculations of this sort. Now that you know how to calculate a percentage, let's try a more involved example involving the previous situation of a factory and some concerned citizens.

Example 7.3 Determine Fraction and Percent

Problem

Out of 1500 residents polled about the construction of a factory, 975 people supported the factory being built. Another 375 were against it, and 150 residents were undecided. Determine the breakdown of the poll by calculating the percentage of residents that are supportive, against, and undecided.

Solution

There are a total of 1500 residents that were polled. Out of 1500 residents there were 975 people in support of the factory. First calculate the fraction and then multiply by 100.

$$\frac{975}{1500} = 0.65$$

$0.65 \times 100 = 65\%$ of the residents polled were supportive of the factory

375 residents out of 1500 were against the construction

$\frac{375}{1500} = 0.25$ and $0.25 \times 100 = 25\%$ of the residents against the construction

150 residents out of 1500 were undecided. We can multiply the fraction by 100 to determine the percentage or we can realize that all percentages must add to 100% and all fractions must add to 1. From parts a) and b), we have 65% and 25%.

$$100 - 65\% + 25\% = 10\%$$

To check your answer, make sure all percentages add to 100 and all fractions add to 1.

$$65\% + 25\% + 10\% = 100$$

$$0.65 + 0.25 + 0.10 = 1$$

Practice Problems

7.4 A student takes a test containing 10 questions with the following points per question:

Question number	Points	Question Number	Points
1	10	6	5
2	5	7	10
3	15	8	10
4	10	9	5
5	25	10	5

a) What fraction and percentage of the total number of questions are worth 5 points?

b) If the student answered questions 3, 8, and 10 incorrectly, what fraction and percentage of the total number of points did she receive on the test? video

7.5 If 232 grams of sodium sulfate, Na_2SO_4, contain 46 grams of sodium and 32 grams of sulfur, what fraction and percent of the compound contains oxygen?

Parts per Million (ppm) and Parts per Billion (ppb)

Let's now consider situations where multiplying fractions by 100 isn't adequate because the resultant percentages are much too small. Suppose we are doing an analysis of drinking water and find that a 1000-gram sample of the water contains 0.0002 grams of lead ion.

To use these numbers in a percentage would yield the following results:

$$\frac{0.0002 \text{ g lead}}{1000 \text{ g water}} = 0.0000002 \text{ (fraction of lead)}$$

$$0.0000002 \times 100 = 0.00002\% \text{ lead in water}$$

To better handle such small numbers, instead of multiplying the fraction by 100, we use 1,000,000 (one million) and by doing, the result is no longer called percentage or parts per hundred, but rather *parts per million*, ppm.

By multiplying the small fraction by 1,000,000, a more manageable number is produced:

$$0.0000002 \times 1{,}000{,}000 = 0.2 \text{ ppm lead}$$

When dealing with chemical pollutants and toxic microorganisms, due to their small concentrations, we normally use parts per million, ppm or even *parts per billion*, ppb. For our example, the fraction of lead in the water is 0.0000002. Multiplying the fraction of lead by 1 billion (1,000,000,000) we obtain

$$0.00000002 \times 1{,}000{,}000{,}000 = 200 \text{ ppb lead}$$

Parts per Million, ppm:
A fraction of the whole multiplied by one million.

Parts per Billion, ppb:
A fraction of the whole multiplied by one billion.

Example 7.4 Calculate Parts per Million, ppm, and Parts per Billion, ppb

Problem

Fluoride is added to public water systems to help reduce tooth decay. A 1000-gram sample of water was found to contain 0.001 g of fluoride ion. Calculate a) the percentage, b) the ppm, and c) ppb of fluoride ion in the water sample.

Solution

a) First find the fraction of fluoride in the water.

$$\frac{0.001 \text{ g fluoride}}{1000 \text{ g water}} = 0.000001 \text{ (fraction of fluoride)}$$

To calculate the percentage, multiply the fraction by 100

$$0.000001 \times 100 = 0.0001\% \text{ fluoride}$$

b) To calculate ppm, multiply the fraction found in part a by 1,000,000.

$$0.000001 \times 1{,}000{,}000 = 1 \text{ ppm fluoride.}$$

c) To calculate ppb, multiply the fraction found in part a by 1,000,000,000

$$0.000001 \times 1{,}000{,}000{,}000 = 1000 \text{ ppb fluoride.}$$

Practice Problems

7.6 The drinking water of a city is tested for cadmium ion. In a 5000-g sample of the water 0.0025 g of the ion is detected. What is the a) percentage, b) parts per million (ppm), and c) parts per billion (ppb) of the ion in the water?

7.7 A 500-g soil sample analysis reveals 50 ppm of radioactive strontium. How many grams of strontium ion are in the soil sample? pencast

7.8 A 180.17 g sample of aspirin, $C_9H_8O_4$, contains 108.09 g of carbon, and 8.08 g of hydrogen. What is the %C, %H, and %O in aspirin? video

Summary Percentages, Parts per Million, and Parts per Billion

- To convert a fraction to a decimal, divide the numerator by the denominator.
- To calculate a percent, first find the fraction, $\frac{\text{part}}{\text{whole}}$, and multiply by 100.
- Parts per million (ppm) is calculated by multiplying the fraction, $\frac{\text{part}}{\text{whole}}$, by one million, 10^6.
- Find parts per billion by multiplying the fraction, $\frac{\text{part}}{\text{whole}}$, by one billion, 10^9.

Worksheet 7.1 Fractions, Decimals, Percent, ppm, and ppb (Section 7.1)

1. Fill in the following table.

Fraction	Decimal	Percent	ppm	ppb
½	0.5	5%	500,000	500,000,000
1/25			40000	
		15%		
			222	
	0.98			
		0.24%		
		48%		
	0.006			
			450	
				580

7.2 Exponents and Exponential Notation

In performing calculations in chemistry, sometimes we will be using very large numbers (e.g., 1,000,000,000,000) or very small numbers (e.g., 0.0000000001). In this section, we will learn a simple way to easily handle such numbers, both by inspection and also with a calculator. As we know, the number one hundred can written as the product of ten times ten, 10×10. Ten multiplied by itself can also be expressed as 10^2. We call this ten squared or ten to the second power. The two in the superscript is called the exponent, and 10^2 is called an **exponential term.**

$$\mathbf{10^2}$$

Exponential Term:
A number raised to a power.

By the same token, the number one thousand, which is the product of tens,

$$10 \times 10 \times 10$$

can be expressed as an exponential term

$$10^3$$

As the numbers get larger, the exponents can be determined by counting the zeros to the right of the one. For example 1,000,000 has 6 zeros to the right of the number 1, and the exponential term is 10^6. But suppose the number is a fraction of ten, such as one tenth, $^1/_{10}$, or one thousandth, $^1/_{1000}$? In such cases we count the zeros as before and place a minus sign before the exponent. For example, the exponential term for $^1/_{100}$, is 10^{-2} and for $^1/_{10}$, is 10^{-1}. The fraction one millionth, $^1/_{1,000,000}$, has six zeros following the one, which yields a 6 for the exponent which gives us the exponential term, 10^{-6}. To get a better feel for expressing whole numbers and fractions in exponential terms, consider **Figure 7.1.**

Figure 7.1

The number scale indicates whole numbers and fractions in exponential terms.

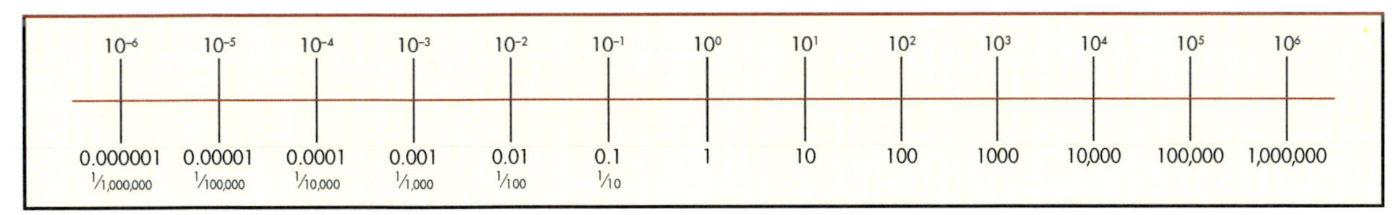

As you can see, when expressing whole numbers, 10, 100, 1000, 10,000, and 100,000, the exponents equal the number of zeros following the 1. For each place the decimal point is moved, the number is divided by 10. We then see that 10^0 is equal to 1. This will be explained a little later in this section. When we look at the fractions, 0.1, 0.01, 0.001, 0.0001, 0.00001, and 0.000001, we see that that each time the 1 moves one decimal place further to the right of the decimal point, the negative exponent increases by one.

Note that the single zero to the left of the decimal is not counted. It is only held as a placeholder to show that there are no numbers in the unit position. For fractions expressed as decimals, we only count the zeros to the right of the decimal point. To see why this is so, consider the two numbers 3.001 and 0.001. We would describe these numbers as three and one thousandth, and one thousandth, respectively. In both these numbers, the decimal part is expressed as one thousandth and both contain two zeros to the right of the decimal point. In the second number, 0.001, the zero to the left of the decimal point only indicates that there are no numbers in the unit position or beyond. So in the number 0.001, only the two zeros after the decimal are counted.

Example 7.5 Converting Numbers to Exponential Terms

Problem

Convert the following numbers into exponential terms.

a) 0.0000001 b) 1,000,000,000,000

Solution

a) Count the number of digits to the right of the decimal point to determine the exponent. There are 7 digits after the decimal point so the exponent is 7. The exponential term is 10^7

b) Count the zeros after the one. There are 12 zeros. The exponential term is 10^{12}

Practice Problems

7.9 Covert the following numbers into exponential terms.

a) 0.000000000000001 ____________ b) 100 ____________

7.10 Convert the following exponential terms to decimals. pencast

a) 10^{-8} ____________ b) 10^8 ____________

Calculations Using Exponents

As we mentioned earlier, exponents make the handling of very large and very small numbers number much easier. Dividing ten million by ten thousand can be accomplished by either long division or with the use of a calculator. By knowing a few simple rules of exponents, this calculation can be done by inspection or very quickly with a pencil and paper.

Both ten million (10,000,000) and ten thousand (10,000) can be expressed as the exponentials 10^7 and 10^4, respectively. Dividing 10^7 by 10^4 is accomplished by subtracting the exponent of the denominator, 4 from the exponent of the numerator, 7, (7 – 4 = 3). The answer to this division is 10^3. The problem is set up as follows:

$$\frac{10^7}{10^4} = 10^{(7-4)} = 10^3$$

Multiplication of exponents is performed in a similar fashion with the exception that we add the exponents instead of subtracting them. Multiplying one thousand, 10^3, by one billion, 10^9, results in

$$10^3 \times 10^9 = 10^{(3+9)} = 10^{12}$$

Example 7.6 Calculations with Exponential Terms

Problem

Divide ten thousand by one millionth.

Solution

Ten thousand and one millionth as exponentials are 10^4 (numerator) and 10^{-6} (denominator), respectively.

$$\frac{10^4}{10^{-6}} = 10^{(4-(-6))} = 10^{10}$$

Recall that when subtracting a negative number, we switch the sign to a positive value.

Practice Problem

7.11 Perform the following calculations and express your answers in exponential terms.

a) 0.0001/1,000 = ________________________

b) 10,000/100 = ________________________

c) 0.000001/0.0001 = ____________________

d) 10,000 × 0.00001 = ____________________

e) 0.01 × 0.000001 = ____________________

In the preceding discussion, it was stated that $10^0 = 1$. Why is this so? We have learned that when dividing numbers such as, 10^6 by 10^4, we merely subtract the exponent in denominator from the exponent in the numerator:

$$\frac{10^6}{10^4} = 10^{(6-4)} = 10^2$$

If we divide a number by itself, say 1000 by 1000, we readily know the answer is 1. Putting this in terms of exponents, the division would read,

$$\frac{10^3}{10^3} = 10^{(3-3)} = 10^0 = 1$$

Calculator Exercises

To apply what we just learned to the scientific calculator, exponents can be inserted by using either the caret key, ^, or 10^X key. The carat key can be used to insert an exponent on any number including 10. It is important to use parentheses around the exponent, 10^(4) or 10^(–6), when using the caret key. The 10^X key is used for inserting exponents on the number 10 and is similar to using the caret key in that the exponent must be put inside parentheses.

Example 7.7 Using a Calculator to Multiply and Divide Exponential Terms

Problem

Use your calculator to multiply $10^4 \times 10^{-3}$.

Solution

Using the caret key

Using the 10^X key. Remember to put the exponent inside parenthesis.

$$10^4 \times 10^{-3} = 10$$

Practice Problem

7.12 Perform the following exercises using your scientific calculator. Express the answer in exponential form.

a) $10^9 \times 10^{12} =$ ______________

b) $\dfrac{10^3}{10^{-8}} =$ ______________

c) $10^4 \times 10^{-10} =$ ______________

d) $\dfrac{106 \times 10^{-3}}{108} =$ ______________

Up until now we have only used exponents with 10. Actually, exponents can be used with any number. For example, if we multiply eight times eight we get sixty–four. This can be expressed as $8 \times 8 = 64$. It can also be written as $8^2 = 64$. If we multiply eight by itself three times, the result is five hundred twelve: $8 \times 8 \times 8 = 512$ or $8^3 = 512$. Whatever number we choose to use an exponent with—that is, the number raised to a certain power—is called the **base.** For all of our calculations we will be using base 10.

Base:
Any number which is raised to a power.

More Uses of Exponents

So far we have used exponentials with nice even numbers such as one thousand, one million, one billion, but in truth, any number, regardless of how large or small, can be expressed in exponential form. Just as 1000 can be expressed as 10^3, and 10,000 can be expressed as 10^4, the number 2000 can be expressed using 3.3010 as the exponent and 10,500 will have an exponent of 4.0211 as an exponent. Two questions that should come to mind are 1) how is this useful?, and 2) how do we find these strange looking numbers?

Let's consider the first question. In the previous section we learned to multiply or divide numbers by either adding or subtracting the exponents of the exponential forms. Using this method, we can divide 10,500 by 2000 by subtracting their respective exponents:

$$\frac{10{,}500}{2{,}000} = \frac{10^{4.0211}}{10^{3.3010}} 10^{(4.0211-3.3010)} = 10^{0.7201} = 5.25$$

OK, but as far as question 2 goes, where do we get these exponents? The answer is simple, we get them from our calculator by using the Log key. Log, short for *logarithm,* is just another name for the exponent to which ten is raised. In other words, the logarithm of one thousand is three, and is expressed as log(1000) = 3. Using your calculator, the operation goes like this:

Example 7.8 Use a Calculator to Calculate a Logarithm

Problem

Use your calculator to find a) log (0.624) and b) $\log\frac{234}{12}$

Solution

a) To find log (0.624) use the following keystrokes.

You can check your answer by calculating $10^{-0.248} = 0.624$.

b) Here we have a fraction. We can divide 12 into 234 and then take the logarithm or enter into the calculator as follows.

Y ou can check your answer by $10^{1.2900} = 19.5$ which is equal to 234/12 = 19.5

Practice Problem

7.13 Use your calculator to find the logarithms of the following numbers.

a) Log(125) = ______________________

b) Log(3,252) = ______________________

c) Log(0.145) = ______________________

d) $\log \frac{17,654}{231}$ = ______________________

Summary of Exponents and Exponential Notation

- When expressing whole numbers such as 10, 100, 1000, etc., the exponent equals the number of zeros following the 1.
- When expressing decimals such as 0.1, 0.01, 0.001, etc., count the zeros in front of the one to determine the negative exponent.
- When multiplying exponential terms with base 10, add the exponents.
- When dividing exponential terms with base 10, subtract the exponent in the denominator from the exponent in the numerator.
- An exponent to which ten is raised is called a *logarithm* (log).

Worksheet 7.2 Exponentials (Section 7.2)

1. Perform the following calculations.

a) $10^2 \times 10^4 =$

d) $\frac{446}{10^2}$

b) $\frac{10^9}{10^6}$

e) $\frac{10^3 \times 10^1}{10^{-3}}$

c) $\frac{124 \times 10^4 \times 10^0}{10^{-3}}$

d) $\frac{10^{-2} \times 10^{-4}}{10^{-3} \times 10^2}$

2. Fill in the table.

Exponential Term	Number
10^6	1,000,000
	0.0000001
10^{-3}	
	0.1
10^0	
	$\frac{1}{100}$

7.3 Scientific Notation

Now that we understand how exponents of ten can be used to express numbers and simplify calculations, let's look at a further use of exponents that is better suited to measurements and calculations in chemistry.

We have just seen that a number such as 125 can be expressed as $10^{2.0969}$, but a much more straightforward way of expressing this number would be as 1.25×100 or as in exponential notation 1.25×10^2. Or instead of expressing 13,000 as $10^{4.1139}$ we write 1.3×1000 or as in exponential notation 1.3×10^3. This notation is called **scientific notation** and consists of a coefficient, the number to the left of the multiplication sign, and an exponential, ten raised to a whole number exponent. We can use this method to avoid fumbling around with long, cumbersome logarithms. The coefficient in **scientific notation** may contain a number of any length, but only a number from one to nine can be placed to the left of the decimal point. For example, to express the number 2341 in scientific notation, we first develop the coefficient by placing the placing a decimal point directly after the two:

$$2.341$$

Scientific Notation:
A number expressed as a decimal with one digit to the left of the decimal point multiplied by 10 raised to a power.

Second, determine what number must be used to multiply 2.341 to give the proper magnitude. In this case, the number is 1000.

$$2.341 \times 1000$$

Using the exponential form of 1000 results in proper scientific notation of 2,341

$$2.341 \times 10^3$$

Another way of looking at this is to observe that when multiplying by 10 to a power, the decimal point is moved the same number of places as the exponent. For our example, 2341, the decimal point is moved to the right of 2 which is our first nonzero number. The decimal point has been moved 3 places to the left of the implied decimal point.

Small numbers can also be put into scientific notation. To express the number 0.0000315 in scientific notation, we first develop the coefficient by placing a decimal point directly after the first nonzero number:

$$3.15$$

Second, determine what exponential must be used to multiply 3.15 to give the proper magnitude. In this case, the exponential is 10^{-5}. The resulting scientific notation is

$$3.15 \times 10^{-5}$$

Note that we move the decimal point 5 places to the right to obtain 10^{-5}. When converting numbers to scientific notation, if the decimal point is moved to the left the exponent is positive, and if it is moved to the right, the exponent is negative.

Example 7.9 Express a Number in Scientific Notation

Problem

Express the number 0.001006 in scientific notation.

Solution

First, develop the coefficient by placing a decimal point directly after the first nonzero number, which in this case is 1.

Second, determine what exponential must be used to multiply 1.006 to give the proper magnitude or move the decimal point to the first nonzero number. The decimal point is moved 3 places to the right and the exponential is 10^{-3}. The resulting scientific notation is

$$1.006 \times 10^{-3}$$

Practice Problem

7.14 Use scientific notation to express the following numbers.

a) 754 ____________

b) 1,256,000 ____________

c) 0.00483 ____________

d) 10,452 ____________

e) 0.000000045 ____________

Calculations using scientific notation can be carried out easily with a scientific calculator without the use of caret, parenthesis or 10^X keys. Scientific calculators contain an EE key which inserts " $\times 10^X$". By using this key, a number such has 3.321×10^3 can be entered as

Remember the EE key inserts $\times 10^X$, and so there isn't any need to enter an $\times$ or a 10. Calculations using the EE key are very easy to perform. To divide 8.263×10^4 by 3.452×10^3 enter them into the calculator using the EE key.

Example 7.10 Use the Calculator's EE Key

Problem

Use the EE key on your calculator to calculate $(1.06 \times 10^{23}) \times 26$. Give your answer in scientific notation.

Solution

We are asked to multiply 1.06×10^{23} by 26. Enter the expression into the calculator using the EE key.

The calculator gives the answer as 2.54E24. Translated to scientific notation we have

$$2.54 \times 10^{24}$$

Practice Problem

7.15 Use the calculator to perform the following calculations.

a) 6.022 × 1023 / 3.67 × 1017 = ______________________

b) 35,841 × 7,415 = ______________________

c) $(3.5 \times 10^{-5}) \times (9.65 \times 10^{15})$ = ______________________

Summary for Scientific Notation

- A number in scientific notation is the product of a number greater than or equal to 1 but less than ten and base 10 raised to a power.
- Use the exponent key for calculations that involve scientific notation.

Worksheet 7.3 Scientific Notation (Section 7.3)

1. Fill in the following table.

Standard notation	Scientific Notation
1651	1.651×10^3
0.000569	
432	
	2.3×10^{-5}
1500008	

2. Use your calculator to perform the following calculations.

a) $(6.6 \times 10^2) \times 423 =$

b) $4.23 \times 10^{-2} + 2.6 \times 10^3 =$

c) $5.67 \times 10^{-23} \times 2.64 \times 10^{24} =$

d) $\dfrac{2.31 \times 10^{-8}}{1.60 \times 10^{-5}} =$

e) $\dfrac{(2.46 \times 10^3) \times (7.2 \times 10^{-2})}{1.3 \times 10^8} =$

7.4 Measurement

Any science—for example, chemistry, physics, or psychology—requires some form of measurement in order to evaluate outcome of an investigation. In chemistry, we need not only a way to measure the amount of products that result in a chemical reaction, but also the amount of reactants, that is, the ingredients that were combined for the reaction. In this section, we will learn about the nature of making measurements, the units involved, the definition and application of precision and accuracy in a measurement. We will also discuss the use of significant figures in recording measurements and the rules for using them in calculations.

Units of Chemical Measurements and Prefixes

Physical properties can be measured in many different units. Recall that physical properties include color, mass, height, etc. For example, we can measure height in inches, feet, yards, meters, or in many other units. For a measurement to be meaningful, it must include both a number and a unit. Without a unit, a height of 44 is meaningless. It could be 44 inches, feet, yards, etc.

Today, most countries employ *metric* units to express measurements. In the United States, the metric system is used primarily in science and medicine. In the metric system, base units are used for length, mass, volume, time, temperature, energy, and the amount of substance. Later we will explain how to modify the base units with prefixes to express very large and small measurements. In 1960, another system called the *International System of Units* was adopted by a group of international scientists in order to keep standard units used when reporting scientific measurements. These agreed upon units are called **SI units** and incorporate metric units (**Table 7.1**). Later we will see that many other units are derived from the metric basic units. For example, *concentration,* the quantity of a substance that is dissolved in a given volume, can have units of g/L.

SI Units:
A set of units that are defined by the International System of Units.

Table 7.1

Metric Base Units and SI Units

Quantity	Metric Base Unit	SI Unit	Equivalents
Length	Meter (m)	Meter (m)	1 m = 3.281 ft
Mass	Gram (g)	Kilogram (kg)	454 g = 1 lb
Volume	Liter (L)	Cubic meter (m3)	3.78 L = 1 gal
Time	Second (s)	Second (s)	same
Temperature	Degrees Celsius (°C)	Kelvin (K)	1°C = 1.8°F
Amount of substance	Mole (mol)	Mole (M)	same

In chemistry, we will not always measure quantities in just grams and liters. Sometimes we will need to measure 10^3 grams of a substance or a volume of 10^{-3} liters. These units of measurement are based on the metric system, which uses either multiples of ten or fractions of ten. In the metric system, prefixes are used to express these differing magnitudes. For example, the amount 10^3 grams would be 1 kilogram, kg. The prefix kilo means 1000. Using another prefix, the volume 10^{-3} liters would be expressed as 1 mL. The prefix milli means 1/1000 or 10^{-3}. The common metric prefixes that most chemists use are shown in **Table 7.2.**

Table 7.2

Common Metric Prefixes

Prefix	Symbol	Multiply Base Unit by	Example
mega	M	1,000,000	Mg = 10^6 g
kilo	k	1000	kg = 10^3 g
deci	d	1/10	dL = 10^{-1} L
centi	c	1/100	cm = 10^{-2} g
milli	m	1/1000	mm = 10^{-3} m
micro	µ	1/1,000,000	µg = 10^{-6} g
nano	n	1/1,000,000,000	nL = 10^{-9} L
pico	p	1/1,000,000,000,000	pg = 10^{-12}g

Aside from just memorizing these prefixes and the amounts they represent, let's do some exercises to get a better feel for what they represent.

Example 7.11 Use Metric Prefixes to Express Magnitude

Problem

How many microliters are in one liter?

Solution

Looking at Table 7.2, we see that one microliter (1 µL) is equal to 10^{-6} L. That is, a microliter is one millionth of a liter. This means that there are one million microliters in one liter. An easy way to see this relationship between grams and micrograms, is to write

$$1\ \mu L = 10^{-6}\ L$$

Now, interchange the number while changing the sign of the exponent, so it becomes positive (divide both sides of the equation by 10^{-6}).

$$10^6\ \mu L = 1\ L$$

We have one million micrograms equal to one liter.

Practice Problems

7.16 How many nanometers are in one meter?

7.17 Express 100 cm as meters.

7.18 How many liters are in 2 ML?

The units of measurement typically used in a chemical laboratory involve mass, volume, length, temperature, and time.

Mass In chemistry, we don't talk about weight, but rather **mass,** which is the amount of matter contained in an object. In chemistry mass is measured in units of grams (g) or kilograms (kg). The **weight** of an object, which is normally stated in units of ounces and pounds, involves the gravitational force of the Earth, Moon, or other large body. For example, a person weighing 150 lb on the Earth would only weigh 25 lb on the Moon, because the gravitational force of the Moon is only 1/6 of the gravitational force of the Earth. Mass, on the other hand, doesn't take the force of gravity into account. Without any gravity involved, the 150-lb person would have a mass of 68,040 g. The mass of this person would be the same, whether on Earth, the Moon, or anywhere else in the universe. In other words, whereas the weight of an object is variable, depending on the amount of gravity, its mass is always constant. In the lab, we use a *balance* to measure mass by comparing the weight of the object to a known reference weight. The SI unit for mass is the kilogram (kg), and it is equivalent to 2.205 lb. In the lab it is more convenient to use grams (g) or milligrams (mg) when measuring mass. Sometimes we need to convert from the English system of units to the metric system. Some common relationships between the English units and metric units for mass are:

$$454 \text{ grams (g)} = 0.454 \text{ kilograms (kg)} = 1 \text{pound (lb)} = 16 \text{ ounces (oz)}$$

$$1 \text{ ounce (oz)} = 28.35 \text{ grams (g)}$$

$$1 \text{ ton} = 2000 \text{ pounds (lb)} = 907.03 \text{ kilograms (kg)}$$

More of these units and their equivalences can be found in Appendix ?

Mass:
A measure of the amount of matter in an object.

Weight:
The measure of the gravitational force exerted on an object.

Length The meter (m) is the standard metric and SI unit for length. One meter is equal to 39.37 inches. Other units that are commonly used are the kilometer (km), which is equal to 1000 m, the centimeter (cm), equal to one one-hundredth of a meter, and the millimeter (mm), which is one one-thousandth of a meter. A few of the more common relationships between the English and metric units of length are

1 meter (m) = 39.37 inches

1 inch = 0.0254 m = 2.54 cm

Volume In the United States, we use pints, quarts, gallons, and other units to measure volumes. **Volume** is a measure of the quantity of space that is occupied by matter. The SI unit for volume is the cubic meter (m^3). In chemistry as well as most other countries, the liter, L, is the unit of volume that is used. Some of the more common relationships are

1 liter (L) = 1.057 quarts (qt)

1 gallon (gal) = 3.7856 L

1 fluid ounce (oz) = 0.02957 L

1 cubic meter (m^3) = 1000 L = 264.2 gal

1 cubic cm (cm^3) = 1 mL = 29.6 mL

1 cm^3 = 1 cc

Volume:
The quantity of space that matter occupies.

Time Around the world, time is measured in seconds, minutes, hours, days and so on. In chemistry, the basic unit of time is seconds, s.

Temperature Most of us are familiar with two temperature scales, the Fahrenheit scale, °F, and the Celsius scale, °C. To convert from °F to °C, we must know that the intervals of degree on the two scales are quite different. Each unit on the Celsius scale is nearly twice each degree on the Fahrenheit scale: 1°C = 9/5 °F = 1.8 °F

A temperature increase of 10 degrees on the Celsius thermometer would equal a temperature change of 18 degrees on the Fahrenheit scale. To convert temperature readings between the two scales, we must not only factor into the calculation the size difference of the degrees, but also the temperature difference for the freezing point of water on both scales, 32 °F verses 0 °C (**Figure 7.2**). When converting from Fahrenheit to Celsius, the 32-degree difference is subtracted, and converting from Celsius to Fahrenheit, the 32-degree difference is added. The two equations for these conversions are:

To convert degrees Fahrenheit to degrees Celsius:

$$°C = (°F - 32) \times 5/9$$

To convert degrees Celsius to degrees Fahrenheit:

$$°F = (9/5 \times °C) + 32$$

Figure 7.2.

Water freezes at zero degrees Celsius, but 32 degrees Farenheit.

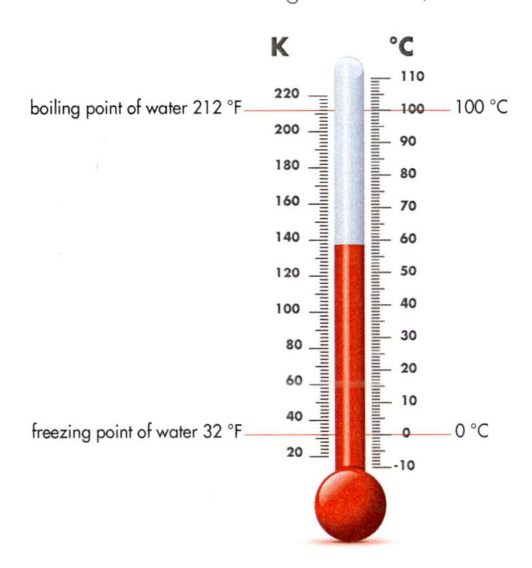

Example 7.12 Converting Degrees Fahrenheit to Degrees Celsius

Problem

A thermometer reads 79 °C. What is this temperature in degrees Celsius?

Solution

Because this problem deals with converting from Celsius to Fahrenheit, we will use the following equation and substitute 79 °C for °C.

$$°F = (9/5 \times °C) + 32$$

$$°F = (9/5 \times 79\ °C) + 32$$

$$°F = 142.2\ °C + 32$$

$$= 174.2\ °F$$

Practice Problems video

7.19 A patient had a body temperature of 104 °F. What is this temperature in degrees Celsius?

7.20 A bread recipe calls for the liquid ingredients to be at a temperature of 45 °C. What is this temperature in degrees Fahrenheit?

There is a third temperature scale, the Kelvin scale, which is measured in kelvins, K. In chemistry, both Celsius and Kelvin scales are used. In laboratories, temperatures are measured using Celsius thermometers. For some calculations, the Celsius measurements are converted to kelvins. In comparing the intervals on the Celsius scale to the Kelvin scale, we find the unit sizes are the same, and this greatly simplifies interconverting between the two scales. A comparison between the Celsius and Kelvin scales is shown in **Figure 7.3.** In the figure, both the freezing point and boiling points of water are indicated. On the Celsius scale the freezing and boiling points are 0 °C and 100 °C respectively. On the Kelvin scale, the same points are 273 K and 373 K. From this we can see that for both temperature scales, the difference between the freezing point and boiling point is 100. This demonstrates that the size of the intervals on both temperature scales are the same. So to convert from degrees Celsius to kelvins, add 273 to the Celsius temperature,

$$K = °C + 273$$

and to convert from kelvins to degrees Celsius, subtract 273 from the Kelvin temperature.

$$°C = K - 273$$

Note that the degree symbol is not used with Kelvin temperatures.

Figure 7.3

Comparison of the Celsius and Kelvin temperature scales.

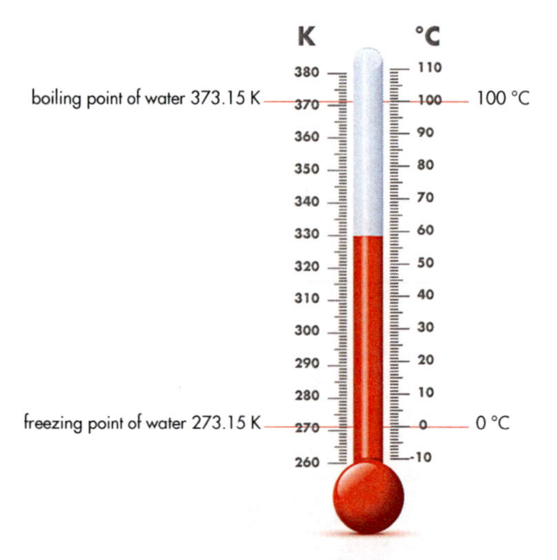

Example 7.13 Conversion from Celsius to Kelvin

Problem

To calculate the volume of a gas, its temperature must be in kelvins. The temperature of a gas was measured at 65 °C. What is this temperature in Kelvins?

Solution

To convert from degrees Celsius to Kelvins, add 273 to the Celsius reading.

$$K = 65\ °C + 273 = 338\ K$$

Answer 338 K

Example 7.14 Conversion from Fahrenheit to Kelvin

Problem

A student measures the temperature of a solution using a Fahrenheit thermometer. She records a temperature of 73 °F. What is this temperature on the Kelvin scale?

Solution

First convert 73 °F to degrees Celsius.

Use the equation °C = (°F – 32) × 5/9 and substitute °F with 73 °F.

$$°C = (73\ °F - 32) \times 5/9$$

$$°C = 41\ °F \times 5/9 = 23\ °C$$

Now that the temperature is in degrees Celsius we can use

$$K = °C + 273$$

$$K = 23\ °C + 273 = 296\ K$$

Answer 296 K

Practice Problems

7.21 A student recorded the temperature of a beaker of water as 98 °C. What is this temperature in kelvins?

7.22 The temperature of a gas was measured as 312 K. What is this temperature in degrees Celsius and degrees Fahrenheit? pencast

Summary of Measurement

- The metric system uses either multiples of ten or fractions of ten, and uses prefixes to express differing magnitudes.
- Mass is a measure of the amount of matter in an object.
- Volume is the amount of space that is occupied by matter.
- Temperature is reported in °F, °C, or K.

Worksheet 7.4 Metric Units and Temperature Conversions (Section 7.4)

1. Give the symbol for each of the following. For example, the symbol for a milligram is mg.

   a) kiloliters ____________ b) megagrams ____________

   c) centiliters ____________ d) nanometer ____________

   e) decimeter ____________ f) cubic centimeter ____________

2. Provide the name for the following units. For example mg is a milligram.

   a) mm ____________ b) dL ____________

   c) pg ____________ d) dm^3 ____________

   e) cc ____________ f) μL ____________

3. Fill in the blanks.

   a) 103 mm = ______ m b) 1 μg = ______ mg

   c) 1 L = ______ mL d) 1 mL = ______ cm^3

   e) 1 Mm = ______ mm f) 10^{-1} L = ______ dL

4. Fill in the blanks.

   a) 1 in = ______ cm b) 1 lb = ______ g

   c) 1 gal = ______ L d) 1 m = ______ in

5. Fill in the following table.

Temperature, °F	Temperature °C	Kelvin
212	100	273
98.6		
		298
−15		
	125	

7.5 Precision and Accuracy of Measurements

In the previous section, we familiarized ourselves with the units of measurement. We can now discuss what measurements are about. In order to proceed, there are two things we must know: 1) measurements are never exact, and 2) we need to know how well a measurement is made in order to use it.

In regard to the first point, saying that measurements are inexact may seem to contradict the idea of scientists as perfectionists. When counting, we have exact numbers. There are twelve items in a dozen; there are 23 people in a room. These, as all counts, are exact numbers. But this doesn't apply to measurements. For example, suppose we have to measure the length of a piece of metal using a meter stick that has centimeter increments (**Figure 7.4**).

Figure 7.4.

When we measure a piece of metal in centimeters, we approximate its length.

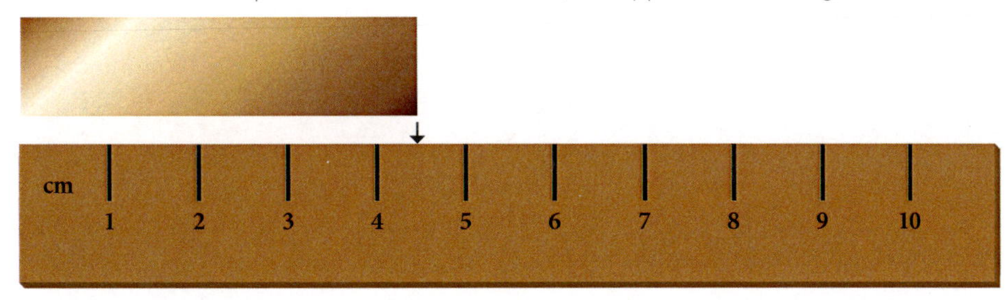

Using this measuring stick, we can see that the piece of metal is definitely greater than four centimeters, but it's less than five centimeters. We have to approximate the distance between the four- and the five- centimeters markings to be halfway or about 0.5 centimeters, giving us a total length of 4.5 cm. As you can see, the 0.5-cm measurement is not exact. It's an approximation.

Now let's try measuring the same piece of metal with a meter stick that has millimeter increments (**Figure 7.5**).

Figure 7.5

When we measure the same piece of metal using a ruler with more increments, we get more precision in our result.

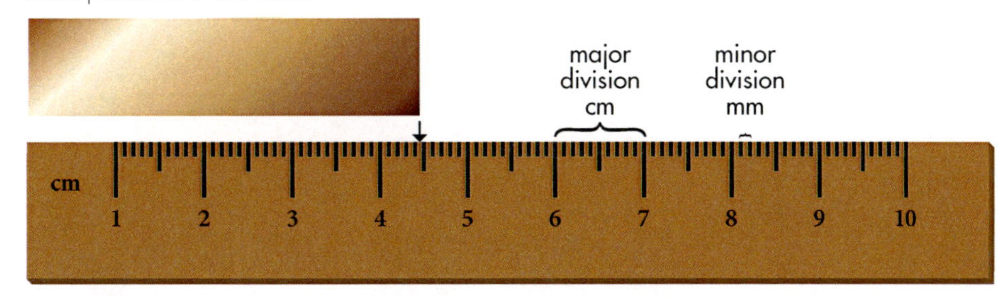

Using more increments in our measuring stick, we can now see that the measurement of the piece of metal is somewhere between 4.4 cm and 4.5 cm. Because it is halfway between we can estimate it as 0.05 cm. The measurement is reported as 4.45 cm. Even though we have more increments to measure with, we still have uncertainty in the last digit of ± 0.01 cm. In the first measurement, it was the 0.5 cm that was estimated and the uncertainty in the measurement is ± 0.1 cm. In the second measurement there is uncertainty in the second decimal place and the uncertainty in the measurement is ± 0.01 cm (**Table 7.3**). The point is, regardless of how many increments we have to measure with, there will always be a point of uncertainty in the measurement. The last digit in a measurement is always uncertain because it is an estimate. The precision of the measuring device is based on the number of increments on the measuring device. The use of the second measuring stick in our example gives a more precise measurement of 4.45 cm compared to that of the first measuring device that gives 4.5 cm.

Table 7.3.

Uncertainty in Measurements

Measurement	Uncertain Digit	Uncertainty in Measurement
437.6 mg	6	± 0.1mg
801 m	1	± 1 m
0.235 L	5	± 0.001 L
0.6245 s	5	± 0.0001 s

Example 7.15 Volume in a Graduated Cylinder

Problem

a) How many milliliters of water does the graduated cylinder contain?

b) Which digit is uncertain?

c) What is the uncertainty?

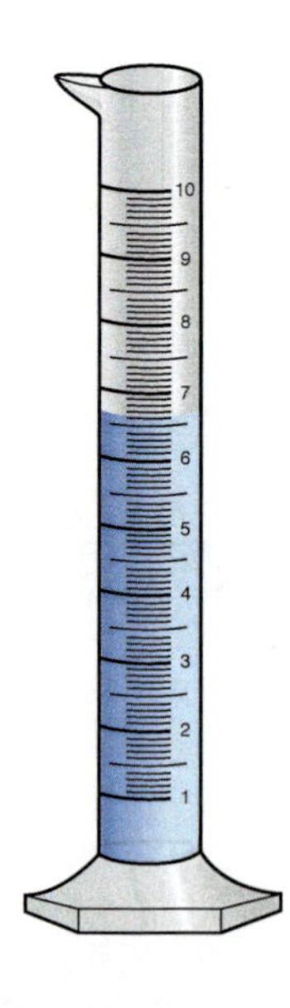

Solution

a) We can see that the volume of the liquid in the graduated cylinder is between 6 mL and 7mL, and using the tenth milliliter increments, the volume is halfway between 6.2 mL and 6.3 mL, about 0.05 mL. The volume of water in the cylinder is

6.25 mL.

b) 6.2 is known for certain and 5 is the estimated or uncertain digit.

c) The uncertainty in the measurement is ± 0.01 mL

Practice Problems

7.23 How many milliliters of water are contained in the beaker?

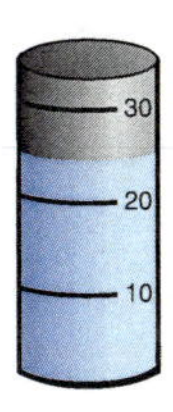

7.24 What is the uncertainty of a measurement using an electronic balance that displays a mass of 25.05 grams?

a) 5 g   b) 0.0 g   c) 0.05 g

What we have been considering in these exercises and problems deals with **precision** in measurements. There are two parts to the precision of a measurement: 1) the "fineness" of the measurement, and 2) how well a measurement can be repeated. By fineness, we mean the number of increments in the unit of measurement. We saw this in the previous examples such as using a meter stick with 1-cm intervals verses using one with 1-mm intervals. For the most part, we can think of the fineness of the measurement as the amount of numbers, including the last uncertain number that the measuring instrument allows us. Repeatability is how close the measurements are to each other. For example, one set of measurements of a volume are 26.01 mL, 25.14 mL, and 23.95 mL compared to a second set of measurements of 25.14 mL, 25.13 mL and 25.12 mL. We can see that the second set of measurements are in much closer agreement than the first set. The second set of measurements have better precision with respect to repeatability. In the following discussion we will address the fineness of the measurement.

precision:
The repeatability and fineness of a measurement.

Accuracy and precision are two terms that many people think as interchangeable. Precision, as we just discussed, is regarded as both the fineness and repeatability of a measurement. **Accuracy,** on the other hand, is how close a measurement comes to an accepted or true value. Under normal conditions, water will boil at 100 °C. But suppose a faulty thermometer measures the boiling water at only 98 °C? The accuracy of the measurement is off. As another example, if a volume of 75 mL is required in an experiment, we can see that using a graduated cylinder with 0.1-mL increments will produce better accuracy than using one with 1-mL increments.

accuracy:
How close a measurement is to a true or accepted value

Summary for Precision and Accuracy of Measurements

- All measurements have an uncertainty.
- Precision is both how closely measurements agree with one another and the "fineness" of the measurement.
- Accuracy is a measure of how close a measurement is to a true or accepted value.

Worksheet 7.5 Precision and Accuracy (Section 7.5)

1. Fill in the table.

Measurement	Uncertain Digit	Uncertainty in Measurement
0.002879 cm		
35.4 mL		
60024 m		
2.6 L		
25.00 mL		
458.004 g		
32.0 g		
1.6578×10^3 K		

2. Record the temperature ______________________

   What is the uncertain digit? ______________________

   What is the uncertainty in the measurement? __________

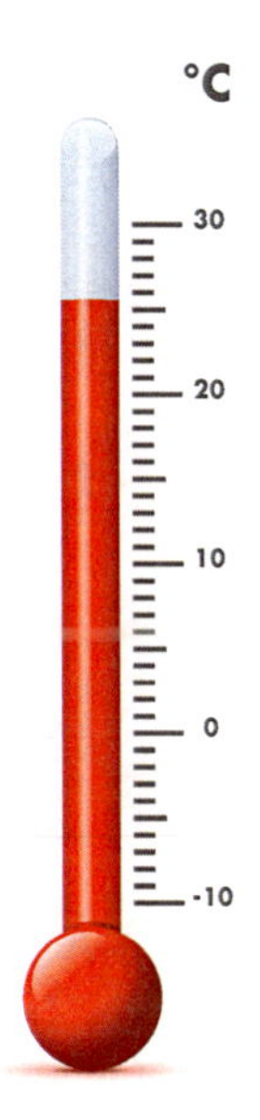

7.6 Significant Figures

We have discussed that in any measurement, the last number is always uncertain. The numbers used in these measurements are called significant figures of which all are certain except for last number. When properly used, **significant figures** represent the precision of a measurement. When we read a volume of 12.5 mL, we know that the increments on the graduated cylinder were in in units of milliliters, and that the last number, 5, reflects an estimation between the 12 mL and 13 mL increments. Identifying significant figures is quite easy when all the numbers in the measurement are non-zero.

> **significant figures:**
> The number of meaningful digits used to express a measurement.

For example, the measurement 25.46 g has 4 significant figures. In chemistry we often use measurements in calculations, and we must report our answers using the correct number of significant figures. We will use the following rules to determine the number of significant figures in a measurement.

Rule 1. Non-zero Numbers: All non-zero numbers in a measurement are significant.

1237 m contains four significant figures.

5.3 g contains two significant figures.

83.25 mL contains four significant figures.

But notice that we did not include zeros in this rule. In measurements such as 1500 cm and 0.0032 g, are the zeros considered significant or not? What we will see now is that sometimes zeros are significant and sometimes they are not considered significant. The following rules will help us to determine when zeros are included in the precision of the measurement, that is, when they are significant.

Rule 2. Sandwiched Zeros: When zeros occur between two non-zero numbers, they are always significant.

305 cm contains three significant figures (two significant non-zero numbers and one significant sandwiched zero)

7001 g contains four significant figures (two significant non-zero numbers and two significant sandwiched zeros)

42,065 L contains five significant figures (four significant non-zero numbers and one significant sandwiched zero)

Rule 3. Leading Zeros: With no non-zero numbers to the left of a decimal point, leading zeros are zeros that come before the first non-zero number. These leading zeros are not significant.

0.035 nm contains two significant figures (two significant non-zero numbers and two non-significant leading zeros)

0.0000328 mL contains three significant figures (three significant non-zero numbers and four nonsignificant leading zeros)

Let's look a little closer at how leading zeros work. Consider the measurement 325 nm. We can readily see that this measurement contains three non-zero numbers, giving it three significant figures. Now, if we want to express this measurement in micrometers instead of nanometers, it would be 0.000325 µm, and in units of millimeters it would be 0.000000325 mm. In this measurement, the precision is retained regardless of which units we use, and it will always carry the same three significant figures.

Rule 4. Trailing Zeros: Trailing zeros occur after the last non-zero number in a measurement, such as 3,500 g, 1.7500 cm, or 30 mL. Trailing zeros can be considered as significant or non-significant. The rule to determine if they are significant is as follows: If the trailing zeros occur after a non-zero number and to the right of a decimal point, they are significant. If there is no decimal point before them, they always are non-significant.

1.7500 cm contains five significant figures (three significant non-zero numbers and two significant trailing zeros that follow a decimal point)

3500 g contains two significant figures (two significant non-zero numbers and two non-significant trailing zeros without a decimal point)

30 mL contains one significant figure (one significant non-zero number and one non-significant trailing zero without a decimal point)

30. mL contains 2 significant figures (one significant non-zero number and one significant trailing zero that occurs after a nonzero number and before the decimal point)

Example 7.16 Determine the Number of Significant Figures in a Measurement

Problem

How many significant figures are in each of the following measurements?

a) 3701 km b) 0.00254 g

Solution

a) 3701 km has three significant non-zero numbers and one significant sandwich zero.

4 significant figures

b) 0.00254 g has three significant non-zero numbers and three non-significant leading zeros.

3 significant figures

Practice Problems pencast

7.25 Determine the number of significant figures in each of the following measurements. Briefly explain your answer.

a) 0.000792 cm ______________________________

Total significant figures ______________

b) 853 L ______________________________

Total significant figures ______________

c) 4038 g ______________________________

Total significant figures ______________

d) 253,000 km ______________________________

Total significant figures ______________

e) 124.4500 kL ______________________________

Total significant figures ______________

Sometimes a measurement without a decimal may contain trailing zeros of which one or more may actually be significant. How do we deal with this? Take for example the measurement 35,000 m. Suppose that this distance was measured in such a way that it contains three significant figures, that is, the first three numbers including the first zero, are all significant, and the remaining two zeros are not significant.

35,0000 m
↓
Significant

The best way to express this measurement with the proper number of significant figures—that is, the correct precision—is with scientific notation. We have learned that a number such as 35,000 can be written as

$$3.5 \times 10^4$$

As it turns out, the numbers in the coefficient, in this case 3.5, are considered as significant. With this in mind, we can express the number 35,000 with three significant figures, as

$$3.50 \times 10^4$$

Example 7.17 Use Scientific Notation to Express the Correct Number of Significant Figures

Problem

Express 75,300,000 km to 5 significant figures.

Solution

The first three non-zero numbers, 7, 4, and 3, are significant figures. In order to complete the needed five significant figures, we need to include the next two zeros. We will place these five numbers into the coefficient of the scientific notation

7.5300

(Remember that only one number can be before the decimal point.)

Now, put the exponential part into the number, and in this case it will be 10^7. The resultant scientific notation will be

$$7.5300 \times 10^7 \text{ km}$$

Practice Problems

7.26 Express 73,000 kg to three significant figures.

7.27 Express 5,200 L to one significant figure.

Using Significant Figures in Calculations

When performing calculations involving measurements, the precision of the measurements may vary. For example, suppose we want to calculate the volume of a rectangle that has been measured as 125.3 cm by 32 cm by 8.016 cm. We can see that the precision of these measurements varies significantly. When we multiply the three measurements together, should the product reflect the most precise measurement or the least? Or should it be an average of the three? As it turns out, the product will reflect the precision of the least precise measurement. In this section, we will learn two rules for handling precision and significant figures in calculations.

Before proceeding, we need to discuss the rules of rounding off numbers in regard to significant figures and decimal places. Let's say that we do a calculation that gives us an answer of 2.30456 and the precision should only have three decimal places. Well, we have to drop the last two numbers, the 5 and the 6. The first rule of rounding off states that if the first number dropped (the 5) is five or greater, we raise the remaining number (the 4) up one unit. That is, the 4 becomes a 5, and the answer now reads 2.305. As another example, suppose we calculate an answer of 123.42, but the answer should only have four significant figures. The answer should read 123.4, because the second rule of rounding off states that if the first number dropped off (in this case 2) is less than five, the remaining number (in this case 4) remains the same. Apply these two rules to the following exercises.

Example 7.18 Rounding Numbers

Problem

Round 0.000060068 to 4 significant figures.

Solution

There are 5 significant figures in 0.000060068. Rounding to 4 significant figures we get 0.00006007. The number that is dropped (the 8 in this case) is greater than 5. The remaining number (6) is raised to 7.

0.00006007 (4 significant figures)

Practice Problem

7.28 Round the following measurements to the indicated number of significant figures.

a) 686.0012 g (5 significant figures)

b) 12.46 mL (3 significant figures)

c) 35,731 m (2 significant figures)

Precision in Addition and Subtraction Involving Measurements with Decimals

When adding or subtracting measurements involving decimals, we don't look at the number of significant figures, but instead we look at the number of decimal places when determining precision. Take for example the simple operation of adding 35.57 mL, 1.2 mL, and 4.028 mL. We can readily see that 4.028 mL is the most precise measurement and the 1.2 mL is the least. Remember that the answer will need to reflect the precision of the least precise measurement, which would be the 2.1 mL. So, how do we handle this?

Rule 1: When measurements involving decimals are added or subtracted, the number of decimal places in the answer will be determined by the measurement with the least number of decimal places.

Let's add 35.57 mL, 1.2 mL, and 4.028 mL.

35.57 mL	2 digits after decimal point
1.2 mL	1 digit after decimal point
4.028 mL	3 digits after decimal point
40.798 mL	

In this case, our answer can only have one digit after the decimal point. The answer is rounded to 40.8 mL. When subtracting measurements, the same rule applies.

6.5002 g	4 digits after decimal point
– 0.201 g	3 digits after decimal point
6.2992 mL	

The answer is rounded to 6.299 g. When adding or subtracting measurements that do not contain decimals, the measurements are just added together. For example, if we add 53 mL to 425 mL we obtain 478 mL.

Example 7.19 Simple Calculations and Significant Figures

Problem

Carry out the following calculation and express your answer using the correct number of significant figures. 2.6391 L – 1.0 L

2.6391 L
– 1.0 L
1.6391 L

Solution

The answer can only have one digit to the right of the decimal point because 1.0 L has one digit after the decimal point. The number 1.6391 is rounded to

1.6 L.

Practice Problem

7.29 Carry out the following calculations. Express your answer using the correct number of significant figures.

a) 225.017 g + 53.687 g + 3.18 g

b) 94.632 mL – 42.8 mL

c) 73.7326 m + 23.715 m

Multiplying and Dividing Measurements

With multiplication and division of measurements, we don't count decimal places. We consider precision by the number of significant figures, and again look for the least precise measure. The answer cannot have more significant figures than the measurement with the least number of significant figures. For example, if we divide 4.2543 g by 25 mL, our answer can only have 2 significant figures, because the measurement with the least number of significant figures has 2 significant figures.

$$\frac{4.2543 \text{ g (5 significant figures)}}{25 \text{ mL (2 significant figures)}} = 0.17 \frac{\text{g}}{\text{mL}} \text{ (answer rounded to 2 significant figures)}$$

Example 7.20 More Calculations and Significant Figures

Problem

Find the area of a rectangle that measures 143.25 m by 15.04 m.

Solution

143.25 m contains five significant figures and 15.04 m contains four significant figures. Counting significant figures, we can see that the second measurement has the least number of significant figures. Carrying out the multiplication, we get a calculator answer of 2,154.48 m^2 that needs to be properly rounded off to two significant figures.

$$143.25 \text{ m (5 significant figures)} \times 15.04 \text{ m (4 significant figures)} = 2{,}154.48 \text{ m}^2$$

Rounding off the answer to four significant figures we have

$$2{,}154 \text{ m}^2$$

Practice Problems

7.30 A bicyclist traveled 20.9 km in 1.2 hours. What was his speed in km/hour? Report your answer using the correct number of significant figures.

7.31 What is the area in cm2 of a piece of gold leaf foil that measures 41.6 cm by 27.04 cm. Report the answer using the correct number of significant figures.

We can add or subtract, multiply or divide the significant figures of measurements, but now let's look at combining these operations. When combining operations, we need to keep track of the number of significant figures that are to be retained from each calculation, but the answer is not rounded until the end. First, we need to know that when mixing multiplication (or division) with addition (or subtraction) in a calculation, the multiplication (division) is performed first followed by the addition (subtraction). To make things easier, certain parts of the calculation are sometimes put into brackets or parentheses, and these operations are performed first. After we obtain our calculator answer, we then go back and determine the number of significant figures to retain in the answer. The best way to understand this is with an example.

Example 7.21 Multiple Operations and Significant Figures

Problem

Perform the following calculation stating the answer with the proper number of significant figures.

$$3.21 \times \left(\frac{3.283+7.03}{2.1} \right) =$$

Solution

We can see that there are three operations: addition, division, and multiplication. To approach this problem, we first perform the operations within the parentheses, the addition, and then the division. The final operation is the multiplication. At this time, do not round off. Keep track of the number of significant figures that can be retained at each step.

The first step will yield, $3.21 \times (\frac{10.313}{2.1})$ =To keep track of the number of significant significant figures for the addition, underline or highlight the last digits that can be retained. In this case we can only have two digits to the right of the decimal point.

The second step will yield, $3.21 \times (4.91095)$ For the division we can retain only two significant figures because 2.1 is the measurement with the least number of significant figures. Again, do not round until the final calculation.

The final step will yield 3.21 × 4.91095 = 15.7641495. Now we can round the final answer. The answer can only have 2 significant figures because 4.9 is the number with the least number of significant figures. The rounded answer with the correct number of significant figures is

16 (2 significant figures)

Practice Problems

7.32 Perform the following calculation, and state your answer with the correct number of significant figures. 5.29 × (32.954 – 27.2)

7.33 Perform the following calculation and state your answer using the correct number of significant figures. pencast

$$\frac{(63.21\ \text{mg} - 34.853\ \text{mg})}{25.97\ \text{L}}$$

Significant figure rules do not apply to exact numbers and defined quantities. An exact number is a number used in counting and has no uncertainty associated with it. For example, we can count 24 books or 150 books. The quantity 1000 mm is defined to be exactly 1 meter and 12 inches is exactly equal to 1 foot.

Summary for Significant Figures

- Significant figures are all of the digits in a measurement that are known for certain plus one estimated digit.
- When a number has trailing zeros, scientific notation is used to express the correct number of significant figures.
- In multiplication and division of measurements, the answer cannot have more significant digits than the measurement with the least number of significant figures.
- In addition and multiplication of measurements, the answer cannot have more digits after the decimal point than the measurement that has the least number of digits after the decimal point.

Worksheet 7.6 Significant Figures, Rounding, and Calculations (Section 7.6)

1. How many significant figures in each of the following?

   a) 0.00430 ________ b) 49.0000006 ________

   c) 0.00000004 ________ d) 4.006×10^{-3} ________

   e) 38,000,000 ________ f) 3.000000×10^{8} ________

2. Round the following measurements to the indicated number of significant figures.

   a) 8.8721 cm (2) ________ b) 126.0041 km (4) ________

   c) 0.0200000 g (4) ________ d) 1.0201×10^{-4} L (3) ________

   e) 3046 kg (2) ________ f) 450,000 mm (2) ________

   g) 20.01 °C (2) ________ g) 2.68900×10^{3} m/s (1) ________

3. Perform the following calculations. Express your answer with the correct number of significant figures.

   a) $42.6 + 7.2 + 3.6663 - 0.689$ b) 0.977×4.1

   c) $\dfrac{2.005 - 0.461 \times 0.321}{0.3000} =$ d) $\dfrac{1.4361 \times (2.0001 + 3.6610)}{26.1000 - 22.00} =$

4. The diameter of Mars is **6794** km.

   The diameter of Mars rounded to 2 significant figures is ________ km.

   The diameter of Mars rounded to 3 significant figures is ________ km.

   Write the diameter of Mars in scientific notation (4 significant figures)

   ________ km.

5. The mass of a gold nugget is 0.00036998 kg.

   Express the mass to 2 significant figures ________ kg.

   Express the mass to 4 significant figures ________ kg.

   Express the mass to 5 significant figures using scientific notation

   ________ kg.

7.7 Conversion Factors and Dimensional Analysis

In problem solving, proper care in setting up calculations is very important, and special attention should always be given to unit cancellation. If the units cancel properly, the problem should solve correctly. This method is called *dimensional analysis* and will be an important part of problem solving in any science course. In this section, we will be putting a lot of practice in learning to use the approach to solving chemical problems.

In order to use dimensional analysis, we must first talk about *conversion factors.* We have been using conversion factors throughout most of our lives without realizing it. By knowing how many dimes are in a dollar, we know that 20 dimes equals two dollars. Conversion factors allow us to convert from one unit (dimes) to another (dollars). The thing about setting up a conversion factor is to know the equivalence of the two units, that is, when the two units equal the same amount. For example, a dime isn't the same amount as a dollar, but ten dimes equals the same amount of money as one dollar. Let's take a closer look using this simple example to determine how many dollars equal 20 dimes. First, set up a conversion factor. In any problem or calculation involving conversions, we need to know the units involved; in this case the units are dimes and dollars. We state the equivalence as

$$10 \text{ dimes} = 1 \text{ dollar}$$

The equivalence can be written in following fractional forms called conversion factors. One of the conversion factors will be used for the calculation. Because the numerator and denominator are equal, the fractions are equal to 1.

$$\frac{10 \text{ dimes}}{1 \text{ dollar}} \quad \text{or} \quad \frac{1 \text{ dollar}}{10 \text{ dimes}}$$

For now we want to concentrate on setting up conversion factors, but as a preview to dimensional analysis, the following calculation shows how the conversion factor is used. In our example, we are asked how many dollars equal 20 dimes. To convert from dimes to dollars, the given (20 dimes) is multiplied by the conversion factor that cancels out the unit dimes.

$$20 \cancel{\text{dimes}} \times \frac{1 \text{ dollar}}{10 \cancel{\text{dimes}}} = 2 \text{ dollars}$$

Notice how the dime units cancel out, leaving the dollar units in the answer. This is the basis for dimensional analysis. As complex as some chemical calculations seem, the dimensional analysis involved remains as simple as the preceding exercise. For now, let's look at the following exercises that deal with setting up the conversion factors. More than one equivalence can be used for a conversion, as you will see in the next example.

Example 7.22 Writing Conversion Factors

Problem

Write conversion factors to convert each of the following.

a) miles to feet   b) grams to pounds   c) millimeters to meters

Solution

a) We need an equivalence to convert from miles to feet. There are 5280 ft in one mile. The equivalence is written as

$$1 \text{ mile} = 5280 \text{ ft}$$

The conversion factors are $\frac{1 \text{ mile}}{5280 \text{ ft}}$ or $\frac{5280 \text{ ft}}{1 \text{ mile}}$

b) to convert grams to pounds, we know that there are 454 g in 1 lb. The equivalence is

$$454 \text{ g} = 1 \text{ lb.}$$

The conversion factors are $\frac{454 \text{ g}}{1 \text{ lb}}$ or $\frac{1 \text{ lb}}{454 \text{ g}}$

We also could have used the equivalence 1 g = 0.0022 lb

c) There are 10 mm in 1 cm. The equivalence is 10 mm = 1 cm

The conversion factors are $\frac{10 \text{ mm}}{1 \text{ cm}}$ or $\frac{1 \text{ cm}}{10 \text{ mm}}$

The equivalence 0.1 cm = 1 mm is also correct.

Practice Problems video

7.34 Write the conversion factors for the following conversions.

a) meters to picometers

b) square inches to square yards

c) quarts to liters

7.35 Write two different equivalences for the conversion of minutes to days.

Knowing how to set up conversion factors, we can now move into setting up calculations using dimensional analysis, which is also known as the factor-label method. When this simple method is used in a calculation, the correct answer is almost guaranteed. The basis for this method is keeping track of the units of the components in the calculations. To determine how many gallons in 24 quarts we first need to set up an equivalence.

The equivalence is written as

$$1 \text{ gallon} = 4 \text{ quarts}$$

The two possible conversion factors are

$$\frac{1 \text{ gallon}}{4 \text{ quarts}} \text{ or } \frac{4 \text{ quarts}}{1 \text{ gallon}}$$

Next, we need to setup the calculation. For this part we need to know the two types of units in our calculation: a) *given units,* or the units that have a given amount and b) *desired units,* the units for which we are solving. In this calculation, the given units are quarts and we are solving for gallons.

Having identified the units and determined the conversion factor, the calculation is set up as follows:

$$\text{Amount with } \cancel{\text{Given Units}} \times \frac{\text{Desired Units}}{\cancel{\text{Given Units}}} = \text{Answer (in desired units)}$$

Notice that the conversion factor used has the given units in the denominator, which allows for proper cancellation of the units—that is, the given units cancel out, leaving only the desired units in the answer.

Back to the calculation

$$24 \cancel{\text{ quarts}} \times \frac{1 \text{ gallon}}{4 \cancel{\text{ quarts}}} = 6 \text{ gallons}$$

In the example we converted 24 quarts to gallons. Most of us could have performed the calculation without setting up equivalences and conversion factors. Many chemistry problems require unit conversions and this is a good method to use regardless of the type of problem encountered. It is important to identify the given and the desired quantities in any problem. This is good practice for the many problems you will encounter in this and future chemistry and science courses.

Example 7.23 Simple Single-Step Conversions

Problem

Use the simple single-step conversion method to convert a) 37.00 yards to inches b) 0.644 lbs to g, and c) the number of drops in 23.5 mL. Assume that one drop is equivalent to one-twentieth of a milliliter, mL.

Solution

First determine the given and desired units.

a) Given units: yd, Desired units: in

Next find an equivalence.

$$1 \text{ yd} = 36 \text{ in}$$

Write the conversion factors.

$$\frac{1 \text{ yd}}{36 \text{ in}} \text{ or } \frac{36 \text{ in}}{1 \text{ yd}}$$

Set up the calculation. The second conversion factor is used in the calculation because the yards are the unit we want to cancel.

$$37.00 \cancel{\text{yd}} \times \frac{36 \text{ in}}{1 \cancel{\text{yd}}} = 1332 \text{ in. (4 significant figures)}$$

b) First determine the given and desired unit.

Given: lbs, Desired: g

The equivalence is 1 lb = 454 g

Write the conversion factors

$$\frac{1 \text{ lb}}{454 \text{ g}} \text{ or } \frac{454 \text{ g}}{1 \text{ lb}}$$

Set up the calculation. To cancel pounds, use the second conversion factor.

$$0.644 \cancel{\text{lb}} \times \frac{454 \text{ g}}{1 \cancel{\text{lb}}} = 292 \text{ g}$$

Note we have 3 significant figures in the answer.

c) Given: mL, Desired: drops

We are told that one drop is equivalent to 1/20 mL. We can write an equivalence from this information

$$1 \text{ drop} = 1/20 \text{ mL or } 20 \text{ drops} = 1 \text{ mL}$$

Write the conversion factors.

$$\frac{1 \text{ mL}}{20 \text{ drops}} \text{ or } \frac{20 \text{ drops}}{1 \text{ mL}}$$

Set up the calculation so that milliliters cancel.

$$23.3 \text{ mL} \times \frac{20 \text{ drops}}{1 \text{ mL}} = 466 \text{ drops}$$

Practice Problems

7.36 Convert 45 in to yards.

7.37 How many micrograms are contained in 0.00264 g? pencast

7.38 A student measures out 750 mL of ethanol in the lab. How many liters of ethanol does this correspond to? pencast

7.39 A piece of tubing measures 18.6 in. How many centimeters is this?

7.40 How many ounces are contained in a 250.-mL bottle of infant formula?

7.41 How many milligrams are contained in 0.02565 g?

Example 7.24 Single-Step Conversions Using Volume

Problem

A cube of metallic silver has a volume of 2.62 in^3. What is the volume in cubic centimeters?

Solution

Given: in^3, Desired: cm^3

To convert the volume we write an equivalence. We know that 1 in = 2.54 cm. But, what about a cubic inch? Obviously we cannot use 1 in = 2.54 cm because volume is given in cubic inches and the desired volume must be in cubic centimeters.

Both sides of the equation must be cubed.

$$(1 \text{ in})^3 = (2.54 \text{ cm})^3$$

$$1^3 \times \text{in}^3 = (2.54 \text{ cm} \times 2.54 \text{ cm} \times 2.54 \text{ cm})$$

$$1\text{in}^3 = 16.387 \text{ cm}^3$$

Now that we have our equivalence, we can write the conversion factors.

$$\frac{1 \text{ in}^3}{16.387 \text{ cm}^3} \text{ or } \frac{16.387 \text{ cm}^3}{1 \text{ in}^3}$$

Set up the calculation.

$$2.62 \text{ in}^3 \times \frac{16.387 \text{ cm}^3}{1 \text{ in}^3} = 42.9 \text{ cm}^3$$

Practice Problems

7.42 How many liters are in 0.56 m^3? pencast

7.43 A cylindrical piece of metal has a volume of 4.82 cm^3. What is this volume in cubic inches, in^3? video

7.44 A fitness ball has a radius,r, of 10.8 in. What is the volume of the ball in cm^3? The volume of a sphere is $\frac{4}{3}\pi r^3$.

Density as a Conversion Factor

Up to this point, we have learned how convert from unit of measurement to another within the same types of units of measurement. For example, converting seconds, s, to microseconds, μs, we are in units of time. Converting milliliters, mL, to nanoliters, nL, we are in units of volume. But it is possible covert from one unit of measurement, for example distance, to a completely different unit of measurement, for example time. In another situation, we can interconvert between mass and volume. How is this possible? Consider driving a car at 35 miles per hour (35 mph). If you drive at this velocity for 2 hours, you can determine the distance you will travel in that time, 70 miles. In other words, with velocity we can interconvert between distance and time. It may not be apparent at first, but 35 mph can written as the conversion factor,

$$\frac{35\text{ mi}}{1\text{ hr}} \text{ or } \frac{1\text{hr}}{35\text{ mi}}$$

As we know, distance and time are totally different units, but they become related through a velocity. Another example of this is density, which is an indication of the mass of atoms, molecules, or ions within a unit of volume. Recall from Chapter 1 that density is defined as the mass of a substance per unit volume. For example, gold has a density of 19.30 g/cm^3. This means that 1 cm^3 of gold has a mass of 19.30 g. On the other hand, lithium, a much lighter substance, has a density of only 0.534 g/cm^3. Just as velocity interrelates distance and time, density allows us to interconvert between mass and volume. As we did with velocity, let's look at the density of gold as a conversion factor.

$$\frac{19.30\text{ g}}{1\text{ cm}^3} \text{ or } \frac{1\text{ cm}^3}{19.30\text{ g}}$$

For example, to find the mass of 4.50 cm^3 of gold we use the conversion factor on the right to cancel out cm^3.

$$4.50\ \cancel{\text{cm}^3} \times \frac{19.30\text{ g}}{1\ \cancel{\text{cm}^3}} = 86.9\text{ g}$$

Recall that density is a physical property of a substance and is useful in determining the identity of unknown substances. For solids, the volumes are stated in cubic centimeters, cm3. For liquids, the volumes are stated in milliliters, mL, and for gases, the volumes are states in liters, L. We can better understand the practical application of densities by doing the following examples and problems.

Example 7.25 Calculations Using Density

Problem

a) A cube of copper measures has a volume of 12.2 cm^3. If the mass of the cube is 109 g, what is the density of the copper? b) Mercury, the only metal that exists as a liquid, has a density of 13.55 g/mL. What is the volume of 25 g of mercury?

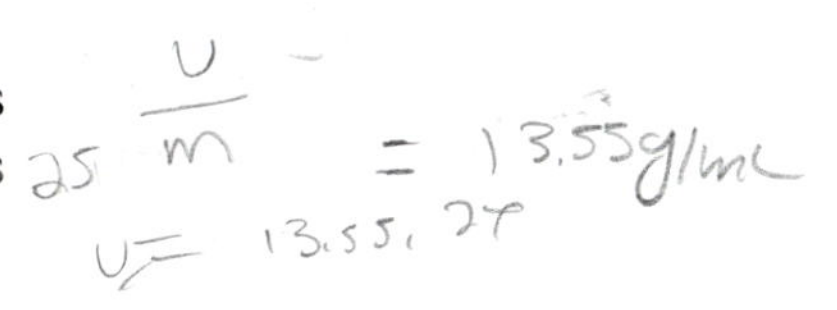

Solution

a) We are given the mass and volume of the copper cube. Density is calculated by dividing the volume by the mass.

$$\frac{109\text{ g}}{12.2\text{ cm}^3} = 8.93\text{ g/cm}^3$$

The density of the copper cube is

$$8.93\text{ g/cm}^3$$

b) Given: g, Desired: mL

Equivalence: 13.55 g = 1 mL

Conversion Factors

$$\frac{13.55\text{ g}}{1\text{ mL}} \text{ or } \frac{1\text{ mL}}{13.55\text{ g}}$$

Set up calculation. Use the second conversion factor for proper unit cancellation.

$$25\text{ g} \times \frac{1\text{ mL}}{13.55\text{ g}} = 1.8\text{ mL}$$

Practice Problems

7.45 A 28-g piece of metal has a volume of 2.47 cm^3. Calculate the density of the metal.

7.46 The density of most substances increases going from the liquid to the solid state, but the density of water decreases as it freezes. At 3.98 °C, 18 g of water occupies a volume of 18.00 mL. When the temperature of the water is reduced to 0.00 °C, its volume decreases to 19.63 mL.

a) What is the density of water at 3.98 °C? b) What is the density of water at 0.00 °C? video

7.47 A student needs 27.45 g of liquid ethanol for an experiment. Ethanol has a density of 0.789 g/mL. What volume, in milliliters mL, will the student need to measure? video

Multistep Conversions and Road Maps

The preceding exercises were based on simple single-step conversions. The following problems will require multistep conversions in the calculations, which means more than one conversion factor and a road map. Road maps are very handy to use in doing calculations. Don't ever think that this approach is beneath you. Taking the time to sketch out the calculation will ensure the right answer. In the following example, we'll show how to use a road map in the calculation. Keep in mind that each type of problem can be done with as many or as few conversion factors as you can write. The number of conversion factors used for each problem will depend on the types and number of equivalences that you use.

Example 7.26 Multistep Conversions

Problem

a) A student needs 2,360 μL of ethyl alcohol for an experiment. He will use a graduated cylinder that reads in milliliter gradations. How many milliliters of ethyl alcohol will he measure? b) A student needs 2.82 kg of sodium chloride for a chemical demonstration. She has access to a scale that measures ounces. How many ounces of sodium chloride should the student measure?

Solution

a) Identify the given units and the desired units:

Given: 2360 μL, Desired: mL

If it's not a single step calculation, develop a road map. Although there is a way to develop a conversion factor that will give us a one-step calculation, for the sake of this example, let's proceed with a two-step method. Looking back at Table 7.2, two relationships are stated:

$$1\ \mu L = 10^{-6}\ L$$

$$1\ mL = 10^{-3}\ L$$

Looking at the two relationships, we can see the common unit of liters, L, between them. Converting from microliters (μL) to liters,(L), is the first step in the calculation:

$$\mu L \rightarrow L$$

This is the first part of the road map. Now convert from liters (L) to milliliters (mL), which will be the second step of the calculation.

$$L \rightarrow mL$$

In terms of the road map, it would look like this,

Road Map				
μL	→	L	→	mL
Given units				Desired units

In this two-step method, we will convert as follows:

microliters to liters and liters to milliliters

Write an equivalence and conversion factors for the conversion microliters to liters

$$1\ \mu L = 10^{-6}\ L$$

$$\frac{1\ \mu L}{10^{-6} L} \text{ or } \frac{10^{-6}\ L}{1\ \mu L}$$

Write an equivalence and conversion factors for liters to milliliters

$$1\ mL = 10^{-3}\ L$$

$$\frac{1\ mL}{10^{-3} L} \text{ or } \frac{10^{-3} L}{1\ mL}$$

Notice that one equivalence and one set of conversion factors is written for each arrow in the roadmap.

Set up the calculation.

$$2{,}360\ \cancel{\mu L} \times \frac{10^{-6} \cancel{L}}{1\ \cancel{\mu L}} \times \frac{1\ mL}{10^{-3} \cancel{L}} = 2.360\ mL \text{ (four significant figures)}$$

The conversion factors are exact leaving us with four significant figures in the final answer.

b) Given: kg, Desired: ounces

Roadmap

There are many ways to solve these problems depending on what equivalences you remember. In this case we will use the following:

kg → g → lb → oz

There are three arrows, so we need three equivalences for three sets of conversion factors. This problem can just as easily be done with only one conversion factor. Again, it depends on the number of equivalences you can remember. The goal is to use the conversion factors to cancel out units until you obtain your desired unit.

kg → g

1 kg = 1000g

Conversion factors (set 1)

$$\frac{1\text{ kg}}{1000\text{ g}} \text{ or } \frac{1000\text{ g}}{1\text{ kg}}$$

g → lb

Conversion factors (set 2)

$$\frac{1\text{ lb}}{454\text{ g}} \text{ or } \frac{454\text{ g}}{1\text{ lb}}$$

lb → oz

1 lb = 16 oz

Conversion factors (set 3)

$$\frac{1\text{ lb}}{16\text{ oz}} \text{ or } \frac{16\text{ oz}}{1\text{ lb}}$$

Set up the calculation. Use the conversion factors to cancel units. Follow your roadmap. First we convert from kg to g, g to lb, and lb to oz.

$$2.82\text{ kg} \times \frac{1000\text{ g}}{1\text{ kg}} \times \frac{1\text{ lb}}{454\text{ g}} \times \frac{16\text{ oz}}{1\text{ lb}} = 99.4\text{ oz}$$

The student would have to measure out 99.4 oz of sodium chloride.

Practice Problems

7.48 How many mg are contained in 36.2 ounces? pencast

7.49 The mass of one carbon atom is 1.99×10^{-23} grams. How many carbon atoms are in 275 μg of carbon? video

7.50 How many seconds are in 2.6 years? pencast

7.51 A container can hold 18.9 cm^3. How many gallons of water can the container hold? video

Summary for Conversion Factors and Dimensional Analysis

- Conversion factors are written from equivalences.
- Dimensional analysis uses conversion factors to convert from one unit to another.
- Density is a physical property of a substance and is defined as the ratio of a substance's mass to volume.

Worksheet 7.7 Writing Conversion Factors and Roadmaps (Section 7.7)

1. Fill in the table.

Units	Equivalence	Conversion Factor
feet to miles	5280 ft = 1 mi	$\frac{1\text{ mi}}{5280\text{ ft}}$
grams to pounds		
L to μL		
μL to L		
	36 in = 1 yd	$\frac{1\text{ yd}}{36\text{ in}}$
mg to g		

2. Write the equivalences and appropriate conversion factors for the following conversions. Provide a roadmap, and set up the problem to show how the units cancel.

   a) months to seconds

   Roadmap:

   Equivalences:

   Conversion Factors:

   Problem Setup:

   b) fluid oz to cm^3

   Roadmap:

   Equivalences:

   Conversion Factors:

   Problem Setup:

Worksheet 7.8 Single-Step Conversions (Section 7.7)

Make the following conversions. Round your answer to the correct number of significant figures and include the unit.

a) Convert 45 in to yards, yd.

Given Units: Desired Units:

equivalence:

conversion factors: ________ or ________

set up: × ________ =

answer (include unit)

b) Convert 18.2 pints to gallons, gall.

Given Units: Desired Units:

equivalence:

conversion factors: ________ or ________

set up: × ________ =

answer (include unit)

c) Convert 75.9 μg to grams, g.

Given Units: Desired Units:

equivalence:

conversion factors: ________ or ________

set up: × ________ =

answer (include unit)

d) Convert 17.93 gal to liters, L.

Given Units: Desired Units:

equivalence:

conversion factors: ________ or ________

set up: × ________ =

answer (include unit)

e) Convert 2.46 mL to cubic centimeters, cm^3.

Given Units: Desired Units:

equivalence:

conversion factors: ________ or ________

set up: $\times$ ________ =

answer (include unit)

f) Convert 452 mg to grams, g.

Given Units: Desired Units:

equivalence:

conversion factors: ________ or ________

set up: $\times$ ________ =

answer (include unit)

g) Convert $1.06 \times 10–7$ m to millimeters, mm.

Given Units: Desired Units:

equivalence:

conversion factors: ________ or ________

set up: $\times$ ________ =

answer (include unit)

h) Convert 36.2 in to centimeters, cm.

Given Units: Desired Units:

equivalence:

conversion factors: ________ or ________

set up: × ________ =

answer (include unit)

i) Convert 10460 g to kilograms, kg.

Given Units: Desired Units:

equivalence:

conversion factors: ________ or ________

set up: × ________ =

answer (include unit)

j) Convert 1.3600 × 105 pg to grams, g.

Given Units: Desired Units:

equivalence:

conversion factors: ________ or ________

set up: × ________ =

answer (include unit)

Worksheet 7.9 Density (Section 7.7)

1. Fill in the table (pay attention to units). Identify the substance using the table of densities.

Mass	Volume	Density	Identity of Substance
22.356 g	29.08 mL	0.70 g/mL	gasoline
57.4 g	6.84 cm3		
22.65 g		19.3 g/cm3	
625 mg		2.7 g/cm3	
6.46 g	5.00 L		

Densities for Several Common Substances

Substance	Density
Water	1.00 g/mL
Air	1.293 g/L
Gasoline	0.70 g/mL
Gold	19.3 g/cm^3
Copper	8.4 g/cm^3
Platinum	21.4 g/cm^3
Ice	0.92 g/mL

2. Calculate the density of the metal bar in g/cm^3.

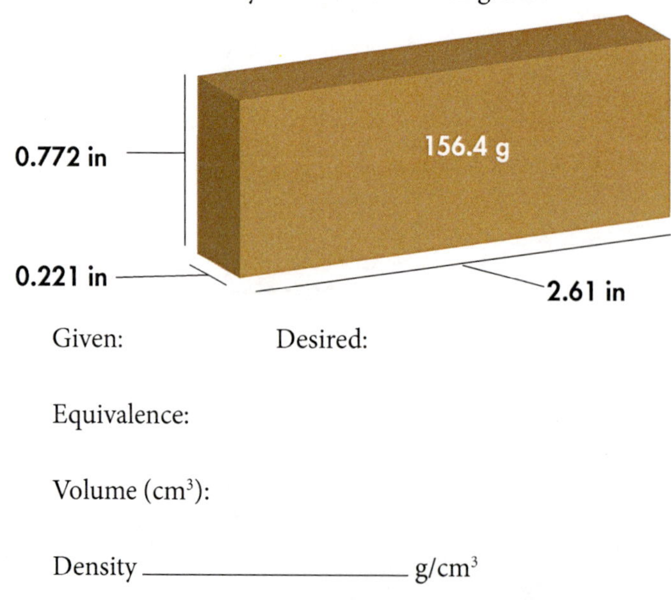

Given: Desired:

Equivalence:

Volume (cm^3):

Density ______________________ g/cm^3

Worksheet 7.10 Multistep Conversions (Section 7.7)

Make the following conversions. Round your answer to the correct number of significant figures and include units. Keep in mind that some problems might require more or less steps than others. This will depend on the number of equivalences you write.

a) **2.4 × 104 mL = ? pints** Given Units: Desired Units:

Road Map

Given Units Desired Units

equivalences:

conversion factors for step 1: ________ or ________

conversion factors for step 2: ________ or ________

conversion factors for step 3: ________ or ________

set up: × ________ × ________ × ________ = **answer**

b) **502.32 cm = ? yards** Given Units: Desired Units:

Road Map

Given Units Desired Units

equivalences:

conversion factors for step 1: ________ or ________

conversion factors for step 2: ________ or ________

conversion factors for step 3: ________ or ________

set up: × ________ × ________ × ________ = **answer**

c) **10.056 seconds = ? years** Given Units: Desired Units:

Road Map

Given Units Desired Units

equivalences:

conversion factors for step 1: ________ or ________

conversion factors for step 2: ________ or ________

conversion factors for step 3: ________ or ________

set up: × ________ × ________ × ________ = **answer**

d) **2.35 qts = ? mL** Given Units: Desired Units:

Road Map

Given Units Desired Units

equivalences:

conversion factors for step 1: ________ or ________

conversion factors for step 2: ________ or ________

conversion factors for step 3: ________ or ________

set up: × ________ × ________ × ________ = **answer**

e) **5.3 gal = ? cm^3** Given Units: Desired Units:

Road Map

Given Units Desired Units

equivalences:

conversion factors for step 1: ________ or ________

conversion factors for step 2: ________ or ________

conversion factors for step 3: ________ or ________

set up: × ________ × ________ × ________ = **answer**

f) **2.46 km = ? miles** Given Units: Desired Units:

Road Map

Given Units Desired Units

equivalences:

conversion factors for step 1: ________ or ________

conversion factors for step 2: ________ or ________

conversion factors for step 3: ________ or ________

set up: × ________ × ________ × ________ = **answer**

g) **2,497 cm^3 = ? L** Given Units: Desired Units:

Road Map

Given Units Desired Units

equivalences:

conversion factors for step 1: ———— or ————

conversion factors for step 2: ———— or ————

conversion factors for step 3: ———— or ————

set up: × ———— × ———— × ———— = **answer**

h) **2,045,000 pg = ? kg** Given Units: Desired Units:

Road Map

Given Units Desired Units

equivalences:

conversion factors for step 1: ———— or ————

conversion factors for step 2: ———— or ————

conversion factors for step 3: ———— or ————

set up: × ———— × ———— × ———— = **answer**

7.8 Counting by Mass in Chemistry

The convenience of counting by weight or mass is an important method for determining the number of atoms or molecules that are required for a chemical reaction. We have already emphasized the smallness of atoms, but to fully appreciate why counting atoms isn't an option, consider that a small handful (12.011 g) of graphite (carbon atoms) would be in the neighborhood of 6×10^{23} atoms, that is, six hundred billion trillion carbon atoms. Now imagine that you could count one carbon atom a second. How long would it take to count a handful of carbon atoms? It would take 1.9×10^{16} years, that is, nearly twenty thousand trillion years to count the handful of carbon atoms! This may not be a concern if chemists were all paid by the hour, but practically speaking, we need a more convenient way to count atoms. In the following examples and problems, we will either determine the number of carbon atoms by mass or vice versa.

Note: the equivalence for the conversion factors is

$$6.022 \times 10^{23} \text{ carbon atoms} = 12.011 \text{ g}$$

Example 7.27 Determine the Number of Carbon Atoms in a Given Mass

Problem

a) How many carbon atoms are contained in 24.02 g of carbon? b) What is the mass in grams of 2.4×10^{24} carbon atoms?

Solution

a) First determine the given and desired units

Given: 24.02 g, Desired: Number of carbon atoms

Equivalence 6.022×10^{23} carbon atoms = 12.011 g

Conversion factors

$$\frac{12.011 \text{ g C}}{6.022 \times 10^{23} \text{ C atoms}} \quad \text{or} \quad \frac{6.022 \times 10^{23} \text{ C atoms}}{12.011 \text{ g C}}$$

Set up

$$24.02 \cancel{\text{g C}} \times \frac{6.022 \times 10^{23} \text{ C atoms}}{12.011 \cancel{\text{g C}}} = 1.204 \times 10^{24} \text{ C atoms}$$

b) Determine the given and desired units. Write equivalence and the conversion factors.

Given 2.4×10^{24} C atoms, Desired: g

Equivalence 6.022×10^{23} carbon atoms = 12.011 g

Conversion factors

$$\frac{12.011 \text{ g C}}{6.022 \times 10^{23} \text{ C atoms}} \quad \text{or} \quad \frac{6.022 \times 10^{23} \text{ C atoms}}{12.011 \text{ g C}}$$

Set up

$$2.4 \times 10^{24} \cancel{\text{C atoms}} \times \frac{12.011 \text{ g C}}{6.022 \times 10^{23} \cancel{\text{C atoms}}} = 48 \text{ g C}$$

Practice Problem

7.52 Fill in the missing information in the following table. The answers from Example 7.22 are inserted in the table.

Mass of Carbon Atoms	Number of Carbon Atoms
24.02 g	1.204×10^{24}
48 g	2.4×10^{24}
0.1201 g	
180.15 g	
	3.01×10^{23}

7.53 How many carbon atoms are in 350. mg of carbon?

Unit of Counting Atoms: The Mole

In the preceding section, we determined the counts of carbon atoms by mass, and one thing for certain is that numbers, such as 6.022×10^{23}, are large and cumbersome to work with. Fortunately, special units of counting exist that make things much easier.

You have probably encountered units of counting such as a dozen, a gross, or a ream. A dozen is twelve, a gross is 144, and a ream is 500. In chemistry the unit used for counting small particles such as protons, atoms, and molecules is the **mole** abbreviated as mol. Recall the smallness of particles such as atoms, and the astronomical number of carbon atoms that fit in the palm of a hand. The number of atoms and molecules for a reaction in a typical research chemistry laboratory requires a unit representing an incomprehensible magnitude, and that unit is the mole. And, how big is a mole?

$$1 \text{ mole} = 6.022 \times 10^{23}$$

The amount 6.022×10^{23} is also called Avogadro's number, N_A.

mole:
An amount that corresponds to 6.022×10^{23} units.

6.022×10^{23} what? It doesn't matter. It's just a number. Yes, it is based on the number of particles in 12.000000 grams of the carbon-12 isotope, but it's still just a number. You could have a mole of electrons or baseballs, but let's face it, the mole is used for very small particles. A mole of carbon atoms will fit in the palm of your hand. A mole of baseballs? Well, you better find an empty back lot in the universe.

In the section of units, we used prefixes to account for multiples or fractions of units, e.g., milliliter, kilogram, and nanosecond. The same prefixes are used with the mole unit in the following examples.

$$1 \text{ millimole} = 10^{-3} \text{ moles}$$

$$1 \text{ nanomole} = 10^{-9} \text{ moles}$$

$$1 \text{ kilomole} = 10^{3} \text{ moles}$$

6.02×10^{23} carbon atoms equals one mole of carbon atoms.

6.02×10^{24} carbon atoms equals ten moles of carbon atoms.

3.01×10^{23} carbon atoms equals one half of a mole of carbon atoms.

In the following example we will convert the number of carbon atoms to number of moles of carbon.

Example 7.28 Convert the Number of Carbon Atoms to Moles of Carbon Atoms

Problem

How many moles of carbon are contained in 6.02×10^{25} carbon atoms?

Solution

First, determine the given and desired units.

Given: 6.02×10^{25} C atoms, Desired: moles

Equivalence: 1 mole = 6.02×10^{23} atoms

Conversion Factors

$$\frac{1 \text{ mole carbon}}{6.022 \times 10^{23} \text{ C atoms}} \quad \text{or} \quad \frac{6.022 \times 10^{23} \text{ C atoms}}{1 \text{ mole carbon.}}$$

Set up: We use the first conversion factor to cancel out atoms

$$6.02 \times 10^{23} \text{ C atoms} \times \frac{1 \text{ mole C}}{6.022 \times 10^{23} \text{ C atoms}} = 0.75 \text{ moles carbon}$$

Practice Problem

7.54 Fill in the missing information in the following table. The answer from Example 7.23 is inserted in the table.

Number of Carbon Atoms	Moles of Carbon
6.02×10^{25}	0.75
	2.5
1.51×10^{23}	
180.15	
	0.499

These preceding sections and related exercises on counting by mass and use of the mole unit are extremely important to students studying chemistry. For this reason, we will work more of these types of exercises before moving on to the next section. Up to this point, we have only dealt with carbon atoms in the conversion exercises. We know that

6.02×10^{23} carbon atoms equals one mole of carbon atoms
and has a mass of 12.011 grams.

But, what about other atoms? Just as one dozen green grapes will have a different weight than one dozen Golan apples, one mole of carbon atoms will have a different mass then one mole of nitrogen atoms.

One mole of any atom will equal 6.02×10^{23} atoms:

One mole of nitrogen atoms equals 6.02×10^{23} nitrogen atoms

One mole of oxygen atoms equals 6.02×10^{23} oxygen atoms

One mole of magnesium atoms equals 6.02×10^{23} magnesium atoms

However, each type of atom (e.g., carbon, nitrogen, oxygen, magnesium, etc.) has its own mass, therefore

One mole of carbon atoms will have a mass of 12.011 g.

One mole of nitrogen atoms will have a mass of 14.00 g.

One mole of oxygen atoms will have a mass of 15.99 g.

One mole of magnesium atoms will have a mass of 24.31g.

You may notice that the mass of 1 mole of carbon is 12.011 g and is the same number as the atomic mass of carbon on the periodic table. Looking at the periodic table you will also see it is the same for nitrogen, oxygen, and magnesium. As it turns out, the numbers for atomic mass on the periodic table can be stated as either atomic mass units (amu) or g/mol. This relationship will be discussed in detail in Chapter 8.

Knowing the above relationships, you can convert grams to moles and moles to grams as shown in the next examples.

Example 7.29 Convert Grams of an Element to Moles

Problem

How many moles of nitrogen atoms are in 0.035 g of nitrogen? One mole of nitrogen has a mass of 14.00 g.

Solution

Referring to the given information, one mole of nitrogen atoms has a mass of 14.00 g, we can obtain an equivalence of

1 mole nitrogen atoms = 14.00 grams

The conversion factors are

$$\frac{1 \text{ mole N atoms}}{14.00 \text{ g N}} \quad \text{or} \quad \frac{14.00 \text{ g N}}{1 \text{ mole N atoms}}$$

The given units are 0.035 g and the desired units are moles

According to the calculation, we know that grams of nitrogen (given unit) need to cancel out, so we will use the first conversion factor.

$$0.035 \cancel{\text{g N}} \times \frac{1 \text{ mol N atoms}}{14.00 \cancel{\text{g N}}} = 0.0025 \text{ moles N}$$

Practice Problems

7.55 How many moles of carbon atoms are in 15.00 g of carbon? One mole of carbon atoms has a mass of 12.011 g.

7.56 A student requires 0.25 moles of magnesium for a chemical reaction. How many grams of magnesium does the student need to weigh out for this reaction? One mole of magnesium has a mass of 24.31 g.

Example 7.30 Convert the number of moles of an element to grams.

Problem

How many grams are contained in 4.68 moles of oxygen?

Solution

From above we see that 1 mole of oxygen atoms is equal to 15.99 g. The equivalence is

$$1 \text{ mole O} = 15.99 \text{ g}$$

The conversion factors are

$$\frac{1 \text{ mole O atoms}}{15.99 \text{ g O}} \quad \text{or} \quad \frac{15.99 \text{ g O}}{1 \text{ mole O atoms}}$$

We are converting from moles to grams.

Set up

$$4.68 \text{ mol O} \times \frac{15.99 \text{ g O}}{1 \text{ mol O atoms}} = 74.8 \text{ g O}$$

The first conversion factor is used to cancel out moles. The answer has 3 significant figures.

Practice Problems

7.57 Calculate the number of moles in 60.2 g of nitrogen. One mole of nitrogen atoms is equal to 14.00 g.

7.58 How many moles of magnesium are equal to 24.31 g?

Summary of the Mole

- 12.011 g of carbon is equivalent to 6.02×10^{23} atoms of carbon.
- The mole is a unit used to count very small particles such as protons, ions, atoms, molecules, etc.
- Just as one dozen of anything is equal to 12, one mole of anything is equal to 6.022×10^{23}.

Worksheet 7.11 Convert Between Grams, Moles, and Number of Atoms (Sections 7.8)

1. Fill in the table. Recall that 1 mole = 6.02×10^{23}

Element	Mass of one mole of atoms, g	Number of atoms	Grams, g	Moles
Lithium	6.941	8.6×10^{24}		
Iron	55.8			
Sodium	22.99			2.00
Helium	4.00			
Sulfur	32.066		45.62	
Gold	196.97			0.68
Silver	107.87	4.29×10^{23}		

Calculations:

Worksheet 7.12 Problem Solving (Sections 7.7 and 7.8)

1. The diameter of a hydrogen atom is 1.06×10^{-7} mm. What is this in picometers, pm?

   Equivalences:

   Conversion Factors

   Set up

2. A chewable tablet contains 81 mg of aspirin. A patient weighing 125 lbs was prescribed a daily dose of 75 mg of aspirin per kg of body weight. How many tablets will the patient need to take in one day?

   Equivalences:

   Conversion Factors

   Set up

3. A single carbon atom has a mass of 1.99 × 10–23 g. What is the mass of 5.25 × 1025 carbon atoms?

   Equivalences:

   Conversion Factors

   Set up

4. A cube has sides measuring 2.65 in. If the density of the cube is 8.26 g/cm3 what is the mass, in grams, of the cube?

   Equivalences:

   Conversion Factors

   Set up

5. An artist takes a blank piece of drawing paper that weighs 3.0128 g. After rendering a pencil drawing, the mass of the paper is 3.0387 g. How many moles of carbon atoms were used in the drawing? The lead in his graphite pencil is made of pure carbon. The mass of a single carbon atoms is $1.99 \times 10–23$ g.

   Equivalences:

   Conversion Factors

   Set up

6. If one gold atom weighs 3.27×10^{-22} g, how many moles of gold atoms are contained in a 1 cm3 cube of gold that weighs 19.2 g? One mole of gold atoms has a mass of 197 g.

   Equivalences:

   Conversion Factors

   Set up

7. How many moles of nitrogen atoms are in 0.035 g of nitrogen? 1 mole of N atoms has a mass of 14.007 g.

   Equivalences:

   Conversion Factors

   Set up

8. If one mole of nitrogen atoms has a mass of 14.00 g, what is the mass of one nitrogen atom?

   Equivalences:

   Conversion Factors

   Set up

9. Benzene is a liquid organic compound with a density of 0.87901. What is the mass of 25 mL of benzene?

   Equivalences:

   Conversion Factors

   Set up

10. Liquid mercury has a density of 13.6 g/mL. One mole of mercury atoms has a mass of 200.59 g. How many mercury atoms are contained in 1 mL of mercury?

    Equivalences:

    Conversion Factors

    Set up

End of Chapter 7 Problems

Section 7.1 Percentages, Parts per Million, and Parts per Billion

1. Express the following fractions in decimal form

   a) $\frac{4}{5}$ b) $\frac{90}{100{,}000}$ c) $\frac{7}{32}$

2. Express the following decimals as fractions.

   a) 0.256 b) 0.000050 c) 0.180002

3. A total of 876 people voted for a particular school levy. Only 98 people voted against the levy. What percent of voters voted for the school levy?

4. Express the following as fractions.

   a) 65% b) 0.025% c) 245 ppm d) 68 ppb

5. Unleaded gasoline can contain a small percentage of lead. If 2725 g of unleaded gasoline contains 89 g of lead,

   a) what fraction of gasoline can contain lead?

   b) what percent of gasoline can contain lead?

6. Analysis of 98.9 g of the deadly gas phosgene contains 12.01 g of carbon, 70.9 g of chlorine, and 16 g of oxygen.

   a) What are the fractions of carbon, chlorine, and oxygen in phosgene?

   b) What are the percentages of carbon, chlorine, and oxygen in phosgene?

7. A 180.18-g sample of glucose, $C_6H_{12}O_6$, contains 72.06 g of carbon and 12.12 g of hydrogen.

   a) How many grams of oxygen does this sample contain?

   b) What is the percent oxygen in this sample?

8. Drinking water may contain a minimum of 0.0007 g of fluoride per 1,000 g of water.

   a) What is the fraction of the minimal amount of fluoride allowed in drinking water?

   b) Express this amount as a percentage and in parts per million, ppm.

9. A student has to make up a 1000-g solution that contains 10 ppm of lead (II) ion, Pb^{2+}

   How many grams of lead ion will she need to use?

10. The analysis of the gases released from a steel mill has shown to contain the chromium (VI) ion at levels of 0.0000001 g of Cr^{+6} per 1 g of air.

    a) What is the fraction of Cr^{6+} in the air?

    b) Express this amount in ppb.

Section 7.2 Exponents

11. A chemist needs 0.001 g of sodium metal for an experiment. Express this amount as an exponential term.

12. Without the use of a calculator, divide ten thousand by one billion. Express the answer in exponential form.

13. Without the use of a calculator, multiply ten thousand by one hundred. Express the answer as an exponential term.

14. Using a calculator, perform the calculation 100,000,000 / 0.001

15. Using a calculator, perform the following calculation $1{,}000 \times 10{,}000$

16. Use base 10 to find the exponent for the number 10,000

17. Use a base of 2 to determine the exponent for the number 64.

18. Use a base of 9 to calculate the exponent for the number 4,782,969.

19. Use your calculator to determine the log of 23 base ten.

20. Use a calculator to determine the log for 0.00274 base ten.

21. Use the number 10^3 to explain the purpose of a logarithm.

22. Calculate log(234/26).

Section 7.3 Scientific Notation

23. Express the following numbers in scientific notation:

a) 0.0000560 b) 123.62 c) 0.2224 d) 50.0

24. Covert the following to standard form:

a) 2.000×10^3 b) 6.548×10^{-5} c) 8×10^{-4}

25. A chemist needs to use 0.0493 g of calcium thiosulfate in a reaction. Express this amount in scientific notation.

26. A student records a mass as 23.768 g. Express this mass in scientific notation.

27. A student calculates an amount of calcium chloride needed for a reaction as 0.00256 g.

    Express this amount using scientific notation.

28. Express the following numbers in proper scientific notation.

    a) 267.3 ______________________

    b) 458×10^{3} ______________________

    c) 1,278,000 ______________________

    d) 0.000356 ______________________

Section 7.4 Measurement

29. Explain the difference between mass and weight.

30. What are the SI units for:

    a) mass b) volume c) time d) temperature

31. Give the name of each unit:

    a) cm b) pL c) dg d) Mg

32. Which is the smallest unit in each pair?

    a) cL or dL b) µg or ng c) m or km

33. What metric prefixes correspond to each of the following:

    a) 10^{3} b) 10^{-2} c) 10^{-3} d) 10^{-6} e) 10^{-1}

34. The radius of a hydrogen atom is 5.3×10^{-11} m. What is this radius in picometers, pm?

35. The mass of the product from a chemical reaction is 0.0010 g. What is this mass in milligrams, mg?

36. A chemical company produces 100,000 kg of potassium nitrate each year.

    a) What is this amount in megagrams, Mg?

    b) What is this amount in grams, g?

37. A meteorologist in Cleveland wants to report the temperature in Toronto, Canada. The temperature in Toronto is 18 °C. What temperature, in Fahrenheit, will the meteorologist report to his audience?

38. Nitrogen gas condenses into a liquid at 77 K. What is this temperature in °F?

39. At what temperature, in K, does water freeze?

40. On an unusually cold day, the temperature dropped to −18.2 °F. What is this temperature in °C?

41. In a laboratory experiment, a gas is heated from 24 °C to 150 °C. The temperature change needs to be recorded in kelvins. The chemist subtracts 24°C from 150°C and records the temperature change as 126 K. Is she correct in doing this? Explain.

Section 7.5 Precision and Accuracy

42. Explain the difference between precision and accuracy.

43. Using a ruler with 0.10 cm (mm) divisions, which of the following measurements is correct?

    a) 27 cm b) 27.1 cm c) 27.10 cm d) 27.100 cm

44. Using a graduated cylinder, a student properly measures a volume of 18.5 mL. What are the increments on the graduated cylinder?

45. Tert-Butyl alcohol freezes at 26 °C. In a laboratory experiment, a student makes four measurements of the freezing point of this alcohol: 25.1 °C, 24.9 °C, 25.0 °C, and 25.1 °C. What can you say about the student's precision and accuracy?

46. In classroom containing 25 students, an instructor demonstrates how to use a graduated cylinder. She determines the volume to be 45.1 mL. Comment on the nature of the two numbers, 25 and 45.1, stated. How do they differ?

47. A measurement of 1.5200 g is recorded.

    a) Which digit is uncertain?

    b) What is the uncertainty in the measurement?

48. A student measures 23.67 mL in a graduated cylinder. What is the uncertainty in the measurement?

Section 7.6 Significant Figures

49. Determine the number of significant figures in each of the following measurements.

    a) 289 mL

    b) 0.305 g

    c) 10,004 m

    d) 0.000650 s

    e) 1.2800×10^{-5} g

50. Round each of the measurements in problem 48 to 2 significant figures.

51. Round off the following to the indicated number of significant figures.

    a) 75,928 (2 significant figures)

    b) 0.00435 (1 significant figure)

    c) 9074 (3 significant figures)

    d) 12 (1 significant figure)

52. How do the rules for multiplying and dividing significant figures differ from the rules for adding and subtracting significant figures?

53. An electron in a hydrogen atom can travel 4.38×10^7 m in 60 seconds. State the electron's velocity using the correct number of significant figures. Perform the following calculations, rounding off the answers to the correct number of significant figures.

    a) $250 \times 17.3 \times 0.1000$

    b) $7.3\,(53.0976 + 2.376)$

    c) $502 \times \dfrac{8.032}{31.7 - 3.927}$

54. A sheet of copper measures 25 cm by 15 cm by 5 mm. Calculate the volume of the copper sheet. Report your answer using the correct number of significant figures.

55. A student added 45.6 mL of ethanol to 26.00 mL of water. What is the total volume of the mixture?

56. A chemist measured the mass of a piece of metal to be 5.6435 g. He determined the volume of the metal to be 0.76 cm^3. The density of the metal can be calculated by dividing the mass by the volume. Calculate the density of the metal and report the answer using the correct number of significant figures.

Section 7.7 Conversion Factors and Dimensional Analysis

57. Do the following conversions and express your answers using the correct number of significant figures.

a) 55 mi to km b) 1.5 lb to g c) 2.59 gal to L d) 12.4 in to cm

58. Carry out the following conversions:

a) 126 cm to m b) 1.26 pm to mm c) 0.0005 μL to mL d) 4.66 mL to cc

59. Carry out the following conversions:

a) 1.61×10^6 s to yr b) 2.35 qt to mL c) 36.2 oz to g

60. A parking area is 150 ft long and 75 ft wide. Find the area of the lot in m^2.

61. A chromium atom has a calculated radius of 166 pm. What this radius in nm?

62. An electron can travel at a velocity of 7.3×10^5 m/s. What is this velocity in miles per hour, mph?

63. A man is driving in Canada at a speed of 104.6 km/hr. How many miles will he travel in 2.5 hours? (1 km = 0.6214 miles)

64. An electron in the first orbit of hydrogen atom is at a distance of 0.529 angstroms from the nucleus. If one angstrom equals 10^{-10} m, what is this distance in pm?

65. A liquid with a volume of 24.32 mL has a mass of 21 g. Calculate the density of the liquid using the correct number of significant figures in the answer.

66. Iron melts at 1538 °C, at which point its density is 6.98 g/cm3. At that temperature, what would be the volume in cubic centimeters of 2 kg of liquid iron?

67. Chloroform is a colorless liquid that boils at 61.2 °C. At 20 °C, its density is 1.483 g/mL. How many milliliters will 36 g of chloroform occupy at 20 °C?

68. The flow rate through a drainpipe is 15 gallons per minute. In 3.45 hours, how many liters will flow through the pipe?

Section 7.8 Counting by Mass in Chemistry: The Mole

69. How many baseballs are in 1 mole?

70. What is the mass of 3.25×10^{25} carbon atoms?

71. How many carbon atoms are in 0.10 pmol of carbon?

72. How many carbon atoms are in 24.6 ng of carbon?

73. What is the mass, in g, of 1.73×10^{15} magnesium atoms. The mass of 1 mole of magnesium atoms is 24.31 g.

74. A student weighs out 2.50 g of sodium. How many sodium atoms are in this mass?

    The mass of 1 mole of sodium is 22.99 g.

75. How many moles are in 35.2 g of zinc. One mole of zinc is 65.38 g.

76. One mole of glucose, $C_6H_{12}O_6$, has a mass of 180 g.

    a) How many molecules of glucose are in 350 g of glucose?

    b) What is the total number of atoms in 350 g of glucose?

General Problems

77. The following items are needed for a construction job:

    2000 1½ inch nails,

    1500 1¼ inch nails

    500 1 in. nails

    Calculate how many pounds of each type of nail will you need given the additional information.

    840 1 inch nails weigh one pound

    530 1¼ inch nails weigh one pound

    300 1½ inch nails weigh one pound

78. If the density of copper is 8.96 g/cm^3, how many copper atoms are in a copper cube that has a volume of 10 cm^3?

79. The hydrogen atom has a diameter of 1.00×10^{-8} cm. What is the total volume in liters of 6.022×10^{27} hydrogen atoms. (*Hint:* treat the hydrogen atom as a sphere. Volume of a sphere = 4/3 pr^3.)

80. How many years would it take to count 1 mole of carbon atoms at a rate of 1 atom per second?

81. A sheet of gold leaf with a length of 8.0 ft and width of 5.2 ft has a mass of 15.62 g. Find the thickness of the gold leaf in millimeters, mm. (The density of gold is 19.35 g/cm^3.)

82. Concentrated sulfuric acid has a density of 1.764 g/mL. A student is running a chemical reaction that requires 12.5 g of this acid. How many milliliters of sulfuric acid will the student need to transfer to the reaction flask?

83. The chemistry lab that you work in has a small bar of sodium metal that measures 2.5 inches in length, 0.5 inches in width, and 0.5 inches in thickness. You are about to start a chemical reaction that requires 0.1 mole of sodium metal. What length of this bar must you cut? (Density of sodium = 0.97 g/cm^3. One mole of sodium has a mass of 22.98 grams.)

84. You are given three samples of different metals. You must arrange them in order of increasing mass, using only the following information:

a) one inch cube of copper

b) 2.0 cm × 0.50 cm × 0.35 cm rectangular bar of lead

c) cylinder of iron, diameter = 68 mm and height = 30 mm.

density of copper = 8.96 g/cm^3

density of lead = 11.35 g/cm^3

density of iron = 7.87 g/cm^3

85. A chemist needs 0.30 moles of liquid mercury for a reaction. How many milliliters, mL, of mercury will she need? (Density of mercury is 13.5 g/mL and 1 mole of mercury atoms has a mass of 200.59 g.)

Applications and Measurements: Reviewing the Balanced Equation

p2

Chapter 8

In part one of this book, we discussed those topics that lay the foundation for understanding both the physical and chemical properties of substances as well as basic atomic structure that gives substances their chemical identities and reactivity. We learned to categorize these substances as elements or compounds, put them into balanced chemical equations, and observe some of the basic reactions they undergo.

In part two, you will be introduced to the application of these concepts. Chemistry is an empirical science, that is, it is based on laboratory experience and observation. To understand what a chemical equation represents is very important in this science, but the ability to apply the equation in running a chemical reaction is essential.

So where do we start? Up until now, chemical reactions have been discussed at the particulate level, i.e. in terms of atoms and molecules.

Recall the combustion of methane:

$$CH_{4\,(g)} + 2O_{2\,(g)} \longrightarrow CO_{2\,(g)} + 2H_2O_{\,(l)}$$

We considered this reaction in terms of one molecule of methane reacting with two molecules of oxygen to produce one molecule of carbon dioxide and two molecules of water. This is well as far as understanding the mechanism and kinetics of the reaction, but what if we wanted to go into a chemistry lab and not only run this reaction, but also control the amount of product? How do we know the amounts of reactants to mix or how to determine the amount of products obtained? To get there we'll review some math, learn some units of measurement, and gain some proficiency in calculations.

But first, let's review the chemical reaction as stated in the form of a balanced chemical equation.

The Symbolic Language of Chemistry

Learning to read, write, and use chemical equations is essential in the study of chemistry at any level. The level of comprehension of a single equation grows as the student proceeds in the course. For example, take an observed chemical reaction such as the ignition (i.e. combustion) of methane in the presence of oxygen to produce water and carbon dioxide.

To state this reaction symbolically, we first divide the reaction into

Reactants
(chemicals that are combined)

and

Products
(new chemicals that are formed).

The reactants are placed on the left hand side of the equation, and the products are placed on the right hand side of the equation. By convention, reactions are written from left to right, with an arrow placed between the reactants and products, indicating the direction of reaction.

Reactants ⟶ Products

Reactants: methane and oxygen

Products : carbon dioxide and water

methane and oxygen ⟶ carbon dioxide and water

Replace each "and" with a + sign

methane + oxygen ⟶ carbon dioxide + water

To perform the next step in writing a chemical equation, the proper chemical symbols and chemical formulas must be used.

Much of learning chemistry is learning a symbolic language. As you progress, so will your "vocabulary" of symbols and formulas. At this point in the course, this information will be given to you.

Methane = CH_4 (methane is composed of one carbon atom and four hydrogen atoms)

oxygen = O_2 (oxygen in the air is composed of two oxygen atoms)

water = H_2O (water is composed of two hydrogen atoms

and one oxygen atom)

carbon dioxide = CO_2 (carbon dioxide is composed of one carbon atom

and two oxygen atoms)

$$CH_4 + O_2 \longrightarrow CO_2 + H_2O$$

gas (g), liquid (l), solid (s)
(aq) is used to indicate a solid dissolved in water.

Another important piece of information which a chemical equation contains is the physical state (i.e. gas, liquid, or solid) of each substance. This information is stated as a subscript in parentheses, at the lower right side of the chemical formula or symbol.

Since all the chemical substances in this reaction are gases,

$$CH_{4\,(g)} + O_{2\,(g)} \longrightarrow CO_{2\,(g)} + H_2O_{\,(g)}$$

However, this particular equation is still incomplete. To take this chemical equation to its final step, the following piece of knowledge is needed:

This is called the conservation of mass.

In other words, in this particular reaction, the number of carbon atoms on the left side must equal the number of carbon atoms on the right side; the number of hydrogen atoms on the left side must equal the number of hydrogen atoms on the right side; and the number of oxygen atoms on the left side must equal the number of oxygen atoms on the right side.

In a chemical reaction, atoms are neither created nor destroyed.

Let's take an inventory of our reaction:

$$CH_{4\,(g)} + O_{2\,(g)} \longrightarrow CO_{2\,(g)} + H_2O_{\,(g)}$$

1 carbon atom	1 carbon atom
4 hydrogen atoms	2 hydrogen atoms
2 oxygen atoms	1 oxygen atom

It is evident that, with the exception of carbon, the number of atoms on the left are not the same as the number of atoms on the right.

That is, this chemical equation is not balanced.

Balancing a chemical equation is done by placing coefficients, where needed, before a chemical term. This coefficient multiplies all the atoms in the chemical formula:

2 H_2O means two molecules of water which also means
$2 \times H_2O$ = 4 hydrogen atoms and 2 oxygens atoms

For our equation, the end product would be:

$$CH_{4\,(g)} + 2\,O_{2\,(g)} \longrightarrow CO_{2\,(g)} + 2\,H_2O_{\,(g)}$$

1 carbon atom	1 carbon atom
4 hydrogen atoms	4 hydrogen atoms
4 oxygen atoms	4 oxygen atoms

This equation is now balanced.

Note: the number one (1) is never used as a coefficient in chemical equations.

Aside from accounting for all the atoms of this reaction, a balanced chemical equation also makes sense on the molecular level:

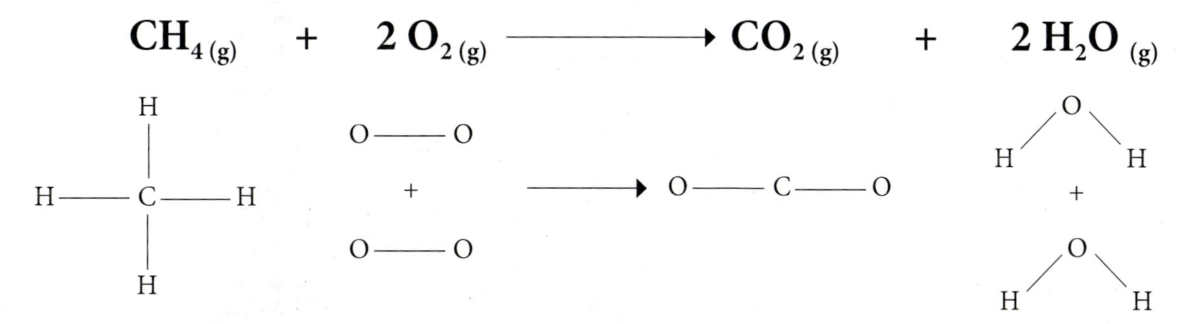

An important observation to make from this is that chemical reactions are, in essence, the rearrangement of atoms to form new chemical species.

Assignment 8.1 Using the given chemical formula, put the following observations into unbalanced chemical equations. Be sure to indicate the physical states of each species.

Magnesium metal (Mg)
Hydrochloric acid (HCl)
Magnesium chloride ($MgCl_2$)
Hydrogen gas (H_2)

1. When solid magnesium metal is placed in a container of hydrochloric acid, (dissolved in water), magnesium chloride (dissolved in water) is formed along with hydrogen gas.

   Reactants:

   Products:

   Unbalanced Chemical Equation with Chemical Formulas

2. Using the unbalanced equation in exercise 1, take inventory:

Reactants	Products
__ magnesium atoms	__ magnesium atoms
__ hydrogen atoms	__ hydrogen atoms
__ chlorine atoms	__ chlorine atoms

3. Balance the equation:

Assignment 8.2 Using the given chemical formula, put the following observations into unbalanced chemical equations. Be sure to indicate the physical states of each species.

Aluminum metal (Al)

Iron (III) oxide (Fe_2O_3)

Iron metal (Fe)

Aluminum oxide (Al_2O_3)

1. The Thermite reaction consists of mixing finely divided aluminum and iron(III) oxide. The highly exothermic reaction (in excess of 2000 °C) produces molten iron and aluminum oxide.

 Reactants:

 Products:

 Unbalanced Chemical Equation with Chemical Formulas

2. Using the unbalanced equation in exercise 1, take inventory:

Reactants	**Products**
__ aluminum atoms	__ aluminum atoms
__ iron atoms	__ iron atoms
__ oxygen atoms	__ oxygen atoms

3. Balance the equation:

Exercises with Polyatomic Ions

Assignment 8.3 Using the given chemical formula, put the following observations into unbalanced chemical equations. Be sure to indicate the physical states of each species.

sodium sulfate (Na_2SO_4)
calcium nitrate ($CaNO_3$)
sodium nitrate ($NaNO_3$)
calcium sulfate ($CaSO_4$)

Mixing aqueous solutions of sodium sulfate and calcium nitrate produces a mixture of soluble sodium nitrate and a precipitate of calcium sulfate.

Reactants:

Products:

Unbalanced Chemical Equation with Chemical Formulas

1. Using the unbalanced equation, take inventory:

Reactants	**Products**
__ sodium ions	__ sodium ions
__ sulfate ions	__ sulfate ions
__ calcium ions	__ calcium ions
__ nitrate ions	__ nitrate ions

2. Balance the equation:

Using the Balanced Equation

From the Particulate to the Macro Scale: How Small is Small?

Having reviewed the balancing act of a chemical reaction, let's discuss laboratory scale up. At the particulate level, it's quite acceptable to discuss a single molecule or atom reacting with another molecule or atom, but in a chemical laboratory we need to deal in measureable amounts, i.e. quantities that can be measured on a balance.

Atoms are incredibly small so small that an eight ounce glass of water contains approximately one trillion trillion atoms. This is approximately the same number as stars in the universe! Or consider the approximate mass of a single atom of carbon:

0.00000000000000000000002 grams,

i.e. 2×10^{-23} grams.

In the following section, **Amounts and Counts**, we will learn how chemists deal with handling such huge numbers of minute particle.

Amounts and Counts

Mass

Mass can be defined as an amount of matter. The larger the amount of matter, the larger its mass, and by the same token, the larger its weight. Is this always true? Can the terms mass and weight be interchanged?

As it turns out, any amount of matter will have a constant mass associated with it. A one inch cube of graphite will have a mass of 2.2 grams. The mass of this graphite cube will remain constant anywhere in the universe.

However, when considering the weight of an object the force of gravity needs to be taken into account. Since the shape of the earth isn't a perfect sphere, there will be very small changes in the cube's weight as you travel from pole to equator, but the approximate weight of the graphite will be 0.08 oz. The weight of the cube on the moon would be about 0.013 oz. since the gravitational field on the moon is 1/6 of that of earth's. Since chemistry is an exact science, it makes sense to describe amounts of matter in terms of mass rather than weight.

Counting Atoms

In chemistry, atoms are the building blocks of matter, and prior to any considerations about atomic structure, for now we will just represent them as very small spheres. Just how small are atoms? If it were possible to count the number of carbon atoms that would fit in the palm of your hand, the number of atoms would look something like

600,000,000,000,000,000,000,000 carbon atoms
or
6×10^{23} carbon atoms

In the investigation and quantifying of chemical reactions, it is imperative to know what quantities of chemicals react together and the quantity of product which results. In dealing with chemical reactions, it is inappropriate to discuss "handfuls" of atoms. We need a more precise method of "counting."

In dealing with very small particles such as atoms there must be a quantitative method for dealing with such astronomically large numbers as:

6×10^{23} atoms.

If there existed an instrument which could count one atom per second, a handful of carbon atoms would take approximately 6×10^{23} seconds to count or about 1.9×10^{16} years. Even if this instrument could count one million atoms a second, it would still take about 19 billion years to tally up the handfull of carbon atoms.

Since it is physically impossible to count each atom we use in a chemical laboratory, we need to resort to another method to "count" atoms.

Counting by weight (or mass) is a method with which we all have some familiarity. At one time or another, we have bought sour balls or candy corn by the pound. Knowing how many sour balls or candy corns were in 1 lb., it was much more convenient for the cashier to weigh them out rather then counting each piece of candy individually. By the same token at the hardware store, you can also buy standard nails of various sizes by the pound:

1 inch nails (two penny nails) 840 nails/pound

1 1/4 inch nails (three penny nails) 530 nails/pound

1 1/2 inch nails (four penny nails) 300 nails/pound

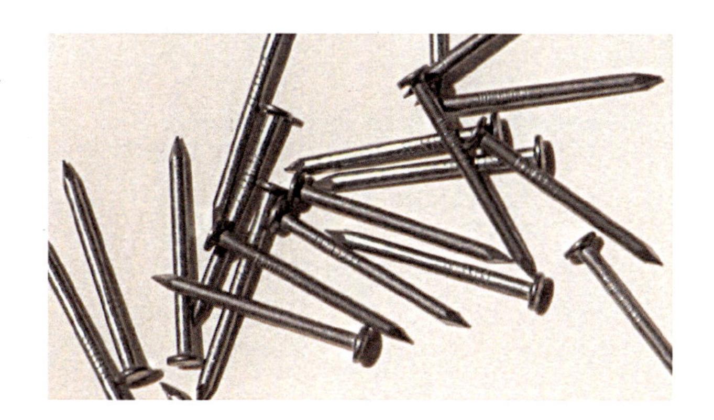

Figure 8.1

The terms two penny, three penny, and four penny goes back to England in the Middle Ages, when nails were fashioned by the village blacksmith.

Conversion Factors

Before attempting the following exercise, which involves converting counts into weights, the use of conversion factors needs to be discussed. In our day to day living, we constantly use conversion factors. Given a one dollar bill, without hesitation, we know its amount in pennies, nickels, dimes, or quarters. i.e. we can easily convert from one unit of currency (one dollar bill) to another (e.g. dimes).

Conversion factors are used to convert from one unit of measurement or count to another.

To know that two one dollar bills is the same as twenty dimes takes very little thought, but it will be very useful to take the time to construct and use the conversion factor your brain took a nanosecond to fire out.

Assignment 8.4

How many dimes equal 2 dollars?

Method

1. state relation between dollars and dimes:

1 dollar = 10 dimes

2. construct the conversion factors between dollars and dimes:

$$\frac{1\text{ dollar}}{10\text{ dimes}} \quad \text{or} \quad \frac{10\text{ dimes}}{1\text{ dollar}}$$

Note: For any relation between two units, two conversion factors can be developed.

3. set up problem with the appropriate conversion factor:

$$2\text{ dollars} \times \frac{10\text{ dimes}}{1\text{ dollar}} = 20\text{ dimes}$$

Note that of the two possible conversion factors, the one which cancelled out the dollar units was chosen. If we needed a conversion factor to calculate the number of dollars in 20 dimes, the other would have been chosen.

Dimensional Analysis

In problem solving, special care should always be given to unit cancellation. If the units cancel properly, the problem should solve correctly. This method is called Dimensional Analysis and will be an important part of problem solving in any science course.

Of course it was much quicker just doing this calculation in your head, but the conversion factors used in chemistry will take some getting used to. But in essence, they will be set up exactly the same way as this simple example.

Assignment 8.5

Complete the table with the appropriate conversion.

Unit	Unit	Conversion factors
miles	feet	$\frac{1\text{ mile}}{5280\text{ ft.}}$ and $\frac{5280\text{ ft.}}{1\text{ mile}}$
days	years	
grams	lbs	
liters	quarts	

yd^2	ft^2	
milliliters	liters	
microseconds	seconds	
yd^3	in^3	

Assignment 8.6

In a construction job you're doing, you are going to need:

2000 — 1 1/2 inch nails,

1500 — 1 1/4 inch nails

500 — 1 in. nails

How many pounds of each type of nail will you need?

1 inch nails (two penny nails)	840 nails/pound
1 1/4 inch nails (three penny nails)	530 nails/pound
1 1/2 inch nails (four penny nails)	300 nails/pound

1 mile = 5280 ft

454 grams = 1 lb

1.057 qts = I L

1 cm^3 = 1 mL

milli = 1/1,000 = 0.001

micro = 1/1,000,000 = 0.000001

Calculations

Scientific Dimensions and Notations

In order to take the conceptual aspects of your chemistry course into a laboratory and apply them to actual chemical reactions, a working knowledge of the units of measurement, accepted by the scientific community*, must be understood. In a scientific laboratory, there are particular units used in dealing with masses and volumes of substances.

Length

The unit of length that is used in the sciences is the meter (m). One meter is nearly 3.5 inches longer than one yard:

By a more scientific definition, one meter is defined as the distance that light travels in a vacuum in $3.33564095198 \times 10^{-9}$ seconds. Just as with Liters and grams, prefixes are used with meters to denote multiples and divisions of this unit.

1 meter = 39.37 inches

Mass

We have already discussed the difference between mass and weight. In the chemical laboratory, amounts of substances are measured in units of kilograms (e.g. grams, milligrams, and micrograms).

Although the mass of an amount of substance is constant whereas its weight is not, using a standard value of g for the gravitational constant, a relationship can be established between mass and weight:

1 kilogram (kg) = 2.205 lbs.

The standard for the kilogram is a cylinder of platinum-iridium alloy which is stored at the International Bureau of Weight and Measures in Sevres, France.

There are one thousand grams in one kilogram

1 kg = 1000 g

There are one thousand milligrams in one gram.

1 g = 1000 mg

There are one million micrograms in one gram

1 g = 1,000,000 mg

There are one billion nanograms in one gram

1 g = 1,000,000,000 ng

There are one trillion picograms in one gram

1 g = 1,000,000,000,000 pg

Note: mg is the symbol for the unit microgram. Since the lower case letter m is already used to indicate the prefix milli, the Greek letter mu, m, is chosen to represent the prefix micro.

Notice in the units milligrams and micrograms, the use of the prefixes *milli* and *micro*. In these instances, the prefixes indicate smaller divisions of the unit grams, whereas in the unit, kilogram, kilo indicates a multiple of the gram unit.

Using the relationships of units of mass on the previous page and what you learned about conversion factors perform the following exercises.

Assignment 8.7

1. What is the mass in kilograms of a 150 lbs man?

2. What is his mass in grams?

3. What is his mass in milligrams?

4. What is the mass of a 15 milligram object in micrograms?

Volume

In our daily routines, we encounter a considerable number of units which deal with volumes: gallons, quarts, pints, fluid ounces, cups, teaspoons, jiggers, and fifths to name a few. In the chemical laboratory, the basic unit of volume is the liter, L.

One liter is close to the volume one quart.

$$\mathbf{1\ liter = 1.057\ quarts}$$

As with the gram, the same prefixes are used to denote either multiples or divisions of this basic unit.

$$\mathbf{1\ kL = 1000\ L}$$

$$\mathbf{1\ L = 1000\ mL}$$

$$\mathbf{1\ L = 1{,}000{,}000\ \mu L}$$

Another unit of volume that is frequently used is the cubic centimeter, cc or cm^3. It is normally used in measurements of volume of solids, but it also has use in some clinical applications.

The relationship of the centimeter to the milliliter is quite simple:

$$\mathbf{1\ cm^3 = 1\ mL}$$

The Liter unit is considered a **derived unit**, in that it is defined as a cube that is 10 centimeters on each side. That is, the Liter is defined by measurements of length. There is no direct standard for the Liter as there are for the kilogram and the meter.

Assignment 8.8 Determine the number of Liters in 2,535 cm^3.

Assignment 8.9 What is the volume in cm^3 of a cube which measures 0.075 meters on each side?

(Note: 1 meter = 100 centimeters)

Density

In measuring the density of a chemical substance, we use a unit which employs both mass (g) and volume (mL or cm^3). The units of density are:

grams per mL (g/mL) for liquids
grams per cubic centimeter (g/cm^3) for solids
grams per Liter (g/L) for gases

The density of a substance can be understood as the number of particles (i.e. atoms, ions, molecules) that are contained in a unit volume (mL, cm^3, or L) of the substance.

There are reference handbooks (e.g. CRC Handbook of Chemistry and Physics) which contain tables of densities of multitudes of chemical substances. With the use of such a handbook, measuring the density of an unknown substance helps in the identification of that substance.

Measuring the density of a substance is quite simple:

1. Measure the mass of the substance in grams.
2. Measure the volume of the substance in mL (or cm^3 or L)
3. Divide the volume of the substance into its mass.

Assignment 8.10 A small metal cube is given to you for identification. With a metric ruler, you determine it to be 1.5 cm on each side. Placing the cube on an electronic balance, its mass is measured at 26.56 g.

1. Determine the density of the cube.

2. Using the following table, identify the substance.

Aluminum	d = 2.70 g/cm^3
Copper	d = 8.96 g/cm^3
Iron	d = 7.87g/cm^3
Lead	d = 11.35 g/cm^3

Answer ______________________

Aside from aiding in the identification of an unknown substance, densities have another practical use. Since the unit of density is actually composed of units of both mass and volume, knowing the mass of a known substance along with its density, we can calculate its volume.

Or, knowing its volume and its density, we can calculate its mass. In other words, the density of a substance can be used as a conversion factor to find a mass or a volume.

Assignment 8.11 At 20 °C, a sample of ethyl alcohol has a volume of 37 mL. Ethyl alcohol has a density of 0.7893 g/mL. What is the mass, in grams, of this sample?

Discussion: The temperature at which a density of a substance is determined is very important to note. Why is that?

Temperature

In the course of physical processes (e.g. dilution of sulfuric acid with water) and chemical processes (e.g. reaction of sulfuric acid and sodium hydroxide solution), energy is transferred in the form of heat, i.e. these processes either give off heat (exothermic processes) or absorb heat (endothermic processes).

For this reason, correct temperature measurement plays an important part in a chemical laboratory. There exist three scales of temperature measurement:

1. The Fahrenheit scale

2. The Celsius scale

3. The Kelvin scale

Aside from being the scale of choice in most countries outside of the United States, the Celsius scale is used in chemistry lab work. The Kelvin scale also appears frequently in the study of General Chemistry, especially in the areas of gas laws and thermodynamics.

In comparing the Celsius scale to the Kelvin scale (Figure 8.2) there are two points to remember:

1. The size of a Celsius degree and a Kelvin "degree" are the same.
2. The zero point (0 °C) on the Celsius scale is defined as the freezing point of water.

The zero point on the Kelvin scale is defined as absolute zero (-273.15 °C) that is, the point at which all atomic motion ceases.

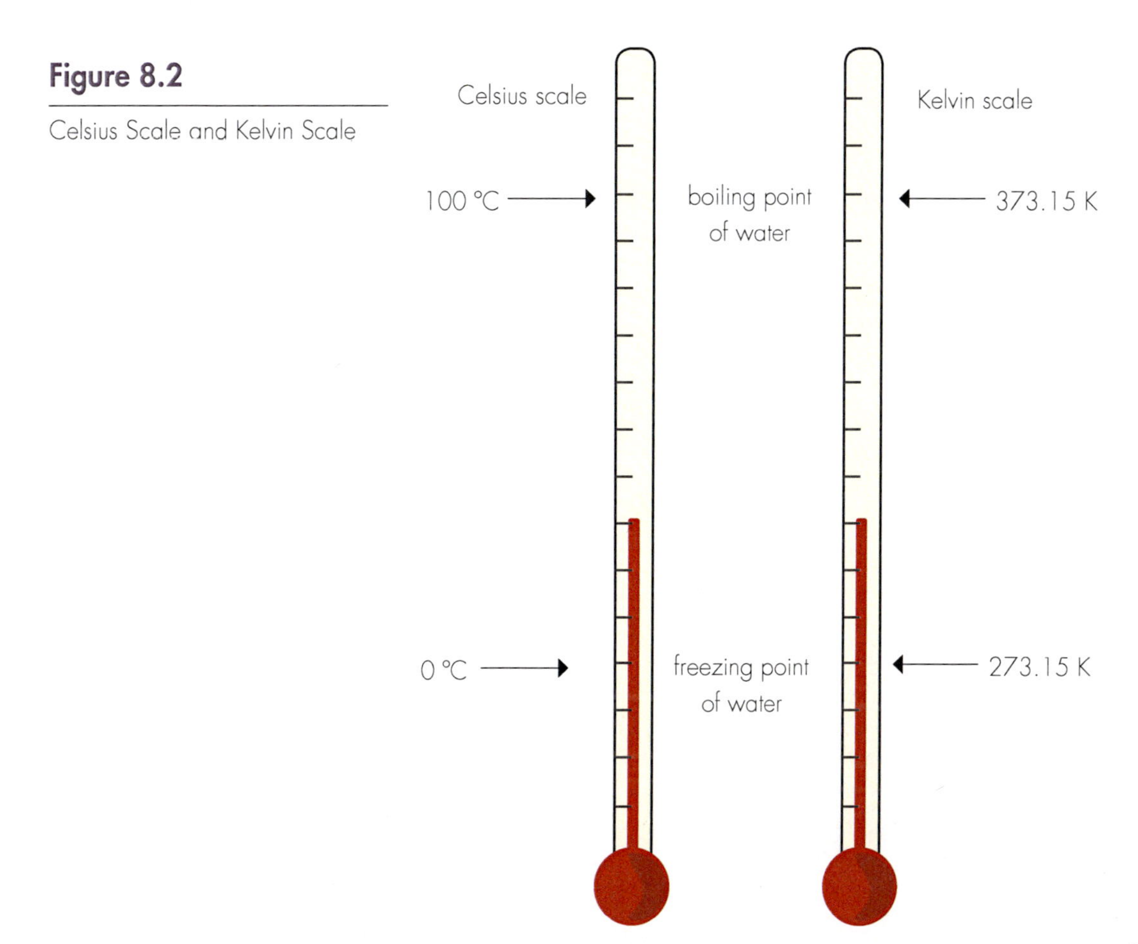

Figure 8.2

Celsius Scale and Kelvin Scale

Assignment 8.12 — Converting between Scales
(Refer to preceding figure)

1) Using the preceding figure, explain what is meant by the statement that the size of a Celsius degree and a Kelvin unit are the same.

2) Elemental bromine boils at 59 °C, on the Celsius scale.
What would this temperature be on the Kelvin scale?

____

3) Write a formula for converting degrees Celsius to units of Kelvin.

____

4) Elemental helium boils at 4 K. What would this temperature be on the Celsius scale?

____

5) Write a formula for converting from the Kelvin scale to the Celsius scale.

____

Converting between the Celsius scale and the Kelvin scale is merely the addition or subtraction of 273.15, depending on which conversion is needed since the degrees or units of division are the same on both scales.

The Relationship of Temperature and Heat

In our daily conversations, heat and temperature are sometimes used interchangeably. By definition, heat is a form of energy that flows spontaneously from one object to another cooler object. The temperatures of both objects determine which direction the heat flows. Heat always flows from a hotter object to a cooler object. Heat is a form of energy, and temperature is a property of matter which determines which direction heat will flow. A thermometer is an instrument which contains a substance with a heat sensitive property. Most lab thermometers contain either alcohol or mercury, both of which expand with heat.

The Metric System and SI Units

Meters, Liters, and grams are all part of the Metric System of measurement which is used in most countries outside of the United States. It is a much more concise and convenient system to use than the English System.

Within the scientific community, seven units from the Metric System are used to denote:

length (meter)

mass (kilogram)

time (second)

temperature (Kelvin)

Electric current (ampere)

luminous intensity (candela)

number of particles (mole)

These seven units are called the SI Units, from the French Systeme International d'Unites.

In chemistry, we will be dealing mainly with meters, grams (more convenient than kilograms), seconds, Kelvins, and moles.

Conversions and Conversion Factors

Simple Single-Step Conversion

Assignments:

Example:

37 yards = ? inches

Given Units: yards Desired Units: inches

equivalence: 1 yd = 36 inches

conversion factors: $\frac{1 \text{ yard}}{36 \text{ inches}}$ or $\frac{36 \text{ inches}}{1 \text{ yard}}$

set up: $37 \text{ yards} \times \frac{36 \text{ inches}}{1 \text{ yard}} = \boxed{\textbf{1,332 inches}}$

Assignment 8.13

45 inches = ? yards Given Units: Desired Units:

equivalence:

conversion factors: ______________ or ______________

set up: × ____________ = [ ]

Assignment 8.14

18.2 quarts = ? pints Given Units: Desired Units:

equivalence:

conversion factors: ______________ or ______________

set up: × ____________ = [ ]

Assignment 8.15

75.9 μg = ? pg Given Units: Desired Units:

May be done in one or two steps

equivalence:

conversion factors: ______________ or ______________

conversion factors: ______________ or ______________

set up: × ____________ × ____________ = [ ]

Assignment 8.16

4.98 pg = kg Given Units: Desired Units:

equivalence:

conversion factors: ______________ or ______________

conversion factors: ______________ or ______________

set up: × ____________ × ____________ = [ ]

Assignment 8.17

2.73 kL = ? L Given Units: Desired Units:

equivalence:

conversion factors: ______________ or ______________

set up: × ____________ = [ ]

Problem 8.1 A student needs 236 mL of ethyl alcohol for an experiment. He will use a graduated cylinder that reads in milliliter gradations. How many milliliters of ethyl alcohol will he measure?

Given Units: Desired Units:

equivalence:

conversion factors: ______________ or ______________

conversion factors: ______________ or ______________

set up: × ____________ × ____________ = [ ]

Problem 8.2 An average drop of water is one-twentith of a milliliter. How many drops are in 3.7 L?

Given Units: Desired Units:

equivalence:

conversion factors: ______________ or ______________

conversion factors: ______________ or ______________

set up: × ____________ × ____________ = [ ]

Problem 8.3 The mass of one carbon atom is 1.99×10^{-23} grams. How many carbon atoms are in 275 nanograms of carbon?

Given Units: Desired Units:

equivalence:

conversion factors: ______________ or ______________

set up: × ____________ = [ ]

Problem 8.4 If the mass of one carbon atom is 1.99×10^{-23} grams, 1.32 grams of carbon will contain how many atoms?

Given Units: Desired Units:

equivalence:

conversion factors: ____________ or ____________

set up: × ____________ = [ ]

Problem 8.5 The radius of a typical atom is one angstrom. How many meters does this equal?

Given Units: Desired Units:

equivalence:

conversion factors: ____________ or ____________

set up: × ____________ = [ ]

Additional Exercises

Simple Multiple-Step Conversion

Example:

3.2 miles = ? inches Given Units: miles Desired Units: inches

Using the conversion table, construct a road map which indicates the conversion steps.

From the conversion table two conversions: one that contains the given units (miles) and another that contains the desired units (inches). Aside from the desired and given units, both conversions need to contain common units (feet).

	Given unit	common unit		common unit		desired unit
equivalence:	1 mile =	5280 feet	and	1 foot	=	12 inches

The road map for the conversion of miles to inches would be:

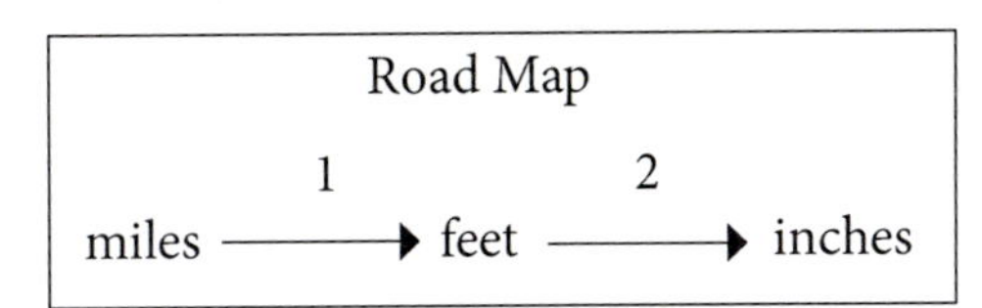

Conversion factors for step 1: $\frac{1 \text{ mile}}{5280 \text{ ft}}$ or $\frac{5280 \text{ ft}}{1 \text{ mile}}$

Conversion factors for step 2: $\frac{1 \text{ ft}}{12 \text{ in}}$ or $\frac{12 \text{ in}}{1 \text{ ft}}$

Combine the two conversion factors as in the following set up:

set up: $3.2 \text{ mile} \times \frac{5280 \text{ ft}}{1 \text{ mile}} \times \frac{12 \text{ in}}{1 \text{ ft}} = \boxed{202{,}752 \text{ in}}$

Assignment 8.18 (two step)

502.32 cm = ? yards Given Units: Desired Units:

Road Map

equivalences:

conversion factors for step 1: ______________ or ______________

conversion factor for step 2: ______________ or ______________

set up: ×______________ ×______________ = [ ]

Assignment 8.19 (two step)

15.93 liters = ? gallons Given Units: Desired Units:

Road Map

equivalences:

conversion factors for step 1: ______________ or ______________

conversion factor for step 2: ______________ or ______________

set up: × ______________ × ______________ =

Assignment 8.20 (three step)

10.056 seconds = ? years Given Units: Desired Units:

Road Map

equivalences:

conversion factors for step 1: ______________ or ______________

conversion factor for step 2: ______________ or ______________

conversion factor for step 3: ______________ or ______________

set up: ______________ ______________ ______________

× × × =

Assignment 8.21 (three step)

2.35 qts = ? mL Given Units: Desired Units:

Road Map

equivalences:

______________ ______________

conversion factors for step 1: ______________ or ______________

conversion factor for step 2: ______________ or ______________

conversion factor for step 3: or

set up: ______________ ______________ ______________

× × × =

Assignment 8.22 (two step)

2,045,000 pg = ? kg Given Units: Desired Units:

Road Map

equivalences:

conversion factors for step 1: ______ or ______

conversion factor for step 2: ______ or ______

set up: × ______ × ______ =

Assignment 8.23 (two step)

$2.3 \times 10^{12} \mu L = ?\ HL$ Given Units: Desired Units:

Road Map

equivalences:

conversion factors for step 1: ______ or ______

conversion factor for step 2: ______ or ______

set up: × ______ × ______ =

Assignment 8.24 (four step) (hint: 1 mL = 1 cm^3)

5.3 gal = ? cm^3 Given Units: Desired Units:

Road Map

equivalences:

conversion factors for step 1: ________ or ________

conversion factor for step 2: ________ or ________

conversion factor for step 3: ________ or ________

conversion factor for step 4: ________ or ________

set up:

×________ ×________ ×________ ×________ =

Assignment 8.25 (two step)

2.46 km = ? miles Given Units: Desired Units:

Road Map

equivalences:

conversion factors for step 1: ____________ or ____________

conversion factor for step 2: ____________ or ____________

set up: × ____________ × ____________ = [ ]

Assignment 8.26 (two step)

2,497 cm³ = ? L Given Units: Desired Units:

Road Map

[ ]

equivalences:

conversion factors for step 1: ____________ or ____________

conversion factor for step 2: ____________ or ____________

set up:

× ____________ × ____________ × ____________ = [ ]

Assignment 8.27 (three step)

2.4×10^4 mL = ? pints Given Units: Desired Units:

Road Map

equivalences:

conversion factors for step 1: ______________ or ______________

conversion factor for step 2: ______________ or ______________

conversion factor for step 3: ______________ or ______________

set up:

× ____________ × ____________ × ____________ =

Problem 8.6 A single carbon atom weighs 1.99×10^{-23} g. How many pounds does 5.25×10^{25} carbon atoms weigh?

Given Units: Desired Units:

equivalences:

conversion factors for step 1: ____________ or ____________

conversion factors for step 2: ____________ or ____________

Road Map

set up:

Problem 8.7. The hydrogen atom has a diameter of 10^{-8} cm. What is the total volume in liters of 6.022×10^{27} hydrogen atoms.

(hint 1: treat the hydrogen atom as a sphere. volume of sphere = $4/3\ \pi r^3$)

(hint 2: given units will be in units of volume)

Given Units: Desired Units:

equivalences:

conversion factors for step 1: ______________ or ______________

conversion factors for step 2: ______________ or ______________

Road Map

set up:

Problem 8.8. If one gold atom weighs 3.27×10^{-22} g, how many moles of gold atoms are contained in a 1 cm^3 cube of gold that weighs 19.2 g?

(hint: The mole is the unit for counting atoms: 1 mole = 6.022×10^{23} atoms)

Given Units: Desired Units:

equivalences:

conversion factors for step 1: ______________ or ______________

conversion factors for step 2: ______________ or ______________

Road Map

set up:

Problem 8.9. If the diameter of a hydrogen atom is 1 angstrom, how many hydrogen atoms need to be placed in a straight line to measure up to a 12 inch ruler?

(hint: 1 angstrom = 10^{-10} meters)

Given Units: Desired Units:

equivalences:

conversion factors ______________ or ______________

Road Map

set up:

Problem 8.10 An artist takes a blank piece of drawing paper which weighs 3.0128 g. After rendering a pencil drawing the mass of the paper is 3.0387 g. How many moles of carbon atoms were used in the drawing?

(hint 1: the lead in his graphite pencil is made of pure carbon)

(hint 2: the mass of a single carbon atom is 1.99×10^{-23} g)

(hint 3: 1 mole = 6.022×10^{23} atoms)

Given Units: Desired Units:

equivalences:

conversion factors ______________ or ______________

Road Map

set up:

Counting by Weight in Chemistry

Just as we can count candy and nails by weight, we also count molecules and atoms by weight (mass).

Just as three hundred 1 1/2 inch nails weigh 1 lb.

6.022×10^{23} (approximately one handful) carbon atoms have a mass of 12.01 g.

By this method, you could determine the number of carbon atoms by the number of grams of carbon atoms.

Determine the number of carbon atoms or the mass of carbon atoms in the following exercise

(Remember to use the proper conversion factors)

Assignment 8.28

Mass of Carbon Atoms	Number of Carbon Atoms
24.02 g	
	2.4×10^{24}
0.1201 g	
	3.01×10^{23}
180.15 g	

Calculations (Assignment 8.28)

Unit of Counting Atoms: The Mole

In the preceding section, we determined the counts of carbon atoms by weight or mass, and one thing for certain is that numbers, such as are large

$$6.022 \times 10^{23},$$

and cumbersome to work with, however special units of counting exit which make things much easier. The unit of counting that is used in chemistry is the mole.

one mole = 6.022×10^{23}

Therefore

6.02×10^{23} carbon atoms equals one mole of carbon atoms.

6.02×10^{24} carbon atoms equals ten moles of carbon atoms.

3.01×10^{23} carbon atoms equals one half of a mole of carbon atoms.

Rather than counts of atoms involving very large exponential values, it is much more convenient to use units of moles rather than large counts.

Assignment 8.29

Number of Carbon Atoms	Number of Moles of Carbon Atoms
6.02×10^{25}	
4.52×10^{23}	
1.51×10^{25}	
3.01×10^{23}	

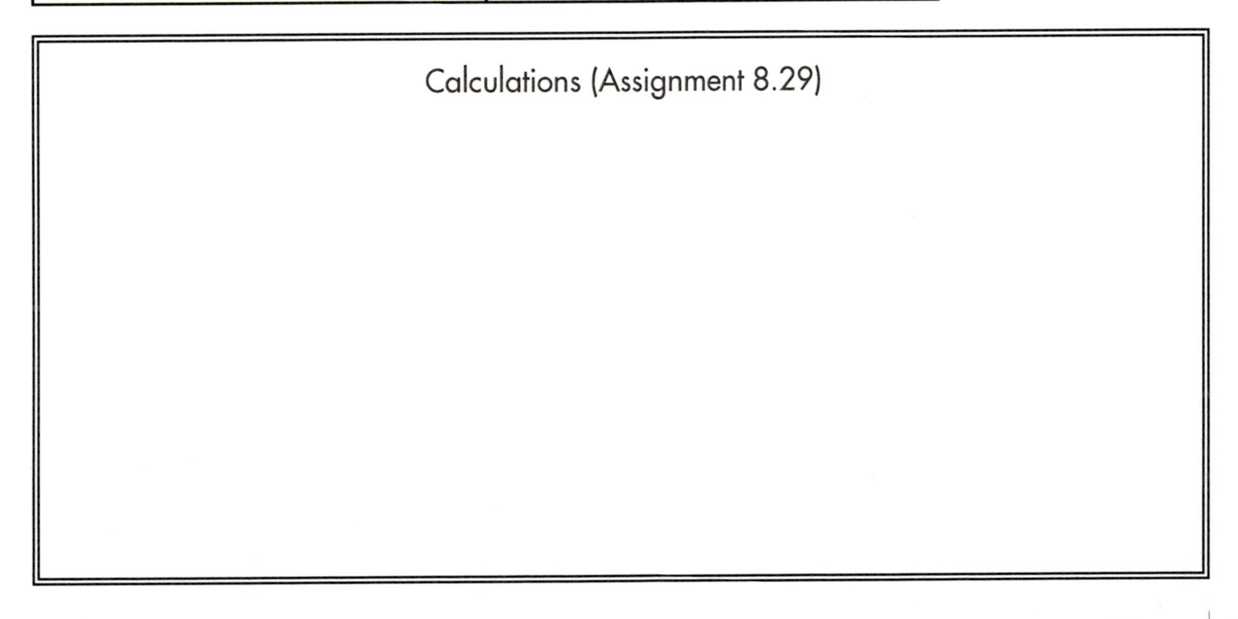

These preceding sections and related exercises on counting by mass and use of the mole unit are extremely important to students studying chemistry. For this reason, we will work more of these types of exercises before moving on to the next section.

Up to this point, we have only dealt with carbon atoms in the conversion exercises. We know that 6.02×10^{23} carbon atoms equals one mole of carbon atoms and has a mass of 12.01 grams. But what about other atoms?

One mole of any atom will equal 6.02×10^{23} atoms:

One mole of nitrogen atoms equals 6.02×10^{23} nitrogen atoms

One mole of oxygen atoms equals 6.02×10^{23} oxygen atoms

One mole of magnesium atoms equals 6.02×10^{23} magnesium atoms

However

Each type of atom (e.g. carbon, nitrogen, oxygen, magnesium, etc.) Has its own mass, therefore

One mole of carbon atoms will have a mass 12.01 grams.

One mole of nitrogen atoms will have a mass 14.00 grams.

One mole of oxygen atoms will have a mass 15.99 grams.

One mole of magnesium atoms will have a mass 24.31 grams.

Work the following problems:

Problem 8.11 How many moles of nitrogen atoms are in 0.035 grams of nitrogen?

Ans.___________

Problem 8.12 How many moles of carbon atoms are in 15.00 grams of carbon?

Ans.___________

Problem 8.13 A student is going to run a chemical reaction which involves 0.25 moles of magnesium. How many grams of magnesium does the student need to weigh out for this reaction?

Ans.___________

Problem 8.14 Four moles of oxygen atoms contains how many oxygen atoms?

Ans.___________

Problem 8.15 How many moles of magnesium are in 2.43 grams of magnesium?

Ans.___________

Problem 8.16 How many atoms of magnesium are in 2.43 grams of magnesium?

Ans.___________

Problem 8.17 One cup of milk contains 285 mg of calcium. How many calcium atoms are contained in one cup of milk?

Ans. ___________

Additional Exercises

Assignment 8.30 Write the conversion factors for the following conversions: (Note: some conversions may involve more than one step)

a) meters to picometers

b) kiloliters to liters

c) square inches to square yards

d) cubic inches to cubic yards

e) cubic centimeters to liters

f) kilograms to nanograms

g) quarts to milliters

h) miles per hour to meters per day

i) milliliters per second to gallons per day

Assignment 8.31 Knowing that one mole of magnesium has a mass of 24.31 grams, complete the following table:

Mass of Magnesium	Number of Magnesium Atoms	Number of Moles of Magnesium Atoms
24.31 g		
		2.5
	2.5×10^{15}	
		15
150 g		
	1.23×10^{3}	
		1.2×10^{4}
4.32×10^{-3} g		
	6.022×10^{28}	
10 kg		

Additional Problems

Problem 8.18 If one mole of nitrogen atoms has a mass of 14.00 grams, what is the mass of one nitrogen atom?

Problem 8.19 The density of carbon (graphite) is 2.25 g/cm^3. How many carbon atoms per cm^3 does this equal?

Problem 8.20 Benzene is a liquid organic compound with a density of 0.87901. What is the mass of 25 mL of Benzene?

Problem 8.21 Nitrogen gas has a density of 1.2506 g/liter. What is the mass of the nitrogen gas when it fills a 10 kiloliter volume?

Problem 8.22 A sample of gold has a mass of 10 grams and a volume of 0.516 cm^3. What is its density?

Problem 8.23 Concentrated sulfuric acid has a density of 1.764 g/mL. A student is running a chemical reaction which requires 12.5 g of this acid. How many milliliters of sulfuric acid will the student need to transfer to the reaction flask?

Stoichiometry

p2

Chapter 9

In part one of this book, we dealt with the nature and structure of matter, i.e. chemical substance. In this final unit, we will discuss the interaction of these substances, i.e. breaking old bonds to form new bonds in the synthesis of new chemical species. Before preceding further, please review balanced chemical equations in section, The Symbolic Language of Chemistry.

Aside from accounting for the conservation of mass in a chemical reaction, as well as much physical description, a balanced chemical equation also provides a reference point for quantitative information. The balanced chemical equation for the combustion of methane is stated as,

$$CH_{4\,(g)} + 2O_{2\,(g)} \longrightarrow CO_{2\,(g)} + 2H_2O_{(g)}$$

In this balanced equation, the coefficients state the ratio of the different molecules involved in this reaction: one molecule of methane reacts with two molecules of oxygen to produce one molecule of carbon dioxide and two molecules of water.

These ratios can also be scaled up to mean: two moles of methane and four moles of oxygen will react to produce two moles of carbon dioxide and four moles of water.

So you can see that within a balanced chemical equation there is a "locked in" ratio between the reactants and products. But what use is this information? In order to use these ratios on a more practical level (i.e. the chemical laboratory) we regard the coefficients of the equation as molar ratios:

$CH_{4\,(g)}$	+	$2O_{2\,(g)}$	$\longrightarrow$	$CO_{2\,(g)}$	+	$2H_2O_{(g)}$
1	:	2	:	1	:	2
1 molecule of methane	:	2 molecules of oxygen	:	1 molecule of carbon dioxide	:	2 molecules of water
1 mole of methane	:	2 moles of oxygen	:	1 mole of carbon dioxide	:	2 moles of water

Regarding these coefficients as moles, a chemist can apply the ratios to predict laboratory yields or to determine the amounts of reagents needed:

Assignment 9.1 How many moles of carbon dioxide and water are produced from 3.5 moles of methane and 7 moles of oxygen?

In dealing with laboratory reactions, chemists must work directly with masses of substances, rather than the number of moles.

Assignment 9.2 Rework the above exercise in terms of grams of carbon dioxide and grams of water produced. *Note: molar mass of carbon dioxide is 44 g/ mol molar mass of water is 18 g/mol.*

154 g CO_2 and 126 g H_2O produced

Assignment 9.3 Using the balanced equation for the combustion of methane, determine how many molecules of methane and oxygen are needed to produce 65 molecules of carbon dioxide and 120 molecules of water?

Actual Yield, Theoretical Yield and Percent Yield

The answer to the above problem, of course, is 3.5 moles of carbon dioxide and 7 moles of water. When the amount of product(s) is calculated from a balanced equation, this amount is referred to as a **theoretical yield**. But in the real world, chemical reactions, for many reasons, do not result in the calculated yield (theoretical yield). The amount of product that results from the actual laboratory reaction is called an **actual yield**. Returning to the above reaction again, the chemist calculates that by reacting 3.5 moles of methane and 7 moles of oxygen, the theoretical yield for the reaction should be 3.5 moles of carbon dioxide and 7 moles of water.

However, upon running the reaction, the chemist recovers only 3.2 moles of carbon dioxide and 6.7 moles of water, the actual yield. Since chemists are very quantitative people, she calculates a **percent yield** for both of the products of the reaction. The calculation is performed as follows:

$$\% \text{ Yield} = \frac{\text{Actual Yield}}{\text{Theoretical Yield}} \times 100$$

Assignment 9.4 Given the preceding information, calculate the % yield of carbon dioxide and the % yield of water.

Stoichiometric Calculations

With the exception of the last exercise which required mass to mole and mole to mass conversions, most of what was calculated was done by inspection. That is, we could pretty much derive the answers by just looking at the equation. We shall now progress into more complicated, and more useful, applications of balanced equations.

Instead of relating the two sides of the equation—the reactant side to the product side—we will now relate only one chemical species to another. Consider the industrial process of generating ammonia from hydrogen gas and nitrogen gas:

$$3\ H_{2\,(g)} + N_{2\,(g)} \longrightarrow 2\ NH_{3\,(g)}$$

Given the 3:1:2 molar ratio of this reaction, the following relations can be established:

3 moles of H_2 requires 1 mole N_2

1 mole N_2 requires 3 moles H_2

3 moles H_2 produces 2 moles NH_3

1 mole N_2 produces 2 moles NH_3

2 moles NH_3 requires 3 moles H_2

2 moles NH_3 requires 1 mole N_2

The one to one relations are very important when calculating amounts of needed reactants or in more complicated calculations involving the determination of limiting reactants and the resultant theoretical yields. As an example, let's try the ammonia reaction.

In order to generate 135 moles of ammonia, how many moles of hydrogen and nitrogen gas are needed?

1. Refer back to the balanced equation:

$$3\ H_{2\,(g)} + N_{2\,(g)} \longrightarrow 2\ NH_{3\,(g)}$$

2. Note the relations between ammonia and hydrogen

$$2\ NH_3 : 3\ H_2$$

and ammonia and nitrogen

$$2\ NH_3 : N_2$$

3. With these relations, two separate sets of conversion factors can be established:

$$\frac{2\ NH_3}{3\ H_2} \qquad \frac{3\ H_2}{2\ NH_3} \qquad \text{and} \qquad \frac{2\ NH_3}{N_2} \qquad \frac{N_2}{2\ NH_3}$$

4. With these conversion factors, the needed number of moles of both reactants can be calculated as follows:

$$135 \text{ moles } NH_3 \times \frac{3\ H_2}{2\ NH_3} = 202.5 \text{ Moles of } H_2$$

$$135 \text{ moles } NH_3 \times \frac{N_2}{2\ NH_3} = 67.5 \text{ Moles of } N_2$$

Assignment 9.5 Using the relations from the previous balanced equation:

a) determine the number of moles of nitrogen gas, N_2, required to react with 223 moles of hydrogen gas, H_2.

b) express the amount of nitrogen gas in grams molar mass of nitrogen is 28.0 g/mol.

Limiting Reactant

Let's move away from the simplistic and take a look at a more real world method of use. As mentioned above, chemical reactions will, in all probability, not result in 100% yields. Think of the products of most chemical reactions (solution reactions) as the result of reactant molecules colliding together. Of course, much more is involved, but in order for the atoms to rearrange to form new compounds, the reactant molecules must first physically "bump into" each other.

In the case of methane and oxygen,

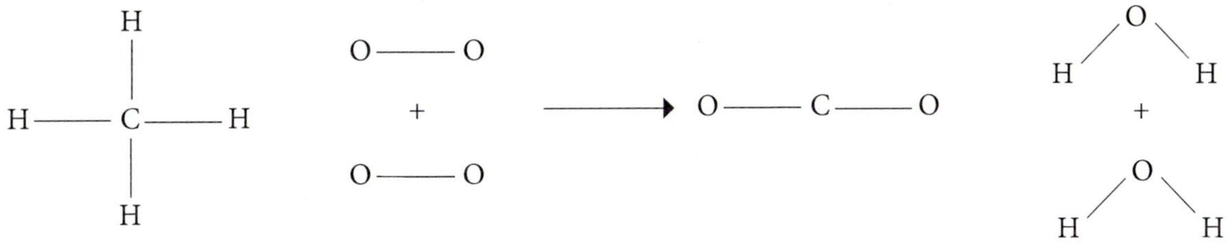

```
     H                                                      O
     |          O——O                                      /   \
                                                         H     H
H——C——H          +      ————>   O——C——O                  +
     |                                                      O
     H          O——O                                      /   \
                                                         H     H
```

You can imagine as more reactant molecules collide and new product molecules are formed, fewer and fewer reactant moles remain.

With fewer reactant molecules, there are fewer collisions and fewer product molecules formed. Thus, from the collision aspect alone, within practical time limitations, to expect a 100% yield of products is highly unlikely. In order to maximize the yield of product (normally, the chemist is interested in only one of the products), one of the reactants will be used in excess to ensure all of the other reactant is consumed. The reactant that is completely consumed is called the limiting reactant. When all the limiting reactant is used up, production stops.

How it works:

Imagine making ham sandwiches. The recipe calls for two slices of white bread and one slice of ham, no mustard. In terms of a balanced equation,

$$\text{2 slices bread}_{(w)} + \text{1 slice ham}_{(nm)} \longrightarrow \text{1 sandwich.}$$

With this ratio, it is obvious, that with 6 slices of bread and 3 slices of ham, 3 sandwiches are produced. Now, in terms of having a limiting reactant, suppose 6 slices of bread and only 2 slices of ham are used. The result is 2 sandwiches with 2 slices of bread left over. As a limiting reactant, the 2 slices of ham determined how many sandwiches could be made. The bread was in excess. Reversing the limiting reactant to the bread,

$$\text{3 slices of ham and 4 slices of bread} \longrightarrow \text{2 sandwiches}$$

the bread determines the amount of product and the ham is in excess.

How does this apply to chemistry? By using one of the chemical reactants in excess, a chemist can push a reaction to maximize the product. Returning to the combustion reaction,

$$CH_{4\,(g)} + 2O_{2\,(g)} \longrightarrow CO_{2\,(g)} + 2H_2O_{\,(g)}$$

If a chemist wants to ensure a high percent yield of product (5 moles of carbon dioxide and 10 moles of water), he can "push" the reaction by using, for example, 5 moles of methane and an excess of oxygen, or 10 moles of molecular oxygen and an excess of methane.

In the former case, an excess of oxygen would enhance collisions of most methane molecules with oxygen. When the 5 moles of methane are used up, 5 moles of carbon dioxide and 10 moles water are theoretically produced and the reaction stops.

With this example in mind, in the following reaction:

Assignment 9.6

a) identify the limiting reagent

b) state the amount of excess reactant that remains

c) state the amount of product yield

4 moles of magnesium are reacted with 4 moles of molecular oxygen

$$2\,Mg_{\,(s)} + O_{2\,(g)} \longrightarrow 2\,MgO_{\,(s)}$$

limiting reactant ____________________

excess reactant remaining ____________

amount of product formed ____________

Rework the previous exercise in terms of grams.

Assignment 9.7 170.1 grams magnesium are reacted with 223.86 molecular oxygen molar mass of magnesium is 24.30 g/mol molar mass of molecular oxygen is 31.98 g/mol

$$2\,Mg_{\,(s)} + O_{2\,(g)} \longrightarrow 2\,MgO_{\,(s)}$$

hint: convert grams to moles to determine limiting reactant

limiting reactant ____________________

hint: find the excess in moles and convert to grams

excess reactant remaining in grams ____________

hint: convert moles of product to grams

amount of product formed in grams ____________

Whenever using balanced equations to determine amounts, always use the ratios in terms of moles and never in grams!

Let's finish this section with a rigorous problem involving limiting reactants.

Iron is obtained from magnetite, Fe_3O_4, in the following reaction:

$$Fe_3O_{4\,(s)} + 4\,CO_{(g)} \longrightarrow 3\,Fe_{(l)} + 4\,CO_{2\,(g)}$$

How many kilograms of iron can be obtained by reacting one kilogram of magnetite with one kilogram of carbon monoxide?

Method

1. **Determine the number of moles of Fe_3O_4 and the number of moles of CO at start of the reaction, and**
2. **determine which of the two is the limiting reactant.**

molar mass of Fe_3O_4 is 231.51 g/mol

molar mass of CO is 28.00 g/mol

$$1000\text{ g } Fe_3O_4 \times \frac{1\text{ mol}}{231.52\text{ g}} = 4.32\text{ mol } Fe_3O_4$$

$$1000\text{ g CO} \times \frac{1\text{ mol}}{28.00\text{ g}} = 35.71\text{ mol CO}$$

Since the molar ratio between the Fe_3O_4 and the CO is 1:4, It's fairly easy to designate the limiting reactant by inspection, however, let's demonstrate the use of conversion factors in the determination of the limiting reactant.

Using the 1:4 molar ratio between Fe_3O_4 and CO, the following conversion factors are obtained:

$$\frac{1\text{ mol } Fe_3O_4}{4\text{ mol CO}} \quad \text{and} \quad \frac{4\text{ mol CO}}{1\text{ mol } Fe_3O_4}$$

It doesn't matter whether this next calculation is performed with the number of moles of magnetite or the number of moles of carbon monoxide. **The limiting reactant can be determined using either one.**

Let's arbitrarily choose the magnetite, Fe_3O_4.

We have determined that there is 4.32 mol of Fe_3O_4. We will now determine if there is enough CO to react with this amount.

$$4.32 \text{ mol } Fe_3O_4 \times \frac{4 \text{ mol CO}}{1 \text{ mol } Fe_3O_4} = 17.28 \text{ mol CO needed}$$

this amount, 17.28 mol CO, is then compared to the amount of CO
we have for the reaction.
We have 35.71 moles of CO on hand.
We only need 17.28 moles of CO.

Therefore the CO is in excess and the Fe_3O_4 is the limiting reactant.

This same conclusion could have been made if the calculation had been done using the 35.71 mole of CO rather than the 4.32 moles of Fe_3O_4.

Assignment 9.8 Confirm the above statement.

Up to this point:

1. **we determined the molar amounts of reactants as**

4.32 mol Fe_3O_4 and 35.71 mol CO,

and

2. **based on this information, we identified the limiting reactant as**

Fe_3O_4.

Since depletion of the limiting reactant will determine how many moles of product are formed, the 4.32 moles of Fe_3O_4 will be used to calculate the amount of iron produced.

Referring again to the balanced equation, we now look for the relation between the magnetite, Fe_3O_4, and the iron, Fe.

$$Fe_3O_{4\,(s)} + 4\,CO_{(g)} \longrightarrow 3\,Fe_{(l)} + 4\,CO_{2\,(g)}$$

The 1 : 3 ratio yields the following conversion factors:

$$\frac{\text{I mol } Fe_3O_4}{3 \text{ mol Fe}} \quad \text{and} \quad \frac{3 \text{ mol Fe}}{1 \text{ mol } Fe_3O_4}$$

Solving for the amount of iron produced,

4.32 mol Fe3O4 × 3 mol Fe = 12.96 mol Fe

1 mol Fe3O4

we proceed to the final step of converting from moles of iron to kg of iron,

12.96 mol Fe × 55.85 g × 1 kg = 0.724 kg Fe

1 mol 1000 g

Summary

1. The molar ratios of the reactants and products of a chemical reaction are taken directly from the coefficients of the reaction's balanced equation.
2. The coefficients of the balanced equations represent the number of moles of each component of both reactants and products.
3. The relations of these coefficients within the balanced equation are called molar ratios and are used to construct conversion factors.
4. To "push" the yield of a reaction, one of the reactants is used as a limiting reactant.
5. When a limiting reactant is used, it will determine the theoretical yield of the reaction.

Review Problems

Assignment 9.9 Reacting benzene, C_6H_6, with Br_2 produces the organic compound bromobenzene, C_6H_5Br, and hydrogen bromide gas, as indicated in the balanced equation:

$$C_6H_{6\,(l)} + Br_{2\,(l)} \longrightarrow C_6H_5Br_{\,(l)} + HBr_{\,(g)}$$

A student mixes 39.0 grams of benzene with 25.0 grams of liquid bromine, and recovers 41.4 grams of bromobenzene.

a. Convert the mass of each reactant into moles.

b. Determine the limiting reactant.

c. Calculate the theoretical yield.

d. Calculate the percent yield.

Assignment 9.10 Hydrogen sulfide and sulfur dioxide react on contact to form sulfur and water.

a. Balance the equation for this reaction:

$$\_\_ H_2S_{(g)} + \_\_ SO_{2\,(g)} \_\_ S_{(g)} + \_\_ H_2O_{(l)}$$

b. If 5.00 g of H_2S gas is mixed with 5.00 g SO_2 gas, how many moles of each are being used?

c. Of the two gases, which is in excess and which is the limiting reactant?

d. How many moles of the excess gas will remain at the end of the reaction?

Assignment 9.11 Aqueous solutions of calcium chloride and sodium phosphate react according to the equation:

$$3CaCl_{2\,(aq)} + 2\,Na_3PO_{4\,(aq)} \longrightarrow Ca_3(PO_4)_{2\,(s)} + 6\,NaCl_{(aq)}$$

a. Identify the limiting and excess reactants in a reaction mixture containing 50.0 g $CaCl_2$ and 100.0 g of Na_3PO_4.

b. How many grams of the excess reactant will remain when the reaction is over?

More Review Problems

Balance the following reactions:

Assignment 9.12

$$P_{4\,(s)} + Cl_{2\,(g)} \longrightarrow PCl_{3\,(l)}$$

__P __Cl   __P __Cl

Assignment 9.13

$$KNO_{3\,(aq)} + NH_4Cl_{(aq)} \longrightarrow KCl_{(aq)} + NH_4NO_{3\,(aq)}$$

(Instead of accounting for each separate atom, recognize the polyatomic ions)

__K^+ __NO_3^- __NH_4^+ __Cl^- __K^+ __Cl^- __NH_4^+ __NO_3^-

Assignment 9.14

$$Ba(OH)_{2\,(aq)} + (NH_4)_2SO_{4\,(aq)} \longrightarrow BaSO_{4\,(s)} + NH_{3\,(g)} + H_2O_{(l)}$$

Assignment 9.15 In the combustion of octane, 2 moles of octane combines with 25 moles of oxygen to yield 16 moles of carbon dioxide and 18 moles of water:

$$2\ C_8H_{18\,(l)} + 25\ O_{2\,(g)} \longrightarrow 16\ CO_{2\,(g)} + 18\ H_2O_{(l)}$$

a. If we combust 8 moles of octane in an excess of oxygen, how many moles of carbon dioxide will form?

b. If we combust 684 g of octane in an excess of oxygen, how many moles of carbon dioxide will form?

c. If we combust 684 g of octane in an excess of oxygen, how many grams of carbon dioxide will form?

Assignment 9.16 According to the reaction,

$$P_{4\,(s)} + 5\ O_{2\,(g)} \longrightarrow P_4O_{10\,(s)}$$

how many grams of tetraphosphorous decoxide, P_4O_{10}, will form by reacting 496 g of phosphorous (MW 124 g/mol) in an excess of oxygen?

Concentration

In dealing with solutions units of concentration such as molarity are very important.

Molarity:

The widely used way of quantifying concentration

$$\text{Molarity} = \frac{\text{solute of moles}}{\text{volume of solution (L)}}$$

Assignment 9.17

Calculate the molarity (M) of a solution that contains 3.65 grams of HCl in 2.0 liters of solution.

Assignment 9.18

Calculate the mass of $Ba(OH)_2$ required to prepare 2.50 liters of a 0.06000 molar solution of barium hydroxide.

Dilution:

A process by which the concentration of a solution is lowered by adding water. In a dilution process the number of moles stays the same.

Moles solute before dilution equals Moles of solute after dilution.

$$M_{initial} V_{initial} = M_{final} V_{final}$$

Assignment 9.19

Calculate the volume of 18.0 M H_2SO_4 required to prepare 1.00 liter of a 0.900 M solution of H_2SO_4.

Assignment 9.20 How many grams of solute are present in 50.0 mL of 1.33 M $CoSO_4$?

Assignment 9.21 An experiment calls for you to use 200 mL of 1.0 M HNO_3 solution. All you have available is a bottle of 6.0 M HNO_3.

How would you prepare the desired solution?

1.0 M KCl solution
means
1.0 M K^+ and 1.0 M Cl^-

1.0 M Na_2SO_4 solution
means
2.0 M Na^+ and 1.0 M SO_4^{2-}

When an ionic compound dissolves, the relative concentrations of the ions introduced into the solution depends on the chemical formula of the compound.

Review Exercises and Problem Solving

Assignment 9.22 Mixing aqueous solutions of silver nitrate and potassium bromide produces a precipitate.

a. Identify the reactants, and identify them as elements, covalent compounds, or ionic compounds.

b. Determine the products, and identify them as elements, covalent compounds, or ionic compounds.

c. Write out the balanced equation for this reaction, include physical states, (s), (l), (g), and (aq).

d. Write out the complete ionic equation for this reaction.

e. Write out the net ionic equation for this reaction.

f. What are the spectator ions for this reaction?

g. Aside from a precipitation reaction could it also be another type of reaction?

h. If 25g of silver nitrate are used in this reaction, how many grams of potassium bromide are needed?

Assignment 9.23 Mixing aqueous solutions of barium hydroxide and sulfuric acid produces a precipitate.

a. Identify the reactants, and identify them as elements, covalent compounds, or ionic compounds.

b. Determine the products, and identify them as elements, covalent compounds, or ionic compounds.

____

c. Write out the balanced equation for this reaction, include physical states, (s), (l), (g), and (aq).

d. Write out the complete ionic equation for this reaction.

e. Write out the net ionic equation for this reaction.

f. What are the spectator ions for this reaction?

____

g. Aside from a precipitation reaction could it also be another type of reaction?

____

h. How many mL of a 1.0 M solution of sulfuric acid are needed to neutralize 0.75 moles of $Ba(OH)_2$?

Assignment 9.24 Aqueous solutions of potassium phosphate and lead(II) nitrate are mixed to produce a precipitate.

a. Identify the reactants, and identify them as elements, covalent compounds, or ionic compounds.

b. Determine the products, and identify them as elements, covalent compounds, or ionic compounds.

c. Write out the balanced equation for this reaction, include physical states, (s), (l), (g), and (aq).

d. Write out the complete ionic equation for this reaction.

e. Write out the net ionic equation for this reaction.

f. What are the spectator ions for this reaction?

g. Aside from a precipitation reaction could it also be another type of reaction?

h. If 3.57 g of lead(II) nitrate are used in this reaction in an excess of potassium phosphate, how many grams of lead(II)phosphate are produced?

Assignment 9.25 Mixing aqueous solutions of calcium chloride and sodium fluoride produces a precipitate.

a. Identify the reactants, and identify them as elements, covalent compounds, or ionic compounds.

b. Determine the products, and identify them as elements, covalent compounds, or ionic compounds.

c. Write out the balanced equation for this reaction, include physical states, (s), (l), (g), and (aq).

d. Write out the complete ionic equation for this reaction.

e. Write out the net ionic equation for this reaction.

f. What are the spectator ions for this reaction?

____

g. Aside from a precipitation reaction could it also be another type of reaction?

____

h. If 0.35 moles of calcium fluoride are produced in this reaction, how many grams of calcium chloride and sodium fluoride were used?

Assignment 9.26 Placing a strip of zinc metal into a solution of copper sulfate results in a precipitation of copper metal and a solution zinc sulfate.

a. Identify the reactants, and identify them as elements, covalent compounds, or ionic compounds.

b. Determine the products, and identify them as elements, covalent compounds, or ionic compounds.

c. Write out the balanced equation for this reaction, include physical states, (s), (l), (g), and (aq).

d. Write out the complete ionic equation for this reaction.

e. Write out the net ionic equation for this reaction.

f. What are the spectator ions for this reaction?

g. Aside from a precipitation reaction could it also be another type of reaction?

h. If mass of zinc is 0.035g, how many gram of copper will precipitate assuming an excess of copper sulfate solution is used.

Problem Solving

Example 1

Up until now, all the calculations have been geared toward the mechanisms of conversion. Just as a music student practices exercises involving scale studies and chords to become proficient in music, we have been performing exercises to strengthen our understanding and abilities in basic calculations.

In this chapter we will now pull everything together in the process of problem solving. Problem solving is not limited to chemistry; we use it daily in our lives to organize our work and routines. In the need to reach a goal, accomplish a task, or even plan a family gathering we evaluate the goal, gather and assess information, work out logistics – in other words, look what needs to be done, check out resources and information, and implement.

The following chemical problems are no different. A goal will be stated, information will be given and assessed, and a method to solve the problem will be established. The following is an example of such a situation.

> In order to run a reaction a chemist needs 0.1 moles of sodium metal. The sodium metal is in the shape of a bar measuring 1.27 cm in width, 1.27 cm height and 6.35 cm in length. Sodium is a soft metal that can be cut with a knife. It has a density of 0.97 g/cm^3 and a molar mass of 22.98 grams. What length of the bar must the chemist cut to obtain 0.1 moles of sodium?

At first glance, the method for solving this problem may not seem apparent. We need to determine the length of the sodium bar needed to give us 0.1 mole of the metal. Since we have done many conversions of grams to moles and moles to grams, just weighing out the sodium would provide a simple solution. But the problem wants an answer in centimeters and not in grams.

(Perhaps the laboratory balance is broken.)

In essence, we want to obtain 0.1 moles of sodium based on volume (Volume = length x width x height) instead of mass.

If we can calculate the mass of sodium needed to obtain 0.1 moles, how can this mass be converted into a volume?

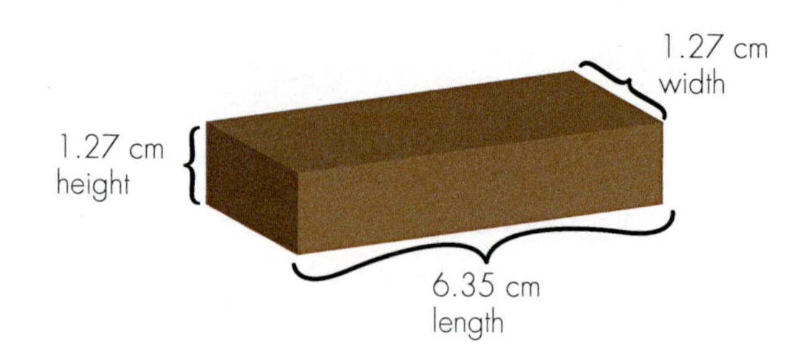

At this point, let's look at the given information:

- The dimensions of the sodium bar
- The density of the sodium
- The molar mass of the sodium
- The number of moles needed

Rather than just jotting down calculations, taking the time to write out a road map can facilitate problem solving.

moles of sodium ⟶ mass of sodium ⟶ volume of sodium

The first step of the solution is accomplished by multiplying 0.1 moles of sodium by the molar mass of sodium, 22.98 g/mol:

$$0.1 \text{ moles} \times \frac{22.98 \text{ g}}{1 \text{ mol}} = 2.298 \text{ g sodium}$$

Converting the mass of sodium into a volume is achieved by division using density, 0.97 g/cm3:

$$2.298 \text{ g} \times \frac{1 \text{ cm}^3}{0.97 \text{ g}} = 2.369 \text{ cm}^3$$

Given the total volume needed to deliver 0.1 moles of sodium, we can plug it into the volume formula:

$$\text{Volume} = \text{length} \times \text{width} \times \text{height}$$

$$2.36\,9 \text{ cm}^3 = \text{length} \times 1.27 \text{ cm} \times 1.27 \text{ cm}$$

Solving yields 1.469 cm for the length of the sodium bar that must be cut to obtain 0.1 moles of sodium.

Problem Solving Example 2

In order to perform an experiment, a student needs to transfer 15 mL of an aqueous solution of iron (III) nitrate, into a reaction flask. The solution is 10% by mass. The density of this solution is 1.0 g/mL. What is the mass of the iron (III) nitrate transferred?

Given

Volume, in mL, of the iron (III) nitrate solution transferred.

Desired

Mass, in grams, of only the Iron (III) nitrate transferred.

Method

1. Convert the volume of solution to mass of solution (mL to g). This can be done by multiplying the volume of solution by the density of the solution:

$$\text{density} = 1 \text{ g/mL}$$

$$15 \text{ mL} \times \frac{1 \text{ g}}{1 \text{ mL}} = 15 \text{ g of solution}$$

2. Convert the mass of the iron (III) nitrate solution to the mass of just the iron (III) nitrate.

Consider:

The iron (III) nitrate solution is a homogenous mixture of iron (III) nitrate and water.

Solution = solute + solvent

(iron (III) nitrate + water) = (iron (III) nitrate) + (water)

Note: A solution is composed of a solute dissolved in a solvent

A 10% (by mass) solution of iron (III) nitrate means that 10% of the solution mass is iron (III) nitrate (the solute).

For example,

$$\text{10\% solution (by mass)} = \frac{\text{10 g iron (III) nitrate}}{\text{100 g solution}} \times 100$$

(iron (III) nitrate and water)

Therefore,

15 g of iron (III) nitrate solution × 0.1 equals the mass of the iron (III) nitrate actually transferred,

i.e.
1.5 grams iron (III) nitrate

Titration Concepts

The process in which a solution of one reactant, the titrant is carefully added to a solution of another reactant, and the volume of the titrant required for complete reaction is **measured**.

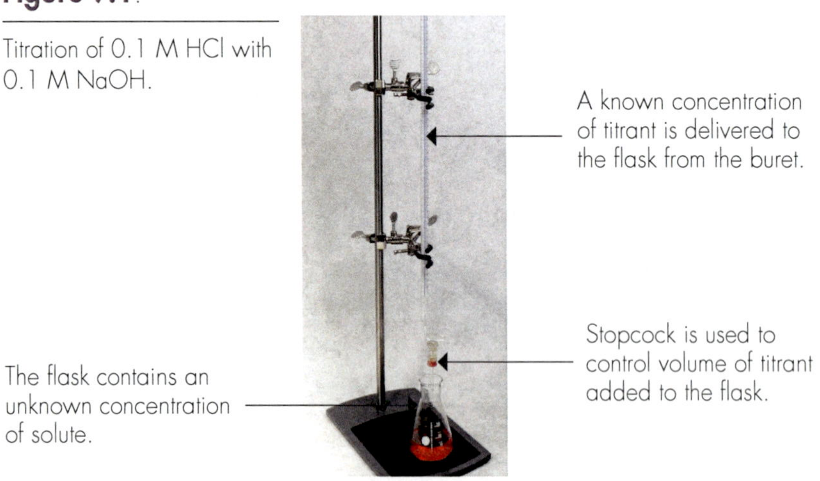
A known concentration of titrant is delivered to the flask from the buret.
Stopcock is used to control volume of titrant added to the flask.
The flask contains an unknown concentration of solute.

Figure 9.1.

Titration of 0.1 M HCl with 0.1 M NaOH.

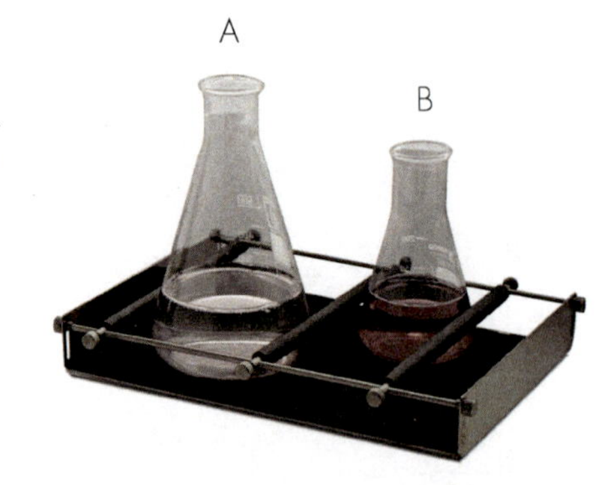
A
B

Figure 9.2.

Before NaOH is added, flask A contains 0.1 M HCl and a few drops of phenolphthalein. Once all of the HCl has reacted with the added NaOH the solution turns pink (Flask B) which indicates the endpoint of the titration.

Standard Solutions:

Solutions of accurately known concentration.

The standard solution undergoes a specific chemical reaction of known stoichiometry with a solution of unknown concentration.

Titration of KOH solution (unknown concentration) with a standard solution of HCl (0.1 M)

$$HCl_{(aq)} + KOH_{(aq)} \longrightarrow H_2O_{(l)} + KCl_{(aq)}$$

Equivalence Point:

When enough HCl is added to neutralize all the KOH. It required 25 mL of HCl to react with all the KOH. Since # mole HCl = # mole KOH and (0.025L)(0.1M) = 0.0025 mole HCl then 0.0025 moles of KOH were in flask. By knowing the volume of unknown solution and the number of moles of KOH, the concentration of the KOH solution can be obtained.

$$\frac{0.0025 \text{ moles KOH}}{0.050 \text{ L solution}} = 0.05 \text{ M KOH}$$

How do you know when the equivalence point is reached?

Indicator

A substance that can exist in different forms, with different colors that depend upon the concentration of H+ in solution (See Figure 6.2).

End Point

The point at which the indicator changes color and the titration is stopped.

Ideally the end point should coincide with the equivalence point.

Indicators

Indicator	color change	acid	base
Thymol blue	1.2 - 2.8	red	yellow
Methyl orange	3.1 - 4.4 r	ed	yellow
Bromothymol blue	6.0 - 7.6	yellow	blue
Phenolphthalein	8.2 - 10	colorless	red

Figure 9.3.

Color chart for selected indicators vs pH.

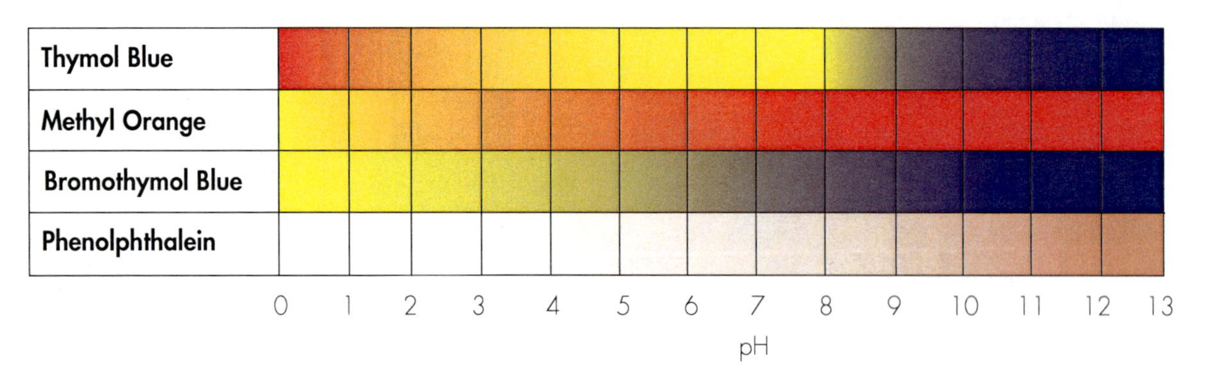

Problems

Assignment 9.27 How many milliliters of 0.155 M HCl are needed to neutralize completely35.0 mL of 0.101 M $Ba(OH)_2$?

Assignment 9.28 In a titration, it takes 43.2 mL of 0.236 M NaOH solution to react with 36.7 mL of an HCl sloution. What is the molarity of the HCl solution?

Glossary

Acid — A substance that ionizes to form hydrogen ions in solution.
Actual Yield — The amount of product that results from a chemical reaction in the laboratory.
Alkali Metal — The metal elements located in Group 1A of the periodic table.
Alkaline Earth Metal — The metal elements located in Group 2A of the periodic table.
Amithuzal Quantum Number (l) — A quantum number that describes the shape of an orbital.
Amorphous Solid — A solid whose atoms are not arranged in an orderly fashion.
Amophoteric Substance — A substance that can act as an acid or a base depending on the reactants involved.
Analyte — The substance that is being analyzed.
Anion — A negatively charged ion.
Aqueous Solution — A solution in which the solute is dissolved in water.
Atom — The smallest particle of an element that still retains all of the characteristics of the element.
Atomic Mass Unit (amu) — A unit that describes the mass of an atom.1 amu = 1.66054×10^{-24} g.
Atomic Number (Z) — The number of protons in an atom.
Atomic Orbital — A region in space around the nucleus where an electron is most likely to be found.
Atomic Weight — The weighted average mass of all the atoms of an element.
Avogadro's Number (NA) — The number of units in one mole of matter. 6.02×10^{23} units.
Base — A substance that increases the hydroxide ion concentration in solution.
Cation — A positively charged ion.
Chemical Change — A change in the chemical composition of a substance.
Chemical Equation — A representation of a chemical reaction where symbols and chemical formulas are used.
Chemical Formula — A representation of a chemical compound that uses symbols for the elements and subscripts to indicate how many atoms of each element are present.
Chemical Reaction — The transformation of a substance into a new substance or substances.
Chemistry — The study of the nature, physical and chemical properties, and the transformations of matter.
Combination Reaction — A reaction in which two or more substances react to form one product.
Compound — A pure substance composed of atoms that are combined in fixed proportions.
Condensation — The transformation of a gas to its liquid state.
Conversion Factors — Equivalences that are used to convert from one unit of measurement or count to another.
Coordinate Covalent Bond — When one atom donates both bonding electrons to the formation of a covalent bond.
Covalent Bond — A bond that is formed by the sharing of electrons between atoms.
Covalent Compound — A compound that is composed of nonmetal species.
Crystalline Solid — A solid with a highly regular repeating arrangement of atoms, ions, or molecules.
Decomposition Reaction — A reaction in which one substance produces two or more substances.
Density — A physical property that measures the mass of a material in a given volume.
Deposition — The transformation of a gas directly to a solid.
Double Covalent Bond — A covalent bond where four electrons are shared.
Electrolyte — A substance that produces ions when dissolved in water.
Electron — A subatomic particle that has a negative charge and negligible mass.
Electron Configuration — The distribution or arrangement of electrons in shells and subshells about the nucleus.
Electronegativity — The ability of an atom to attract electrons in a covalent bond.
Electron–Electron Pairing — When each atom contributes one electron to the formation of a covalent bond.

Element — A pure substance that is composed of one type of atom.
Element Symbol — A one or two letter symbol that represents a given element.
Endothermic — A process or reaction that absorbs heat from the surroundings and has a positive ΔH.
Endpoint — The point at which the indicator changes color and the titration is stopped.
Energy — The ability of a system to do work.
Equivalence Point — The point in the titration when the amount of added titrant is stoichiometrically equivalent to amount of analyte.
Exothermic — A process or reaction that releases heat to the surroundings and has a negative ΔH.
Extranuclear Region — The region that surrounds the nucleus of an atom where the electrons are found.
Formal Charge — The charge on an atom in a Lewis structure. Formal charge is used to determine the most stable structure.
Freezing — The transformation of matter from a liquid to its solid state.
Gas — Matter that has both variable volume and shape.
Group — A column of elements in the periodic table.
Halogens — The non metal elements located in Group 7A of the periodic table.
Heat — A form of energy that flows spontaneously from one object ot another cooler object.
Heterogeneous Mixture — A mixture that has non-uniform composition throughout.
Homogeneous Mixture — A mixture that has a uniform composition throughout.
Hybridization — The mixing of atomic orbitals to form a new hybrid orbital.
Hydrogen Bonding — The attraction between a hydrogen that is bonded to an electronegative O, N, or F atom and another nearby O, N, or F atom.
Indicator — A substance used in a titration that can exist in different forms, with different colors that depend upon the concentration of hydrogen ion in solution.
Inner Transition Elements — The metal elements that are in the two rows below the periodic table. The lanthanides and actinides.
Intermolecular Force — A force or forces that act between molecules to hold them close together in their liquid or solid state.
Ion — An atom or group of atoms that is electrically charged.
Ionic Compound — A compound that is composed of metal and nonmetal species.
Ionic Equation — A chemical equation that indicates explicitly whether dissolved substances are present as ions or molecules.
Isotope — Atoms that have the same atomic number (Z) but different mass numbers (A).
Kinetic Energy — The energy of motion.
Kinetic Molecular Theory — A set of postulates used to describe how gas particles behave.
Lewis Dot Symbol — Dots that represent valence electrons are placed around the element's atomic symbol.
Lewis Structure — A molecular representation that shows how the atoms are connected and the location of the lone pair electrons.
Limiting Reactant — The amount of reactant that determines the amount of product that can be produced.
Liquid — Matter that has a definite volume but variable shape.
Lone Pair — A nonbonding pair of electrons.
Magnetic Quantum Number (ml) — A quantum number that describes the orientation of atomic orbitals.
Magnetic Spin Quantum Number (ms) — A quantum number that describes the electron spin as clockwise or counterclockwise.
Main Group Elements — Elements that are in Groups 1A through 8A in the periodic table.
Mass — The measure of how much matter is contained in an object or substance.
Mass Number (A) — The sum of protons and neutrons in the nucleus of an atom.
Matter — Anything that has mass and occupies space (volume).
Melting — The transition of a solid to its liquid state.
Metal — An element that is malleable, ductile, and is a good conductor of heat and electricity.
Metalloid (semi-metal) — An element that has properties that are intermediate between a metal and nonmetal.
Mixture — A blend of pure substances that retain their chemical identity and have variable composition.

Molar Mass — The number of grams in one mole of substance.
Mole — Corresponds to 6.02 × 1023 units.
Molecular Formula — A formula show shows the number and types of atoms in one molecule of a compound.
Molecule — A group of atoms held together by covalent bonds.
Net Ionic Equation — A chemical equation that only includes ions and substances directly involved in the formation of product.
Neutralization Reaction — The reaction between an acid and a base to produce a salt and water.
Neutron — A neutral subatomic particle that resides in the nucleus of the atom.
Noble Gas — The inert gases located in Group 8A of the periodic table.
Nonmetal — An element that is not a good heat or electrical conductor.
Nucleus — The region of the atom where the protons and neutrons are combined. The nucleus contains most of the mass of the atom.
Octet Rule — Representative Elements undergo reactions that result in 8 valence electrons.
Oxidation — The loss of one or more electrons.
Oxidation Number — The charge on a monotomic ion or the charge an atom would have if the shared electrons in the bond were held by the more electronegative atom.
Oxidizing Agent — The substance that is reduced.
Oxo Anions — Negatively charged polyatomic ions which contain a metal or nonmetal atom combined with one or more oxygen atoms.
Percent Yield — The amount of the actual yield divided by the theoretical yield multiplied by 100.
Periodic Table — A tabular listing of the elements ordered by increasing atomic number (Z) and grouped according to chemical similarities.
Period — A row of elements in the periodic table.
pH — A measure of hydronium ion concentration.
Physical Charge — A change that does not alter the chemical composition of matter.
Polar Bond — A covalent bond where the electrons are not shared equally between atoms.
Polyatomic Ion — An ion composed of two or more atoms.
Potential Energy — Stored energy.
Principle Quantum Number — A quantum number that describes the distance of an electron from the nucleus.
Proton — A subatomic particle that resides in the nucleus of the atom and has a positive charge.
Pure Substance — A substance that has uniform chemical composition.
Radical — A substance that contains an unpaired electron.
Reducing Agent — The substance that is oxidized.
Reduction — The gain of one or more electrons.
Representative Elements — Elements that are in Groups 1A through 8A in the periodic table.
Resonance Structures — An average of two or more Lewis structures that only differ in the position of the electrons.
Saturated Solution — A solution that contains the maximum amount of solute.
Schroedinger Equation — Mathematically describes all the energy levels within an atom.
Single Covalent Bond — A covalent bond where two electrons are shared.
Solid — Matter that has both a definite shape and volume.
Solubility — The maximum amount of solute that will dissolve at a given temperature.
Solute — A substance that is dissolved in another substance and is usually present in the lesser amount.
Solution — A homogeneous mixture that contains small particles the size of ions and small molecules.
Solvent — A substance that dissolves a solute and is usually present in the larger amount.
Space–filling Models — A visual representation of a molecule where color coded spheres are used to represent the atoms.
Specific Gravity — The ratio of the density of a substance to the density of water.
Spectator Ion — An ion that does not participate in product formation.
Stoichiometry — A quantitative measure of reactants and products in a chemical reaction.
Strong Acid — An acid that completely ionizes in solution.
Structural Formula — A representation of a molecule that shows the covalent bonds between atoms by using lines.

Subatomic Particle — Protons, neutrons, and electrons.
Sublimation — The transformation of a solid directly to its gaseous state.
Supersaturated Solution — A solution that contains more than the maximum amount of solute.
Temperature — A property of matter that determines the direction of heat flow.
Theoretical Yield — The amount of product that is calculated from a balanced chemical equation.
Thermal Energy — Heat energy.
Titration — The process in which a solution of known concentration and volume is added to an unknown analyte in a solution of known volume to determine the concentration of the analyte.
Transition Element — The metal elements that are in Groups 1B through 8B in the periodic table.
Triple Covalent Bond — A covalent bond where six electrons are shared.
Valence Electron — An electron that resides in the valence shell.
Valence Shell — The outermost energy level (shell) of an atom.
Vaporization — The transformation of a liquid to its gaseous state.
Weak Acid — An acid that does not completely ionize in solution.
Weight — The measure of the gravitational pull that a body exerts on an object.